Green Vegetable Oil Processing

Revised First Edition

Green Vegetable Oil Processing

Revised First Edition

Editors
Walter E. Farr and Andrew Proctor

Urbana, Illinois

AOCS Mission Statement

AOCS advances the science and technology of oils, fats, surfactants and related materials, enriching the lives of people everywhere.

ISBN 978-0-12-810216-9 (print)
ISBN 978-0-9835072-0-8 (.epub)
ISBN 978-0-9835072-1-5 (.mobi)

Library of Congress Cataloging-in-Publication Data

Green vegetable oil processing / editors, Walter E. Farr and Andrew Proctor. — Revised first edition.
pages cm
Includes index.
ISBN 978-0-9888565-3-0 (hard copy : acid-free paper) 1. Vegetable oils. 2. Separation (Chemistry) 3. Sustainable engineering. I. Farr, Walter E., 1938- editor of compilation. II. Proctor, Andrew, 1953- editor of compilation.
TP680.G74 2014
665'.3—dc23

2013036761

Printed in the United States of America
16 15 14 13 12 5 4 3 2 1

The paper used in this book is acid-free, and falls within the guidelines established to ensure permanence and durability.

Contents

Preface

Two years have passed since the first edition of *Green Vegetable Oil Processing* was published. The *Revised First Edition* includes much of the content of the first edition, but incorporates updated data, details, images, figures, and captions. We are grateful to the many authors that submitted their contributions for this book for the second printing.

The vegetable oil processing industry is a large and profitable sector of the international food business community. Current processing technologies and increased oilseed production developed in the twentieth century have been very successful in the industrial scale production of commercially available plant oils. The technologies were developed to extract and process vegetable oil when energy costs were relatively low and the environmental impact of plant operations and chemical use were not a significant consideration. These processing techniques provided generations of consumers reliable, safe, food oil based foods, at a very reasonable cost. However, for many modern affluent consumers price is not the only consideration when purchasing food and other consumer products. They now wish to purchase 'green' food products that require less energy for production and transport, and are thus perceived as being environmentally responsible. This change in consumer perception is an opportunity by the vegetable oil industry to reduce costs of energy and materials, by increasing processing efficiency through development of new innovative technologies.

Alternative green food processing technologies have gained much technical and industrial attention in recent years as a potential means of reducing costs and promoting consumer awareness of corporate environmental responsibility. However, utilizing green principles is now becoming an effective business approach to enhance vegetable oil processing profitability. This can be done by reduced use of fossil fuels, organic solvents, and other chemicals in processing operations. Such practices will increase the sustainability of the vegetable oil processing industry and reduce the environmental footprint produced. The book addresses alternative green technologies at various stages of oilseed and vegetable oil processing. This includes oil extraction technologies such expeller, green extraction processes, including aqueous and supercritical methods; and green modifications of conventional unit operations, such as degumming, refining, bleaching, hydrogenation, winterizing/dewaxing, fractionation, and deodorization (physical refining). A range of technologies are described, from those already in commercial use to those that are still in development stages. While most chapters describe soy oil processing, the techniques described are equally applicable to oils and fats in general.

This book is important in documenting the current situation with respect to "green" oil processing technology developments and evaluation of their commercial potential. It is hoped that this book will provide useful insights and perspective to industry leaders and researchers in industry, academia and government as they creatively engage to meet the business, technical and consumer challenges of the twenty-first century.

Walter E. Farr
Andrew Proctor

1

Extrusion/Expeller® Pressing as a Means of Processing Green Oils and Meals

Vincent J. Vavpot, Robert J. Williams, and Maurice A. Williams
Anderson International Corp, Cleveland, Ohio, USA

Introduction

Today, there is much interest in "green processing." This catchy term has a specific meaning, which can be obtained from the Internet. It means "any process that is safe, economical, and environmentally friendly" (http://www.eng.uwo.ca/chemical/green-new.htm). Put into different words, it means any process that has little or no impact on the environment and also any process that optimizes the use of energy for processing. Another catchy term is "organic food." "Organic foods" (from Wikipedia) means those foods "that are produced using environmentally sound methods that do not involve modern synthetic inputs such as pesticides and chemical fertilizers, do not contain genetically modified organisms, and are not processed using irradiation, industrial solvents, or chemical food additives" (http://www.wikipedia.org/wiki/Organic_food).

Vegetable oils are, today, obtained by either solvent extraction or by mechanically crushing the oil-containing seeds that come from the farmlands. Solvent extraction does impact the environment because of the effluent streams of unrecoverable solvent vapors, and solvent extraction also brings food materials into contact with the solvent, which, today, is undesirable. All manufacturers of oil processing equipment should be aware of both the green processing and the organic foods expectations of those who buy machinery for oil processing.

Mechanical screw presses have been manufactured for more than 100 years. Since mechanical screw pressing is an alternative to solvent extraction, if someone wishes to produce oils for human foods that comply with the expectations of organic, a mechanical screw press would be a proven piece of equipment to use. The title of this chapter coincides with redesign criteria that some manufacturing companies have been considering for the past several years. It is the authors' pleasure to discuss here what is being done to better adapt screw presses and extruders for green processing and organic foods production. Since the authors are already on a committee doing this work, we have a great deal to say about the subject. To be fair to other manufacturers of screw presses and extruders, we have also invited them to consider adding their own input on how their companies are responding to the challenges presented by green processing and organic foods.

Mechanical Screw Presses

The mechanical screw press was developed at a time when labor was inexpensive and capital equipment costs were of concern (Shahidi, 2005). Today, labor costs are very high, and any simplification of time-consuming maintenance procedures would, of course, make mechanical screw presses much more attractive. Also, when the mechanical screw press was first introduced, concerns about the environment were minimal. Today, however, these concerns are a major consideration when operating modern processing plants. So, with these considerations in mind, we would like to discuss what some manufacturers are doing to upgrade their mechanical screw press and extrusion systems to address both green processing and organic foods concerns.

The major objectives of green processing are optimization of the use of energy and having minimal or no impact on the environment. One could also specify reduction in labor costs, particularly in replacement of wear parts, especially the replacement of worn choke parts. Optimization in the use of energy would entail consuming less electrical energy by operating with better prepared oilseeds, by using motors that are more efficient, and by using fewer motors. Having less effect on the environment would entail preventing escape of oily mists and pungent vapors from the processing rooms. These concerns will be discussed in greater detail showing what is being done by different manufacturers.

The mechanical screw press accepts a continuous stream of oil-bearing material, compresses it under very high pressure exerted by a worm shaft rotating within a slotted-wall cage with the flights on the worm shaft propelling the oilseed forward as the shaft exerts pressure. The pressure releases the oil, most of which flows out through the slotted-wall cage. The deoiled solids flow through an adjustable port, like a movable cone, at the discharge end of the press.

In some cases, a second worm shaft is used to force feed the main worm shaft. A few screw presses have been designed to eliminate the need for a force feeder. These screw presses are fed by free-falling input streams discharged from variable speed screw conveyors. The metered feed stream falls directly into the feed worm on the main worm shaft. Flighting is designed so that a gravity-fed feed worm is capable of accepting the designed input mixed with air as it falls onto the feed worm. Having only one worm shaft eliminates the need for a motor-driven forced feeder and also avoids potential problems with balancing motor loads pulled by two worm shafts. This optimizes the use of energy for processing and also lowers costs for the screw press.

Some screw presses available today are supplied without adjustable chokes. An adjustable choke presents an advantage because it provides a means to assure

Fig. 1.1. Victor 600™ Press (Courtesy Anderson International Corp).

obtaining minimum residual oil even though the worms undergo some wear. Without a means of adjustment, there would be a steady reduction of pressure exerted by the shaft as the worms wear, and this reduction in pressure would result in a gradual rise in residual oil. In addition, the ability to open the choke makes it possible to reduce the compaction exerted by the press just before shutdown to avoid leaving hard cake within the cage. Worm shafts in most screw presses can be supplied with internal bores for water cooling. Some manufacturers also supply water-cooled cages. Other manufacturers supply a cooled oil flushing system to both cool the cages and flush away any solids clinging to the cages.

Some screw presses are constructed onto cast steel housings containing speed reduction gears and thrust bearings; others are constructed on channel iron bases and are equipped with off-the-shelf inverter duty motors, gear reducers, (sometimes using V-belt drives if more reduction is desired), and couplings (Fig. 1.1). The availability today of reasonably priced VFD drives permits the use of worm shaft RPM as a variable during screw press operation. Varying worm shaft RPM affects the compaction exerted by the worm shaft and, therefore, serves as an alternate way of applying choke. Varying worm shaft RPM now becomes a valuable way to optimize screw press performance with a greater range of capability than is achievable with fixed RPM screw presses.

A problem with many screw presses is that they emit a mist of oil droplets along with the oil, and this mist easily escapes through the shroud surrounding the barrels and enters into the press room, causing environmental problems in the processing plant and in the effluent air stream leaving the plant. All screw press manufacturers are looking into this. Anderson International's Victor presses are now being equipped with much improved shrouds (Fig. 1.2) that prevents escape of oily mist into the press room. Duct connections on the shroud allow for an air stream to draw vapors through ductwork through a deodorizer scrubber to diminish odors escaping into the atmosphere. These shrouds perform much better than shrouds on previous Anderson screw presses.

Extruders

Extruders were originally developed for extrusion of cereal grains and pet foods, but were later used to process other materials such as rice bran prior to solvent extraction to improve extractability, and to process other oilseeds to deactivate enzymes that impair oil quality. Extrusion as a preparation for mechanical screw pressing has been described in earlier publications (Gunstone & Padley, 1997).

An extruder consists of a rotating worm shaft within a cylindrical barrel. Material to be extruded is fed into one end of the barrel, and the worm shaft forces it out through the opposite end. The worm shaft flighting is usually not a continuous wrap; in some cases it is composed of partial wraps and unflighted segments. Stationary pins protruding from the barrel wall or shaft-mounted steam locks quickly masticate the oilseed and elevate the temperature within the extruder.

Fig. 1.2. Shrouds (Courtesy Anderson International Corp).

The oilseed within the extruder is subjected to high pressure and is pressure-cooked at optimum moisture and temperature to convert the protein into a tacky, elastic-like condition. The injected steam, if steam is being injected, releases its heat of vaporization, which helps to heat the oilseed. Additional heat is generated by friction from the rotating shaft. Friction and shear can be optimized to rupture oil cells and release oil for easy removal in mechanical screw presses.

The experimentation of Nelson et al. in 1987 led to the use of extrusion as a means to pre-treat oilseeds for subsequent full pressing. Coarsely ground soybean was processed through a high-shear extruder producing a hot, foamy mixture of small particles, liberated oil, and vaporizing moisture. The product upon discharge was passed into a screw press and pressed to 6% residual oil. Nelson's work was done on an Insta-Pro extruder (Nelson et al., 1987).

Unlike traditional atmospheric vessels that require 30 or more minutes residence time to achieve desired pretreatment, all extruders operate under pressure, creating a high-temperature short-residence time environment while cooking the oilseed, and an extruder also generates shear against the oilseed. The shear ruptures the oil cells, which are embedded in the oilseeds, making the oilseed better prepared for oil release in a screw press. The high pressure reduces required residence time sixty-fold, resulting in 30 seconds required residence time. Short residence time permits a much smaller and less expensive vessel, which reduces both the size and cost of the pretreatment vessel, and short residence time provides for better quality oils and meals. Today oilseed extruders are produced by several manufacturers, and extruders have been used for many years as better devices to prepare oilseeds for subsequent mechanical screw pressing (Farnsworth et al., 1993).

Manufacturers of Screw Presses and Extruders

Anderson International

Anderson International (Cleveland, Ohio, USA) has inaugurated an improved line of mechanical screw presses called Victor presses. Three models are envisioned. The first is a 6" diameter, 55" long drainage cage model powered by a 125 HP variable speed motor: the Victor-600™ screw press (see Fig. 1.2). Target capacity for canola is 50 MT/d pressing to 5–7% residual oil, when used in conjunction with preparation in Anderson's Dox-Hivex™ extruder. Early field performance already indicates up to 55 MT/d pressing canola to 5–6% residual oil and 40 MT/d pressing soybean to 4.2%. The second model is a 12" diameter, 88" long drainage cage model powered by a 400 HP variable speed motor: the Victor-1200™ screw press. Target capacity for canola is 200 MT/d pressing to 5–8% residual oil when used in conjunction with Anderson's Dox-Hivex™ extruder and 150 MT/d with traditional preparation. The third model is a small pilot model, 6" diameter with a 33" long drainage cage, powered by a 20 HP variable speed motor: the Victor-60™ press.

All three models have floating worm shafts that are supported at the discharge end (and in the center on the largest model) by hardened wear sleeves allowing the oilseed being processed to serve as a lubricant. This does away with bearings that are difficult to align properly at the discharge end and are also difficult to replace. Anderson has a long history of supplying equipment with floating shafts. Victor-600™ presses are already in operation in three commercial oilseed plants.

The Victor series screw presses have simplified choking devices that do not require any disassembly of the mechanical screw press to replace worn choke parts. The choke consists of a rotating cone on the worm shaft with replaceable wear faces: the proximity of the cone to a retaining ring provides choke adjustment (Fig. 1.3). The cone rotates with the shaft, and its position is fixed on the shaft; the non-rotating retaining ring is mounted on the barrel, and its position is movable. The cone, mounted on the rotating shaft, absorbs much of the forward thrust applied through the material passing through the press. All rearward thrust against the worm flights is absorbed on the feed end of the same worm shaft. This application of thrust in both directions onto the same solid piece of metal greatly reduces the structural requirements of presses that have all forward thrust absorbed by the choke mechanism that is mounted on the barrel and all rearward thrust absorbed by the worm shaft.

The wear faces of the cone, as well as the retaining ring, are easily replaced without disturbing the barrel cages. This results in considerably less maintenance time to replace worn parts. Anderson International is also marketing a Combo-System™ that combines an extruder with a screw press. This will be described elsewhere in this chapter.

Fig. 1.3. Quadrant Choke (Courtesy Anderson International Corp).

Desmet Ballestra

Desmet Ballestra (Zaventem, Belgium) offers the Rosedowns line of screw presses known as the Sterling Series (see Fig. 1.4) with bore diameters of 200 mm (8") up to 400 mm (15.75"). The Sterling Series presses have one main drive motor, varying in size from 45 kW (60 HP) to 700 kW (900 HP) that does all the crushing work. The largest press in the series is the 800 Series. This press will process about 750 MT/d of canola as a pre-press and up to 175 MT/d of canola as a full press. When configured as a pre-press, a highly extractable cake, typically with 18% oil content, is produced. When configured as a full press, consistent operation at 7–8% is possible.

Proper oilseed preparation is essential for press performance both for expected oil yields and for minimum power consumed by the motor. For smaller soft seeds, such as canola, flaking is an essential step for pre-pressing, as the larger clearances in a high-capacity press do not guarantee that every seed will be ruptured. However, when processing more fibrous materials such as sunflower, simple breaking is often sufficient to give good performance. Similarly, when used in a double pressing plant, even the flaking step can be successfully omitted. This would reduce some of the capital equipment required for preparation. The ability to operate in this manner is a result of modern low pressure, multistage compression technologies. In essence, the feed material is subjected to a series of gentle squeezes rather than one hard one. This prevents the cake from becoming too compacted and closing up the fissures in the cake that are essential for the drainage of oil from it.

The multiple compression approach means the press is less sensitive to the variations in the feed, and this allows successful operation of the press without an

Fig. 1.4. Sterling Press (Courtesy Desmet Ballestra).

adjustable choke mechanism. The choke essentially becomes the final stage of compression and is integral to the worm assembly, meaning its maintenance is no more complicated than that of any other worm assembly component. This is a significant simplification of the press that makes it much easier to maintain and greatly enhances the service reliability of the press. Service reliability is a key issue because plants that operate with a few high capacity presses need each press to be quickly put back into operation if any press should go down for repair.

All Rosedowns presses come complete with their own feed arrangement, which consists of a horizontal variable speed feed conveyor and fixed speed vertical conveyor. The purpose of the vertical feed conveyor is to give a more even and more efficient flow of feed into the press. The maximum power installed on these vertical conveyors is 5.5 kW (7.5 HP). These feeder screws are not designed to force material into the worm assembly; instead they are designed to regulate the feed and to facilitate a free flow of material into the press. All the pressing work is done by the main drive and the worm assembly, which is designed by the manufacturer for the required duty.

All Sterling Series presses come complete with state-of-the-art shrouds to contain any oily mist produced through the drainage cages. The shrouded enclosure surrounding the drainage cages can be vented through a scrubber system to ensure minimum effluent flow into the atmosphere and surroundings of the plant.

Maintenance costs today are very high. When designing a plant with large presses, attention to proper location of presses goes a long way in making the plant easier to run and the presses easier to maintain. Bigger machines mean bigger and heavier components, making access for lifting an important consideration. While the design of the press itself can improve access for maintenance, it can only do so much. When installing these large presses in modern plants, providing lifting beams, or even an overhead crane, is a desirable feature to ensure less downtime for maintenance.

Dupps

The Dupps Company (Germantown, Ohio, USA) offers five models of Pressor® presses (Fig. 1.5) with capacities up to 59 MT/d pressing to residuals as low as 5–7%. The diameters of the bores range from 7" to 12". The Pressor® has a simplified feed design that features a large feed hopper that eliminates the need for a down feeder and its drive. This ability to use a single motor to drive all mechanical parts of the press contributes to green processing expectations by eliminating the horsepower that would have been used for down feeders.

These presses use hydraulically operated chokes that automatically exert uniform choking pressure. The position of the choke "floats" in balance to the force exerted on the hydraulic side of the piston and the force exerted by the material exiting the press. Hydraulic oil pressure is selected by the pressroom operator and is held constant by the hydraulic system. Any deviation of force exerted by the material against the choke is compensated for by a shifting of the choke position because the oil pressure is held constant. The result is that the choke "floats" so as to permit uniform oil pressure, and, at the same time, the back pressure against the choke is

Fig. 1.5. Dupps Pressor® Press (Courtesy The Dupps Company).
(Pressor® is a registered trademark of The Dupps Company.)

also held uniform. This helps to prevent swings of motor load as conditions in the incoming feed material change. Keeping the motor load uniform, while the feed flow remains constant, also helps to keep a more uniform product discharging from the screw press and reduces operator attention to prevent motor overload.

The largest Dupps screw press is Model 12-12-6 (having a 12"diameter feed hopper with a 12" diameter main drainage cage that is 6 feet long) equipped with a 300 HP motor, and is rated for full pressing soybean at 59 MT/d pressing to approximately 6% residual oil.

The smaller Model 10-10-6 screw press, with a 200 HP motor, can press canola that has been prepared through a slotted-wall extruder at capacities of up to 45 MT/d, pressing to 9% residual oil. The Model 10-10-6 press is also running successfully on camelina seed prepared through a slotted-wall extruder at 45 MT/d, pressing to 8% residual oil.

Dupps presses are equipped with shrouds to prevent oily mists from escaping into the press room. The captured mist can be directed thorough scrubbers before discharge into the atmosphere, which will greatly reduce the plant's impact on the environment.

The French Oil Machinery Company

The French Oil Mill Machinery Company (Piqua, Ohio, USA) manufactures a broad line of screw presses and other machinery for processing a wide variety of oil bearing materials. Already well known for their ruggedly built machines that can achieve high performance and long service life, French Oil is also addressing green oil processing concerns.

Most of their presses already feature water-cooled main worm shafts and water-cooled inserts on the main cages to prevent overheating the oilseed during pressing;

this helps assure higher quality cake and oil. Their presses utilize motorized force feeders to expel air from the feed stream, prevent bridging in the feed hopper down-spout, and ensure 100% feed flow entering the feed hopper and cage.

French Oil has a newer line of presses, the Achiever Press, with five models ranging in capacities from 12 to 130 MT/d. These presses can produce press cake with residual oils ranging from 2.5 to 9.0%, depending upon the size of the press in operation and the seed processed. At the other end of the French spectrum, French Oil offers a Laboratory Screw Press, the Model L-250, with a processing range of 65–120 kg/hr.

The French® Achiever 66 (Fig. 1.6) is the largest full press presently offered by French Oil. It is capable of processing up to 130 MT/d of canola, producing press cake with residual oil in the 5–9% range. The Achiever 66 press is typically installed with a 500 HP main drive motor to ensure maximum capacity and oil extraction, and consumes 400 HP when pressing 130 MT/d canola to 5–9% residual oil. This press is equipped with a 25 HP vertical force feeder, and it has water-cooled main cages and a water-cooled main shaft to help raise the organically beneficial nutrition levels of the meal. (French® is a registered trademark of the French Oil Mill Machinery Company.)

The Achiever presses are equipped with sealed cage covers surrounding the cage drainage area, and the covers can be supplied with aspiration vents over the main cage and the discharge cone (choke) mechanism to allow for aspiration of oily mist away from the screw press and press room. The oily mist can be directed to a vapor scrubber system to ensure low environmental impact by the crushing plant.

French Oil also provides high efficiency motors on their presses, and endeavors to design their press worm shaft assemblies to obtain desired capacity and lowest

Fig. 1.6. French Oil Achiever screw press (Courtesy The French Oil Mill Machinery Company).

possible residual oil in the press cake at the lowest possible HP/ton ratio so as to minimize energy consumption in obtaining the oil.

French Oil has also simplified time-consuming maintenance procedures for replacing wear parts by designing the split cages, the cone mechanism, and the worm shaft to permit easy maintenance on these assemblies.

Harburg-Freudenberger

Harburg-Freudenberger Press+LipidTech, the division of Harburg-Freudenberger Maschinenbau GmbH (Hamburg, Germany) that deals with oilseed processing equipment, offers several models of screw presses. The largest full press is a newer version of the Model EP-16 full press designated as the SP 280 F Press (Fig. 1.7). The Model SP 280 F press can be used in a standard single-step pressing operation on flaked and conditioned seed. It can be used in combination with seed pre-treated by extruders or as the final press in a Harburg-Freudenberger two-step pressing process. This unique Harburg-Freudenberger two-step process consists of two consecutive conditioning and pressing steps without expensive and maintenance intensive flaking of the seed or press cake prior to pressing. This system is very flexible and can be applied to many oilseeds without the requirement of mechanical preparation equipment.

Depending on the preparation process, the seed capacity of the press ranges up to 125 MT/d, producing cake with residual oil content of 5–8%. Typical main drive is 250 kW. All Harburg-Freudenberger screw presses feature specially designed fixed throttle rings that serve as the choke in contrast to the mechanically adjustable chokes in their earlier presses. The throttle rings alternate with the pressure worms that build

Fig. 1.7. Model SP 280 F Press (Courtesy Harburg-Freudenberger GmbH).

up pressure followed by alternating conical throttle rings that contribute to relaxation of pressure between the worms.

Harburg-Freudenberger presses are designed for easy access to all wear parts. There is no support bearing at the discharge end of the shaft. This "free floating" shaft is supported by the oilseed material that exits the press. This results in a continuous balancing of the radial forces active on the cage support structures. These free-floating shafts, as well as shafts supported by bearings at the discharge end, do not require force feeders: they are designed to accept desired capacity entirely by free-falling feed material dropping onto the feed worm of the shaft. The absence of force feeders contributes to less cost, less maintenance, and to less down time to repair force feeders. And the absence of a force feeder motor complies with green processing expectations for the use of less energy.

To minimize oily mists escaping from the press and, therefore, have less impact on the environment, Harburg-Freudenberger SP 280 F presses have inspection doors that are equipped with labyrinth seals to help keep the press environment clean. These presses also have flange connections to operate in combination with an aspiration system, which is recommended to be connected to the corresponding flanges at the press frame.

While the earlier EP 16 is still considered state-of-the-art pressing equipment, Harburg-Freudenberger is continuously striving for further improvement, not only in mechanical details but also seeking new, more energy efficient direct drive solutions. The newer SP 280 F can be driven by a standard motor that is flanged directly onto the press housing through a dampened coupling element and a planetary gear reducer. It can also be driven through an innovative low-noise CMG drive which consists of four three-phase asynchronous motors that directly form a unit with the gear reducer. The CMG drive lends itself to green processing expectations. The SP 280 F is a press aimed not only at better and more efficient screw pressing but also at addressing green processing expectations.

Insta-Pro

Insta-Pro International (Urbandale, Iowa, USA) was founded in the 1960s. The company began with a line of extruders to cook soybean. Later they developed a line of screw presses, which, when combined with their extruders, are offered as their ExPress® oil extraction system (ExPress® is a registered tradename of Insta-Pro International). New company ownership in mid-2008 spurred increased efforts in R&D, leading to improved products and processes that are chemical free and environmentally friendly. Today Insta-Pro offers their 600 Series extruders, which are their smaller extruders, and their larger 2000 Series extruders.

They also offer several lines of screw presses: the Model 1000, Model 2000, and Model 5005 presses. The Model 5005, when combined with two Model 2000 Series extruders, can extrude and press oilseeds such as soybean, cottonseed, and canola at capacities of 4,000 lbs/hr, pressing to as low as 5–8% residual oil.

Optimizing Energy Use

A major consideration in green processing is optimization of energy used in processing plants. Anderson International has for a number of years been concerned with this issue. Ongoing improvements in equipment and procedures for preparation led to a newly developed Combo System™ (Fig. 1.8 and Fig. 1.9). This system uses a specially modified extruder built on the same base as the Victor presses, and uses some

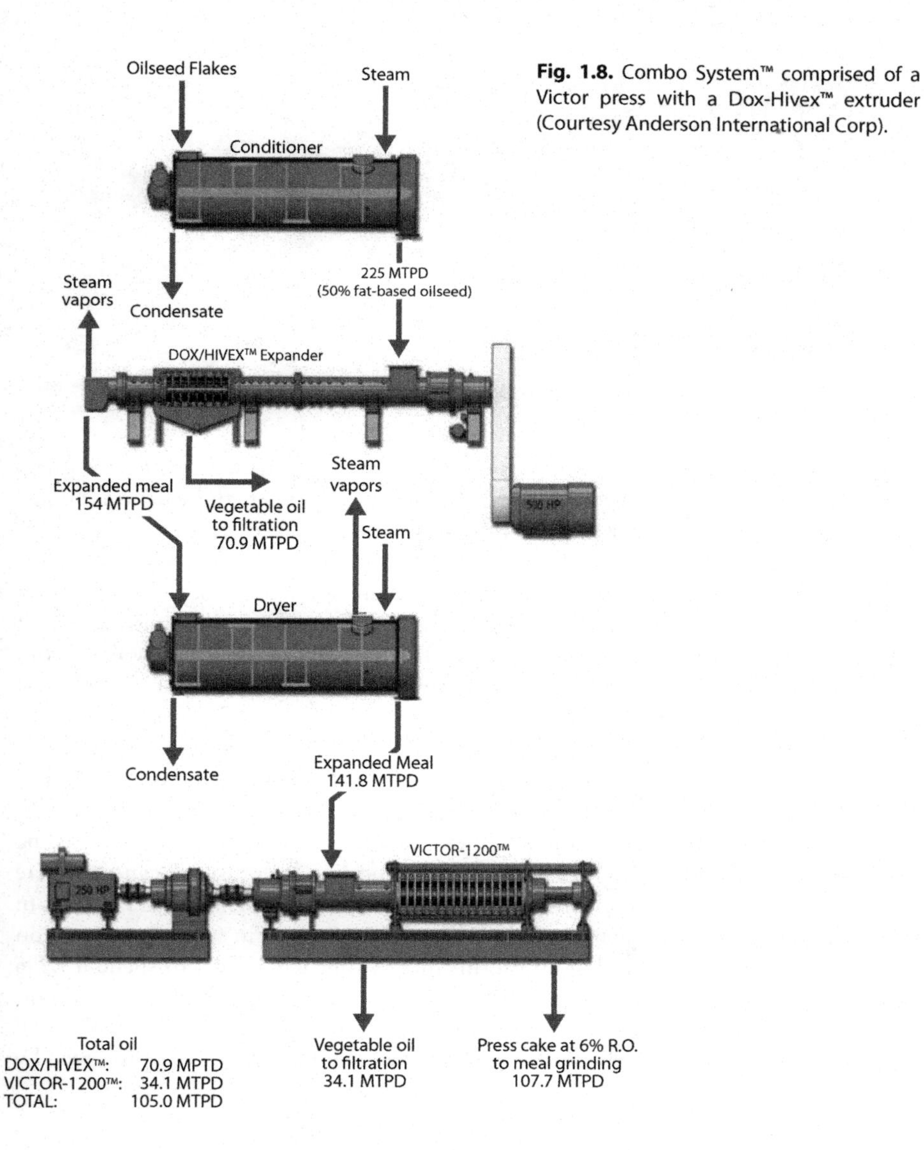

Fig. 1.8. Combo System™ comprised of a Victor press with a Dox-Hivex™ extruder (Courtesy Anderson International Corp).

Fig. 1.9. Dox-Hivex™ extruder (Courtesy Anderson International Corp).

of the same components, such as feed barrels and drainage cages, as the Victor press. These extruders come in three sizes: 12", 10", and 8" diameter, and each can be retrofitted as a solid-walled extruder, a slotted-walled extruder, or can contain a shear inducing head using a conical discharge port (Fig. 1.10).

Anderson International has studied the overall cost benefit of using extrusion as pretreatment rather than conventional atmospheric cookers. A big advantage of extrusion is the additional attrition (shearing) of the oilseed afforded by extrusion. The results are tabulated in Table 1-A, which summarizes the benefits for a 200 MT/d processing line on canola.

The costs are based on purchase price and energy consumption of multiple pieces of equipment in the processing line. Some of the equipment will be used in the conventional line; others will be used in the Combo line, as can be readily seen in Table 1-A on page 16. There is a large difference in capital cost: $4,405,060 for the conventional line versus $2,480,230 for the Combo line for crushing canola. There is more connected horsepower in the Combo line (1,462.5 versus 1,142.5) than in the conventional line, but there is a very substantial reduction in steam consumption using the Combo line. Summing up the advantages, one can expect to spend 43.7% less in capital equipment costs for a Combo line, and reduce steam consumption by 58.8% in a Combo line even though the Combo line requires 21.4% additional connected horsepower. Taken together, this will reduce the cost per ton processed by 29.7%.

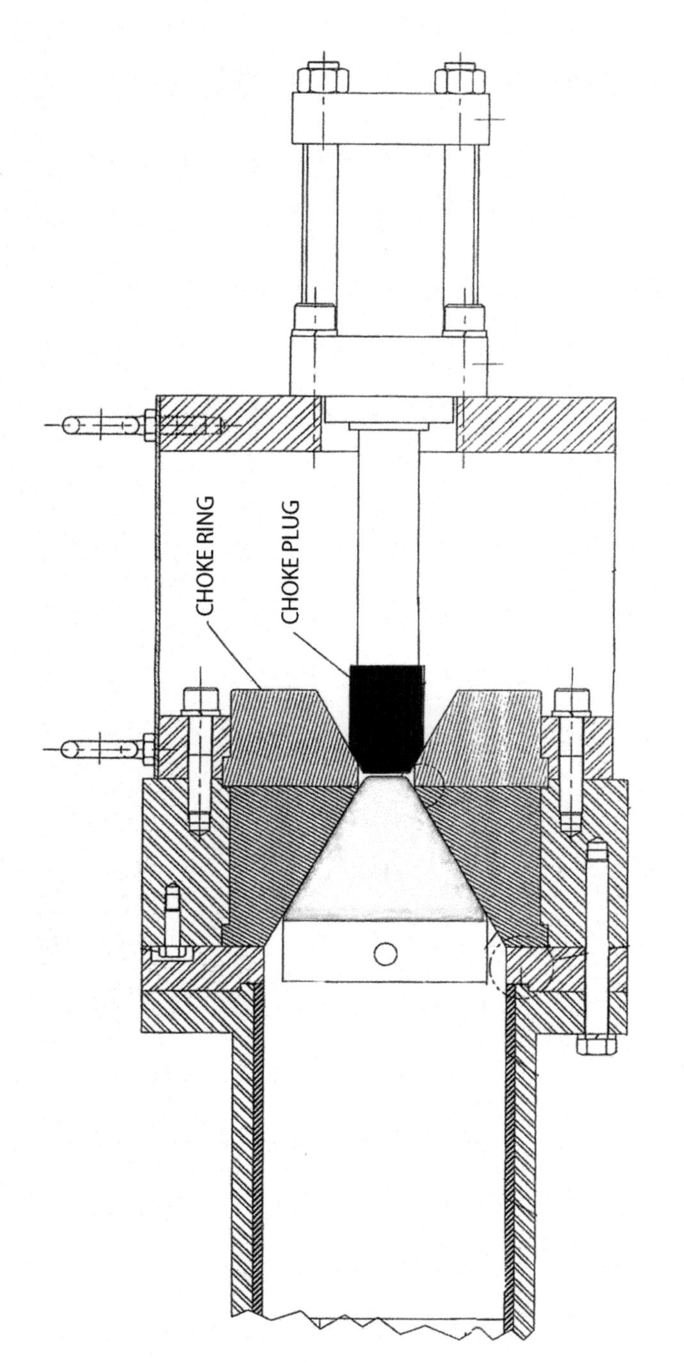

Fig. 1.10. Shearing head (Courtesy Anderson International Corp).

Table 1-A. Canola Plants (Courtesy Anderson International Corp).

		200 Tons per Day Canola							
		Capitol Costs and Energy Consumptions—May, 2007							
		Conventional Process				Dox/Hivex™ Process			
				Energy Consumptions				Energy Consumptions	
Item	Description	Qty	Total Price	HP	Steam	Qty	Total Price	HP	Steam
1	Cleaners	1	$ 45,300.00	2.0		1	$ 45,300.00	2.0	
2	Preconditioners	2	$ 106,000.00	17.0	800	2	$ 106,000.00	17.0	800
3	Flaking Rolls	1	$ 254,000.00	125.0		1	$ 254,000.00	150.0	
4	Cracking Rolls	0	$ —			0	$ —		
5	Cookers	3	$ 159,000.00	25.0	1200	0	$ —		
6	Dryers	5	$ 265,000.00	42.5	2000	0	$ —		
7	Conditioners	8	$ —	40.0	1600	0	$ —		
8	Expanders	0	$ —			4	$ 92,500.00	800.0	
9	Preconditioners	0	$ —			3	$ 159,000.00	25.5	1200
10	Expellers	8	$ 2,320,000.00	648.0		3	$ 645,000.00	273.0	
11	Screening Tanks	1	$ 55,000.00	2.0		1	$ 55,000.00	2.0	
12	Filter Presses	1	$ 91,710.00	0.0		1	$ 91,710.00	0.0	
13	Heat Exchanger	1	$ 16,950.00	0.0		0	$ —		
14	Cooling Tower	1	$ 27,500.00	7.5		0	$ —		
15	AN-195 Cooler (w-F&C)	2	$ 99,320.00	20.0		2	$ 99,320.00	20.0	
16	AN-197 Fan & Cycl.	2	$ —			2	$ —		
17	Boiler	1	$ 176,400.00	5.0		1	$ 149,900.00	5.0	
18	Lot Screw & El Conv.	1	$ 256,700.00	80.0		1	$ 256,700.00	80.0	
19	Metering Bin	1	$ 70,800.00	7.5		1	$ 70,800.00	7.5	
20	Overflow Bin	1	$ 23,400.00	3.0		1	$ 23,400.00	3.0	
21	Cake Bn	1	$ 47,500.00	5.0		1	$ 47,500.00	5.0	
22	Cooling Water Pump	1	$ 2,740.00	20		0	$ —		
23	Circ. Oil Pump	1	$ 2,740.00	20		0	$ —		
24	Oil Unloading Pump	1	$ 1,900.00	8		1	$ 1,900.00	8	
25	Cake Grinder	1	$ 36,200.00	40.0		1	$ 36,200.00	40.0	
26	Air Compressor	1	$ 9,800.00	25.0		1	$ 9,800.00	25.0	
27	Oil Meter	1	$ 4,200.00	0.0		1	$ 4,200.00	0.0	
28	Filter Press Platform	1	$ 52,000.00	0.0		1	$ 52,000.00	0.0	
29	Oil Work Tanks	3	$ 32,200.00	0.0		3	$ 32,200.00	0.0	
30	Expeller® Discharge Chutes	8	$ 7,200.00	0.0		7	$ 6,300.00	0.0	
31	Lot Elect. Equip.	1	$ 140,000.00	0.0		1	$ 140,000.00	0.0	
32	Lot Process Pipes	1	$ 32,000.00	0.0		1	$ 32,000.00	0.0	
33	Engineering Services	1	$ 69,500.00	0.0		1	$ 69,500.00	0.0	
			$ 4,405,060.00	1142.5	5600		$ 2,480,230.00	1462.50	2000

Capital Investments			**Steam Energy**		
Difference ($)	$1,924,830.00	Savings	Difference (LBS/HR)	2612	Savings
% Difference	43.7%		% Difference	58.8%	
Electrical Energy			**Cost/Ton Production**		
Difference (KWH)	207.4	Additional Energy	Difference ($/Ton)	$8.49	Savings
% Difference	21.4%		% Difference	29.700%	

Conclusions

In conclusion, this chapter addressed concerns for operating extruders and screw presses within constraints of green processing and within the constraints of organic foods. The green processing goal of having little or no impact on the environment should favor mechanical crushing and discourage solvent extraction because of solvent losses in effluent air and water streams. Today, most vegetable oils serving the food market are separated by solvent extraction because solvent extraction permits large capacity plants and maximum separation of oil. However, concerns about foods exposed to hexane and other chemicals may encourage food processors to favor some method other than solvent extraction where no hexane or other chemical solvent will be introduced to the food material being processed. Mechanical screw presses can meet this requirement, but screw presses have traditionally been low capacity machines compared to solvent extractors and consume more horsepower per ton of oil extracted than solvent extractors. The authors believe that concerns related to green processing and organic feeds might become an incentive for equipment manufacturers to design screw presses for higher capacities, pressing to lower residual oils and consuming less horsepower. Significant work is already being done in these areas.

Today's higher capacity full-presses operate at around 130 MT/d and produce residual oil levels of approximately 8%. A modest first step goal for some manufacturers is to supply full presses that can achieve 150 MT/d on canola pressing to 3% residual oil. The authors believe that achieving this goal is just around the corner.

Note: "Expeller" is a registered worldwide trademark of Anderson International Corp. since 1900. Also there are some names that use the suffix™. They are not registered the same as trademarks but are used in business or trade to designate the business of the user and to which the user asserts a right to exclusive use.

References

Farnsworth, J.; Johnson, L.; Wagner, J.; Watkins, L.; Lusas, E. Enhancing Direct Solvent Extraction of Oilseeds by Extrusion Preparation. Oil Mill Gaz. 1986, 91, 30, 32–35.

Green Process, University of Western Ontario http://www.eng.uwo.ca/chemical/green-new.htm.

Nelson, A.; Wijeratne, I.; Yeh, W.; Wei, T.; Wei, L. Dry Extrusion as an Aid to Mechanical Expelling of Oil from Soybeans. JAOCS 1987, 64, 1341–1347.

Wikipedia, Organic Foods. http://www.wikipedia.org/wiki/Organic food.

Williams, M. Extraction of Lipids from Natural Sources; Gunstone, F. and Padley, F., Eds.; Lipid Technologies and Applications. Marcel Dekker, Inc.: New York, 1997; pp. 113–135.

Williams, M. Preparation of Oilseeds to Improve Extraction of Fats. Extrusion Communique, 1993, 6, 12–14.

Williams, M. The Recovery of Oils and Fats from Oilseeds and Fatty Materials; Shahidi, F., Ed.; Bailey's Industrial Oil and Fat Products, Sixth ed., Vol. 6; John Wiley & Sons: New York, 2005; pp. 2589–2678.

2

Modern Aqueous Oil Extraction—Centrifugation Systems for Olive and Avocado Oils

Marie Wong[1], Laurence Eyres[2], and Leandro Ravetti[3]
[1]Institute of Food, Nutrition & Human Health, Massey University, Auckland, New Zealand, [2]ECG Consultants, Auckland, New Zealand, and [3]Modern Olives, Lara, Victoria, Australia

Introduction

Olive oil is extracted from the flesh of the olive fruit. The flesh at maturity contains 20–25% oil (by fresh weight), while the pit typically contains less than 1% oil (by fresh weight). Similarly, for avocado fruit (e.g., "Hass" cultivar), the flesh at ripe maturity contains up to 32% oil (by fresh weight), and the seed or pit has approximately 2% oil (by fresh weight). Avocados have primarily been grown for consumption of the edible fruit; the lipid material from the flesh was not normally extracted. Once extracted, both olive and avocado oils can be consumed with little further processing. This results in the retention of color, distinctive flavors, vitamins, and nutrients.

History of Olive Oil Extraction

Extraction of oil from the olive fruit for use as food dates back to 2000 BC (Amouretti, 1996) when oil was extracted by mechanical means involving crushing, pressing, and decantation (separation of the oil and vegetable water phases). Olives were crushed by hand using simple stone implements, the paste obtained was further pressed to expel the oil and then hot water was added to aid the separation of the oil from the vegetable matter. It has been well reported that during the Bronze Age stone rollers/millstones were built and used for crushing the olives. After crushing, the olives were placed under large mechanical presses in order to use pressure to expel the oil from the flesh. To aid with the decantation process water was added to the extracted oil as the oil would float and could be separated from the top of the earthenware jars (Amouretti, 1996). These three steps are still the key principles used in olive oil extraction to this day. There have been continual improvements to each step of the process to ensure oil of high quality not necessarily greater oil yields, though this is still important (Amouretti, 1996; Kiritsakis, 1998). Examples of improvements include the implementation of rapid processing soon after harvesting or improvement of press designs. More recently, improvements to the pressing and oil separation step and decantation/centrifugation have been implemented due to new equipment and technology.

Crushing results in the formation of a paste containing plant and oil cells. The oil is released from the oil cells by mixing and the application of pressure. The resulting oil is then separated from the vegetable material by decantation and in more recent times with the application of centrifugation to separate the various phases, making the process continuous and less labor intensive (Petrakis, 2006). The extraction process must produce oil that is of the same composition and sensory attributes as found in the fruit. To achieve this, oil must be extracted at no higher than 27°C to avoid any thermal degradation, and no additives can be used (European Commission, 2002). The key objective during virgin olive oil processing is to extract the maximum oil yield from the fruit while maintaining oil quality and important sensory and composition characteristics.

The recovery of oil from the olive pomace produces oil of lesser quality compared to the first press in terms of sensory attributes and quality, with higher free fatty acids. Pomace oils can be distinguished from virgin oils by their chemical composition as the second extraction process results in the extraction of previously unrecovered minor components. This oil is either sold as olive pomace oil or is further refined using standard oil refining techniques (Fedeli, 1996; Kiritsakis, 1998; Petrakis, 2006).

Development of Avocado Oil

In contrast to olives, avocados are primarily grown for fresh fruit consumption. Avocado production worldwide has increased by 52% since 1999. With this growth in production there is a concurrent supply of reject grade fruit for which growers would like to find a market. There is growing competition for this reject fruit for culinary oil and for guacamole products made possible with the development of high pressure processing technologies. Over 90% of the world's production of avocados is of the cultivar "Hass." The flesh of a "Hass" avocado can contain as much as 32% oil (based on fresh flesh weight) depending on the time of harvest. This is a maximum, and the average at the peak of the processing season is approximately 18%.

Avocado production in New Zealand has increased from 1,000 tonnes in 1987 to 26,766 tonnes in 2007 (Requejo-Tapia, 1999; Plant & Food Research, 2008) with considerable potential for future increases in production. Most first grade avocados are exported overseas while the New Zealand market absorbs the export rejects, rejects only in the sense of cosmetic appeal (size, shape, and color). Production levels in New Zealand vary year to year depending on the season (some trees bear fruit biennially), weather (wind and storms can damage fruit, which are unsuitable for fresh fruit export), and export markets. Over the 2008/2009 season, New Zealand processors produced more than 150,000 liters of extra virgin avocado oil, with production expected to increase in the 2009/2010 season. Extra virgin avocado oil is also being produced in Mexico, Chile, South Africa, and Kenya (Wong et al., 2010).

The use of avocado oil for skin and cosmetic uses has been reported as early as the 16th century, but it was not until the 21st century that commercial operations began to extract avocado oil for culinary uses. Early attempts to extract avocado oil involved dehydration first of the flesh and also the use of solvents to maximize oil yields (Woolf et al., 2009). Early in the 21st century two commercial producers in New Zealand began extracting avocado oil using a process similar to the extraction of olive oil. Earlier, olive oil extraction involved crushing and pressing but the pressing step was made continuous with the introduction of the three-stage decanter in 1970 by Alfa Laval (Alba et al., 2011). For large operators, the use of horizontal decanting centrifuges is now standard and final polishing of the oil was achieved with vertical disc centrifuges. Avocado oil processors in New Zealand adapted the process for olive oil extraction by taking advantage of the continuous process afforded by horizontal decanters. The process requires minimal addition of water and no other additives are required. The goal of this extraction process for culinary avocado oil is the same as for olive oil: maximum oil yields are desired while maintaining oil quality, sensory and compositional characteristics with no subsequent processing or refining required.

Olives and avocado fruit contain high levels of water, 50 and 65% (by fresh weight), respectively (Kiritsakis, 1998; Woolf et al., 2009). Advances in the extraction process for olives led to the development of continuous processing with minimal energy usage and reductions in waste streams. Where oil extraction from other oil bearing fruits and seeds may require pre-drying or digestion steps prior to oil extraction, the oil extraction process from olives and avocados uses the presence of water to aid with separation during the decantation steps and does not require high temperatures to encourage the release of oil from the oil bearing cells. However, too much water can lead to emulsion problems, which will be discussed later in this chapter.

Standards for Virgin Oils

The International Olive Council (IOC) was set up in 1959 under the auspices of the United Nations. The IOC standards for olive oil, also adopted by the European Commission regulations and Codex Alimentarius for food, define the various grades for olive oil. Virgin olive oil is oil extracted solely from the olive fruit by physical or mechanical means, involving the use of washing, crushing, decanting, centrifugation, and filtration. The process does not alter the oil found in the fruit, and the process must be controlled thermally without the use of solvents, esterification, or other oils (International Olive Council, 2011). Virgin olive oils are fit for consumption immediately after extraction. There are various classes of olive oils ranging from extra virgin olive oil (the highest quality) to ordinary virgin olive oil. For oils to fall into each class, they must meet standards based on chemical quality, chemical composition, and sensory evaluation. After virgin olive oil, there are also standards for olive pomace oil, extracted from the olive pomace, both refined and unrefined (International Olive

Council, 2011). Suggestions that the IOC standard does little to prevent old and stale oils being sold on export markets led to the development of improved standards by the USA in 2010 and by Australia in 2011 (USDA, 2010; Standards Australia, 2011).

Guided by the standards for virgin olive oil, standards have been proposed for avocado oil produced by the equivalent cold pressed extraction procedure (Wong et al., 2010). For avocado oil, the procedure allows for washing, grinding, and separation by decantation. The maximum temperature for avocado oil extraction is 50°C which does not lead to any alterations in the oil, and still allows the oil to be classified as extra virgin or virgin as long as it meets the standards proposed for these classes. For each class the avocado oil must meet chemical quality, composition, and sensory standards. It will be discussed subsequently in this chapter, but, due to the different attributes of avocado oil, it can be extracted at higher temperatures than olive oil and results in no chemical alteration of the oil found in the fruit.

This chapter will focus on the production of virgin oils from primarily olive and avocado by using only aqueous extraction methods which use mechanical and physical means to release the oil from the fruit as briefly described above. The goal of the extraction processes for olive and avocado oils is to produce high quality oils ready for consumption requiring no further processing. Any poor quality oils produced are generally sent for further refining, bleaching, and deodorizing and sold as lower grade oil. These subsequent refining processing steps will not be covered in this chapter.

Centrifugation Systems for Olive Oil

Overview

Virgin olive oil is the oily juice obtained from olives, which is separated from the other components of the fruit solely by mechanical means. When obtained by using appropriate processing techniques and from high quality fresh fruit that have no defects or alterations and which are at the suitable stage of ripeness, the oil can have exceptional chemical and organoleptic characteristics. It is one of the few vegetable oils that can be consumed in its raw state, preserving its fatty acid composition and minor component content, which are extremely important for health and nutrition, particularly its vitamins and antioxidant content.

Unfortunately, not all virgin olive oils produced in the world fulfill the above conditions. Significant amounts of this product will be refined as a result of their unpleasant organoleptic characteristics, and/or high acidity and other chemical and physical parameters. Experience has shown that virgin olive oil spoilage occurs almost exclusively as a result of deficient fruit handling and poor processing methods.

Olive oil represents approximately 2% of the total world oil and fat production by volume and 11.2% by value. This higher value is in recognition of its unique organoleptic, chemical, and physical characteristics. It is precisely due to these characteristics that virgin olive oil is considered a true fruit juice, and that is how it needs to be treated throughout the entire process: from olive harvest through processing and storage. Each step must be performed with the utmost care in order to produce the highest possible quality extra virgin olive oil.

Process Flow Diagram

Virgin olive oil quality is impacted by a number of factors and circumstances, the most significant of these include olive variety, climate, soil type, crop age, and management systems. The oil extraction technology is a fundamental step to obtaining good quality virgin or extra virgin olive oil. The main operations in the virgin olive oil manufacturing process are shown in Fig. 2.1.

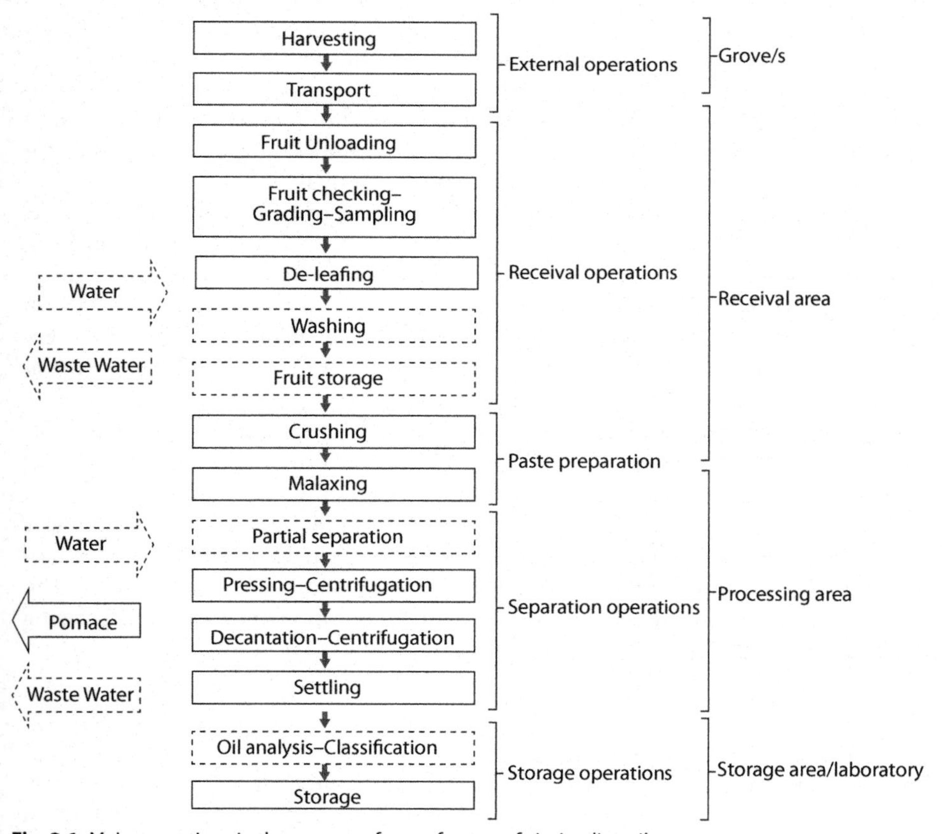

Fig. 2.1. Main operations in the process of manufacture of virgin olive oil.

External Operations

Although picking or harvesting is an independent operation from the manufacture of olive oil, it has a significant impact on oil yields and characteristics. From the processor's point of view, two factors need to be taken into consideration at harvest time: timeliness and the harvesting system to be used.

Many aspects need to be considered simultaneously while deciding when to harvest the fruit. To determine the optimal harvesting time, the following points need to be evaluated: the olives must have the maximum weight of oil; the quality of the oil must be optimal; fruit and tree damage must be minimal; next year's crop must not be affected; and harvesting must be as cheap as possible. The moment at which the olives have the maximum amount of oil of the highest quality varies according to environmental conditions, varietal characteristics, and the amount of fruit per tree. It is commonly said that this critical point is reached once there are no more green olives on the tree and the majority have changed color. However, this rule can lead to harvesting the fruit at other than the optimal time when implemented under different environmental conditions. Continuous monitoring of oil and moisture content in the fruit as well as average fruit retention force and natural fruit drop are key management tools to precisely determine the optimal time to harvest. The utilization of more than one variety with different ripening periods allows the grower and processor to extend their season without compromising yields or quality.

Regarding the harvesting system, it is important to bear in mind at all times that olive oil is like a "fruit juice," so that the chosen system must not spoil the olives, cause any lesions, bruising, breakage of branches, or soft shoots. Almost all forms of harvesting, particularly mechanical, lead to some form of damage to the olives. In these circumstances, it is critical that the harvested fruit is processed within a very short time frame. The ideal processing plant will extract the oil at the same time as fruit is being harvested, as this will achieve the highest olive oil yield with the same characteristics as in the fruit at the time it is received at the mill. This perfect coordination is very difficult to obtain and is only possible for small mills or some of the large scale modern groves where mechanical harvesting and processing occur 24 hours a day, nonstop, and where processing equipment and harvesting capacity have been designed to match, leading to a fully coordinated single operation. If olive harvesting and processing are not synchronized correctly, some of the fruit will need to be stored. The storage will vary according to the conditions in each olive growing area.

Receipt Operations

Fruit Receipt

In order to maintain the quality of the oil present in the fruit, the processing plant needs to manage the receipt of the fruit in an effective and careful way. It is important that the processing plant receipt area is designed with sufficient room and capability to receive, classify, and, if necessary, separate the different olives that arrive for crushing.

Any type of production control requires that fruit delivered is weighed either by utilizing a weighing bridge or an in-line set of scales. Each batch of fruit must be carefully sampled and evaluated by a responsible and trained receipt area manager. Some of the elements that should be evaluated in order to properly classify the fruit are: fruit ripeness, level of damage, incidence of pests, diseases, or other disorders; fruit temperature at depth in the bin; and presence of foreign materials or materials other than olives.

De-leafing and Washing

Olives arriving at the mill carry a variable percentage of foreign matter of very diverse origins, including soil, stones, leaves, timber, weeds, and steel. In order to obtain good quality oil and to prevent the impact of these products on the oil's organoleptic characteristics and to minimize machinery wear and tear or damage, it is important to eliminate as much as possible, if not all, of this foreign matter. For this purpose, air cleaners are used to remove objects that are lighter than the olives and water washers are used to dissolve and/or remove heavier materials.

When olives have been harvested directly from the tree and do not present significant quantities of dirt or other heavy materials, it is possible to bypass the washer. The main advantages of bypassing the washer are: minimizing the addition of moisture that could lead to emulsion problems in the extraction phase, maintaining higher levels of polyphenols, and improving overall processing plant efficiency without compromising oil quality. In the case of olives harvested mechanically, the incorporation of metal detectors and/or in-line magnets will be required to protect crushers when olives are not washed.

Fruit Storage

As discussed previously, oil should ideally be extracted at the same time as the fruit is being harvested. As this requires perfect coordination, which is very difficult to obtain, some of the fruit will need to be stored prior to its processing. Fruit storage can be performed before or after de-leafing and washing. When it is performed before, it is common to place the olives in bins, bags, or in piles for hours or even days until they are finally processed. When fruit storage is performed after the washing process, the most common storage method is placing the olives in large above ground stainless steel hoppers that will then feed the olives into the crushing systems for 12 to 24 hours continuous operation. New modern processing plants only have small clean fruit hoppers for buffering purposes so the continuous process is not affected by small issues associated with fruit delivery. Those hoppers would only hold enough fruit for a maximum of 2 to 3 hours of processing.

Preparation of the Paste

The purpose of olive oil extraction techniques is to separate the continuous oily phase from the other olive fruit components, without altering its composition or its organoleptic characteristics. The steps in the oil making process itself are described below.

Crushing

The first step in obtaining olive oil is the crushing of the olives to tear apart the cellular structure of the olives in order to release the oil accumulated inside them. The cutting action applied during this process ruptures membranes, and frees oil droplets contained inside the fruit cells. These freed oil droplets gradually come together, coalescing to form droplets of many different sizes, which come into direct contact with the aqueous phase in the paste. The aqueous phase consists of the vegetable water from the olive and from the water left over from the washing process. Lipoproteic membranes usually form with the dissolved proteins in the water, giving the oil droplets considerable stability and sometimes forming emulsions.

Crushing plays a crucial role in the oil extraction process because the type of crusher or grinder used and the degree of crushing will have a clear influence on the rest of the processing phases, extraction efficiencies, and on oil quality and characteristics. For centuries, this step was performed by stone mills. Today, modern plants have replaced this traditional system with different kinds of mechanical grinders, with hammer mills being the most popular. These grinders consist of a set of high-speed rotating hammers that smash the fruit through a grid, turning the fruit into a uniform paste. This grid comes with different sized apertures, typically ranging from 5 to 7 mm. The choice of the aperture size is usually related to the size of the fruit, the ripeness of the olives, and their moisture content.

Malaxing

All crushing systems used will require a complementary malaxing or mixing process in order to form a larger continuous oil phase from the oil droplets found throughout the paste before proceeding to the olive oil separation phase. By malaxing the olive paste slowly, the oil droplets merge more easily into larger droplets and form a continuous oil phase. Table 2-A shows the changes in oil droplet size before and after malaxing (Di Giovacchino, 1988). It must be noted that there are always drops of oil in an emulsion state or "occluded" among the paste solids that cannot be separated (Alba et al., 1982).

Temperature is a significant factor to be taken into account during the malaxing process. During this phase, malaxers usually employ a heating system consisting of a

Table 2-A. Changes in Oil Droplet Size During Malaxing (Di Giovacchino, 1998).

	Diameter of olive oil droplets in microns					
	<15	15–30	30–45	45–75	75–150	>150
Before malaxing (%)	6	49	21	14	4	6
After malaxing (%)	2	18	18	18	19	25

double water heating jacket. Many positive changes are triggered by higher malaxing temperatures; oil viscosity decreases and higher extraction efficiencies can be achieved during the separation process. Nonetheless, heating the oil, particularly in excess, can cause alterations in oil quality. Various authors mention temperatures ranging between 20 and 35°C as a normal range, but it is important to remember that oxidation processes double their rate with every 12°C increase in paste temperature, and the volatile compounds that give olive oil its aroma are easily degraded at high temperatures. Higher malaxing temperatures typically lead to oils with higher polyphenol content, with increased bitterness and darker color as a consequence of a higher content of chlorophyll and carotenoids (Uceda et al., 2006). Within the European Union, designations of "first cold pressing" and "cold extraction" can only be used for virgin olive oil obtained below 27°C (European Commission, 2002). Some researchers have questioned the relevance and accuracy of such a definition (Boselli et al., 2009).

Time is another important aspect of the malaxing process. Although it is generally accepted that longer malaxing times can lead to higher extraction efficiency, various authors indicate that malaxing times between 45 and 90 minutes would be a good compromise between yield and oil quality, depending on fruit type, the style of oil to be produced, and the presence of emulsions (Solinas et al., 1978; Uceda et al., 2006). An excessive malaxing time may lead to a drop in polyphenol levels and a change in the volatile profile of the oil (Angerosa et al., 2001; Di Giovacchino et al., 2002).

Malaxers are usually classified into two types according to their mode of operation within the extraction system: continuous malaxers allow olive paste to continuously overflow from one vat to the next until processed through to the appropriate separation system; batch malaxers allow separate olive batches to be processed separately as there is no continuous connection between the vats. The first system is preferred for large operations working continuously with similar fruit cultivars and when the oil obtained from each fruit batch will ultimately be combined. The second system is more suited to operations processing smaller batches of olives when the oil from the malaxers is kept separate for different cultivars or individual growers.

Use of Processing Aids

Technological co-adjuvants added in the malaxing stage help separate the oil from the other components of the olive paste and thus increase extraction efficiency. In order to capitalize on the above mentioned advantages, it is important that adjuvants are used correctly and in appropriate quantities.

Processing aids have been used in the olive oil industry for more than 30 years to improve the oil's extractability from the olive paste. The introduction of processing aids in the oil extraction process was due to the difficulty found in extracting oil from the paste of certain olive cultivars that led to high oil losses in the pomace. Among these aids, talc and microtalc powder and enzymes have been the most commonly used and studied in past years in Spain, Italy, and other Mediterranean countries (Alba, 1982; Ranalli, 2003; Moya et al., 2010). Furthermore, some new processing aids and techniques like common salt, calcium carbonate, and hot water dipping have also been evaluated in recent works (Perez et al., 2003; Garcia et al., 2005; Cruz et al., 2007). Extraction efficiency improvements ranging between 10 to 30% have been recently reported with the individual and combined usage of these processing tools for olive pastes in those countries (Ranalli, 1997; Linares et al., 2006; Moya et al., 2010). In addition, the appropriate use of talc powder and enzymes has been reported to reduce the pollution potential of the processing waste water stream up to 30% (Ranalli et al., 2003). The regulations related to the use of the different co-adjuvants vary from country to country.

Separation Operations

Oil-paste Separation

The separation of the oil phase from the rest of the fruit components is the most critical phase of the oil extraction process. During this phase, operators try to separate the liquids contained in the olive paste, particularly the oily phase, from the solid components. This separation process may use different systems: selective filtration, pressing and/or paste centrifugation in three or two-phase decanters. Even though the traditional mat pressing system is still used in a number of mills around the world and selective filtration is utilized in small scale boutique mills, the vast majority of the olive oil manufactured at the present time is recovered by two or three-phase decanter centrifugation. When compared with the classical manufacturing method of mat pressing, centrifugation has advantages and disadvantages. Among the advantages that determine the dominance of this system in the industry are: compact set-up using a smaller area with significantly higher throughput; no mats are used; shorter assembly time; less labor intensive; allows for automation of the process; and produces consistently higher oil quality. Some of the disadvantages of this modern system that could be mentioned when compared with the traditional extraction process are: higher energy consumption; requires water to operate (although this water need is significantly reduced in the two-phase decanter); and greater financial investment in the overall installation (but not on a per tonne of capacity basis). Within the olive oil industry, olive paste centrifugation using centrifugal force is considered to be the most modern system to separate the solid phase of the fruit paste from the lighter liquid phases, particularly the oil.

The main device for any continuous oil centrifuge extraction operation is the horizontal centrifugal decanter (Fig. 2.2). This decanter consists of a cylindrical-conical rotating bowl and a helical hollow-axis screw rotating coaxially inside it and at a slightly different speed to the bowl. The olive paste is injected inside the decanter through an opening in the screw axis, and when centrifugal force is applied, the solids are displaced toward the wall of the bowl, and are slowly dragged toward the end by the differential speed between the auger and the bowl. The lighter oily phase forms an internal ring around the axis, and is drained through outlets placed at the opposite end of the pomace outlet. In the three-phase continuous centrifuge system, the water ring formed between the oil ring and the pomace ring is drained separately, while in

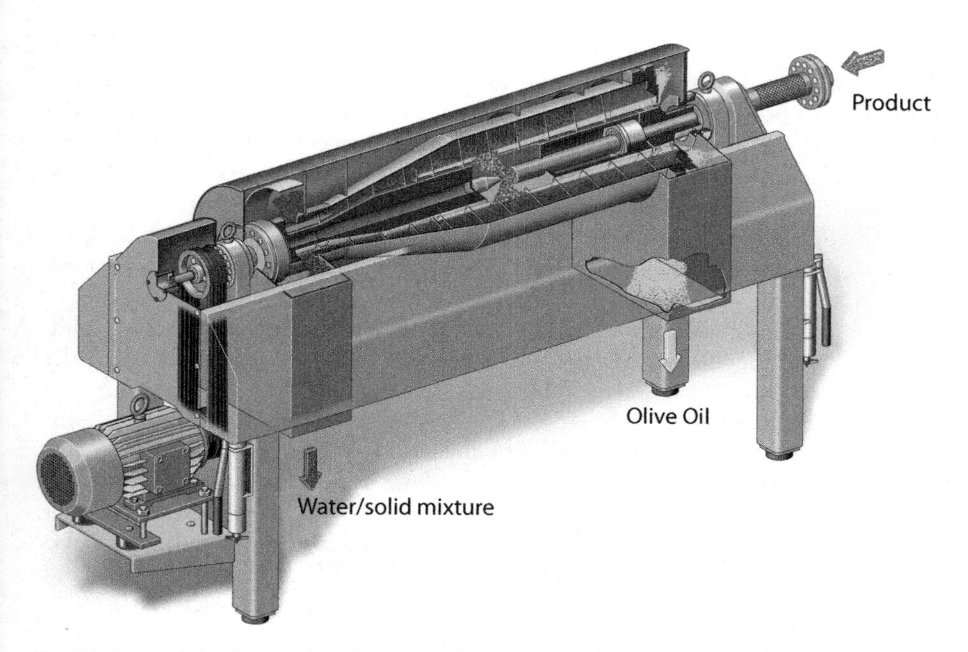

Fig. 2.2. An example of a two-phase horizontal decanting centrifuge (Image courtesy of GEA Westfalia).

the two-phase centrifuge system there are only two outlets, and both pomace and vegetable water come out together.

Table 2-B compares the aqueous phase production in the various manufacturing systems. The three-phase outlet centrifuge system was the first type of centrifuge to be used in the olive industry. It produces approximately 1.2 liters of final vegetable water per kilogram of olives, creating an effluent with a particularly high chemical oxygen demand, and if released into natural waterways, has a negative impact on the

Table 2-B. Comparison of Aqueous Phase Production in a Press, Two- or Three-Phase Horizontal Decanter.

	Aqueous phase produced		
		Centrifugation	
Processes	Press	Three-Phase	Two-Phase
Olive washing (l/kg)	0.04	0.09	0.05
Solid-liquid separation (l/kg)	0.40	0.90	0.00
Liquid-liquid separation (l/kg)	0.20	0.20	0.15
General cleaning (l/kg)	0.02	0.05	0.05
Final effluent (l/kg)	0.66	1.24	0.25

biological development of flora and fauna (Uceda et al., 2006; Alba et al., 2011). In order to mitigate this problem, centrifugation technology has evolved, and new equipment has been designed that is able to function with less fluid paste inside the decanter. In Spain, at the beginning of 1992, a demonstration was held showing the operation of a horizontal centrifuge that was able to produce virgin olive oil without adding water and without producing an aqueous phase in the decanter, drastically reducing effluent pollution (Uceda et al., 2006; Alba et al., 2011). Research in Spain has also demonstrated that alperujo, the solid by-product, can be reused as a fertilizer in organic farming.

It is obvious that both decanters operate with three internal phases (solid, water, and oil). Nonetheless, because the three-phase decanter had three outlets, the new decanter was then popularly and incorrectly known as a "two-phase" decanter by association with its two outlets (Uceda et al., 2006). Since then, the expression "two-phase" decanter has been widely adopted by the industry.

There are significant differences between both types of decanter centrifuges. Apart from the most obvious difference explained above, the significantly higher water usage in the three-phase decanters and the number of outlets, there are also differences related to the average characteristics of the oils produced by each, and in the solid waste (pomace) and liquid waste (vegetable water) characteristics (Alba et al., 1992; Salvador et al., 2003). In general terms, the oils produced with two-phase decanters tend to have higher polyphenol levels, higher bitterness levels, and longer shelf life than those produced with a three-phase decanter. The pomace from the two-phase decanter has higher moisture levels and is therefore more difficult to handle than the three-phase pomace. On the other hand, the larger volume of vegetable water produced by the three-phase process has a significantly higher contaminant level than the two-phase wastewater.

Oil-water Separation

Due to the nature of the extraction system, the liquid oily phase separated in the centrifuge decanter contains mostly oil and diluted water together with fine flesh particles of lighter densities. Consequently, there is a need to separate these three phases (aqueous, oil, and solid matter) in order to obtain a final product with limited amounts of impurities. The existing processes for separation are limited to natural decantation, centrifugation, or a combination of both processes.

Decantation has been the method used to achieve separation of liquids of different densities since ancient times. Oil density typically ranges between 912 and 918 kg m^{-3}, and vegetable water density ranges between 1015 and 1085 kg m^{-3}. This method is still used today in certain mills, particularly smaller mills that cannot justify a vertical centrifuge. This separation of the oil and aqueous phases by decantation is carried out in a set of interconnected vats made of tiled concrete, fibreglass, or stainless steel. Temperature, cleanliness, and time can affect the results obtained with this type of decantation. The use of solely natural decantation requires a considerable number of containers and plant area. It is also important to understand that during the time required for natural decantation to occur, oil remains in contact with the aqueous phase for a long time, leading to fermentation processes and quality deterioration.

Centrifuge separators are increasingly popular in most modern processing plants, particularly those with large capacity, as they allow continuous and fast separation. In most two- and three-phase extraction systems, the liquid phases separated are sieved through a vibrating screen and then processed through a vertical centrifuge. In this centrifuge, in general terms with the addition of water, oil is rinsed and separated from the other components. In order to achieve a good result from this type of separator, it is important that the liquid to be treated is as uniform as possible. Furthermore, the volume and temperature of the added water needs to be well regulated, the feed flow properly adjusted according to the capacity of the vertical separator, and a rigorous cleaning regime must be maintained for the separator.

Most processing plants prefer to supplement this centrifuge with a final decantation or settling process to remove the air added to the oil during the high speed centrifugation and to allow the product quality to be classified before it is placed in its final container.

Storage

For virgin olive oil, the issue of conservation is particularly important as the rate of change in its unique chemical and organoleptic characteristics largely depends on storage conditions. While it is true that virgin olive oil should be consumed as soon as possible and within its first year, it is not uncommon that the industrial mills and trading companies that are responsible for packaging and marketing are forced to store oil throughout the year and to carry over some stock to the following season, making storage conditions even more important.

Whenever possible, olive oil must be kept at a quasi-constant temperature of 15–18°C; sudden temperature changes must be avoided as they could cause freezing when the temperature is too low or encourage oxidation when too high. Oil should be stored away from light sources, and when stored for prolonged periods of time, it is highly recommended that the air in the headspace of the tank be replaced with a modified atmosphere of low oxygen content.

It is important to size the tanks according to the production capacity. Oil storage containers should be built with suitable materials so that the oil does not penetrate or react with the container's surface. Oil that is absorbed and hence resists cleaning will alter and compromise the future use of the container. Stainless steel is the material of choice for the storage of high quality virgin olive oils. Storage tanks should be designed with an inspection door for easy inspection after cleaning, and with a cone shaped or slanted bottom with a bleed valve in order to continue draining settlings as they occur during the storage period.

Storage and processing rooms must be built with walls and flooring made of materials that can be easily and regularly cleaned. It is important that both areas are covered and protected from weather and temperature variations. It is advisable to avoid storing in those areas materials or objects that might affect the aroma of the processed or stored oil.

Olive Oil Organoleptic Characteristics

Olive oil has organoleptic characteristics that most other oils do not possess, such as fruity olive characteristics. The loss of these characteristics can be a consequence of the refining process which is commonly carried out in other oils, and it is one of the factors that have an impact on consumer choice.

The characteristic aroma of olive oil results from a complex set of over 100 volatile components that are found only in very small quantities. Although much progress has been achieved in identifying and researching the compounds responsible for the aroma, color, and flavor of virgin olive oils, it has not been possible to satisfactorily substitute human senses in the organoleptic description and appreciation of this product. Therefore, the assessment of virgin olive oil's quality by a panel of specialized trained tasters is carried out together with the chemical analysis of this product.

Extra virgin olive oil is characterized by the absence of organoleptic defects. The intensity and complexity of the fruity characters, pungency, and bitterness are the most common positive attributes that describe a virgin olive oil. The most commonly detected sensory defects include: winey, musty, muddy sediment, fusty, and rancid; there are also other faults among which the most frequently detected are pomace flavors and vegetable water flavors.

Factors that Impact Olive Oil Quality

There are a wide range of factors that can have an impact on the quality of olive oil. They can be classified as either agronomical factors or processing factors.

Agronomical Factors

Agronomical factors impact the quality of the oil mostly through their direct impact on the olives themselves. Under normal circumstances, neither cultivar nor agricultural environments have a clear impact on oil quality. It is generally accepted that any cultivar and environment has the potential to produce extra virgin olive oils, provided that healthy olives, which have been harvested at their optimal time, are processed correctly. Nonetheless, it is also well documented that there are significant differences among oils coming from different cultivars and agricultural environments. These differences will be mostly associated with the intrinsic chemistry of the oils through parameters such as fatty acid composition, sterol composition, wax content, polyphenol content, etc. Furthermore, genetics and agricultural environments clearly influence the organoleptic characteristics of the oils. Sensory characteristics of virgin olive oils obtained through a careful process may all be considered as extra virgin, but the different varieties and environments will have their own organoleptic profile. There are a large number of publications from researchers in all olive growing regions worldwide illustrating the variation in composition and sensory characteristics due to cultivar and agricultural environments (e.g., Garcia-Gonzalez et al. 2010; Mailer et al., 2010).

Regarding grove management, most crop management practices are known to have a strong impact on tree production and oil yields, but they are not relevant with respect to the chemical quality of the oil. Neither pruning nor fertilizing has a significant impact on the quality of virgin olive oils, but fertilization will increase fruit production (Chouliaras et al., 2009). Irrigation does not have an impact on regulatory quality indices, but it does influence the polyphenol content of the oils (Servili et al., 2007), with irrigated trees displaying lower levels of these compounds and associated lower levels of bitterness and pungency in the oil. Similar to many other crops, plant health treatments are crucial to obtaining good quality oils, particularly related to the control of certain pests and diseases such as olive fly and anthracnose that are well known to have a detrimental impact on oil quality.

As discussed earlier in this chapter, olive harvesting is a critical operation not only for the grove's economic success, but it also has a major impact on the quality of the oil produced.

Processing Factors

There is no doubt that olive oil quality is born in the grove as a combination of environmental aspects (climate and soil), genetics (variety), and agronomic factors (cultural practices). There is an old saying that affirms: "The olive has its own potential and all you can do is stay out of the way." It is well documented that the environment, the cultivar, and all those cultural practices that ensure the grower will produce well-developed and healthy fruit will certainly provide the opportunity to produce oil of the highest quality. However, recent research results in Australia and elsewhere would indicate that the extraction process can play a more important role in terms of defining the characteristics of the produced oils than what has been previously assumed. These results do not contradict the fact that quality already exists in the olive fruit, and the subsequent acts of harvesting, storage of fruit, transport, processing, settling, storage of oil, filtration, and bottling can individually or in combination result in the addition of defects and the consequent reduction of oil quality.

During the oil extraction process, plant managers can have a decisive influence on the character of the oils. Most of these actions are mainly related to achieving more balanced and complex oils, trying to maximize desired aspects and minimizing those aspects that may lead to an unbalanced final product. One of the first processes when preparing olives for the extraction process is to remove twigs, leaves, stones, and dust from the fruit. In modern continuous plants, the presence of high levels of leaves and twigs due to mechanical harvesting may intensify the flavor characteristic described as "green leaves" in the oil. An excessive amount of leaves can lead to the production of unbalanced oil, which will be inferior from the organoleptic point of view. Recently, many processors in Spain have decided to bypass the washing process, especially when the olives are clean and picked directly from the tree. The main purpose of bypassing the washing process before crushing the fruit is to avoid the mechanical action of the washing machine, which may cause flesh separation and consequent oil losses, as well as to avoid increasing the fruit moisture content; these processors have discovered that this action has also some positive effect on the oil characteristics as well. Bypassing washing has no impact on free fatty acids, peroxide value, or ultraviolet coefficients, but it increases total polyphenols content, oil shelf life, and oil fruitiness.

As previously discussed, the main objectives of crushing the olives is to tear the flesh of the fruit and to release the oil that is entrapped in the vacuoles. From the organoleptic point of view, it is well known that oil obtained using aggressive actions, such as hammer mills, tends to have a more bitter note, with a stronger and more pungent taste due to the higher amount of polyphenols that are released. However, this does not mean higher or lower quality in the oil; the oil extracted has different characteristics that may be desirable under certain circumstances or undesirable in other situations.

After being crushed, the olive paste has to be stirred slowly and continuously in malaxers using horizontal or vertical shafts. The stainless steel malaxing chambers are fitted with a heating system using circulating hot water. As discussed previously, the combination of time and temperature must be calibrated with each variety of olives that is processed and with each kind of oil that is desired in order to ensure maximum efficiency without affecting quality. Angerosa et al. (2001) and Uceda et al. (2006) have shown that malaxing at higher temperatures and for longer periods will decrease polyphenol levels with a dramatic change in the volatile compounds present in the oils.

Centrifugation Systems for Avocado Oil

Commercial extraction of avocado oil based on continuous virgin olive oil extraction was developed in New Zealand (Eyres et al., 2001), and avocado oil was commercially produced by two processors, long after the process had been in use by the olive oil industry, primarily around the Mediterranean oil producing countries (Woolf et al., 2007). There are unique characteristics of avocados that require some simple modifications to the oil extraction process, but primarily the process uses only mechanical or physical means of extraction including washing, grinding, malaxing, and centrifugation. The temperature during processing is maintained below 50°C, and the only additive that is sometimes needed is purified water. The use of enzymes and natural pH adjustment has also been used in trials. A process adopted in New Zealand for continuous extraction of cold pressed avocado oil is shown in Fig. 2.3. The avocado extraction process uses a procedure similar to that used for olive oil extraction; such a process is shown in Fig. 2.4.

Oil Accumulation in Fruit

As mentioned earlier, the majority of the oil is present in the flesh of the avocado, up to 32% (by fresh flesh weight), while only 2% can be found in the stone. The proportion

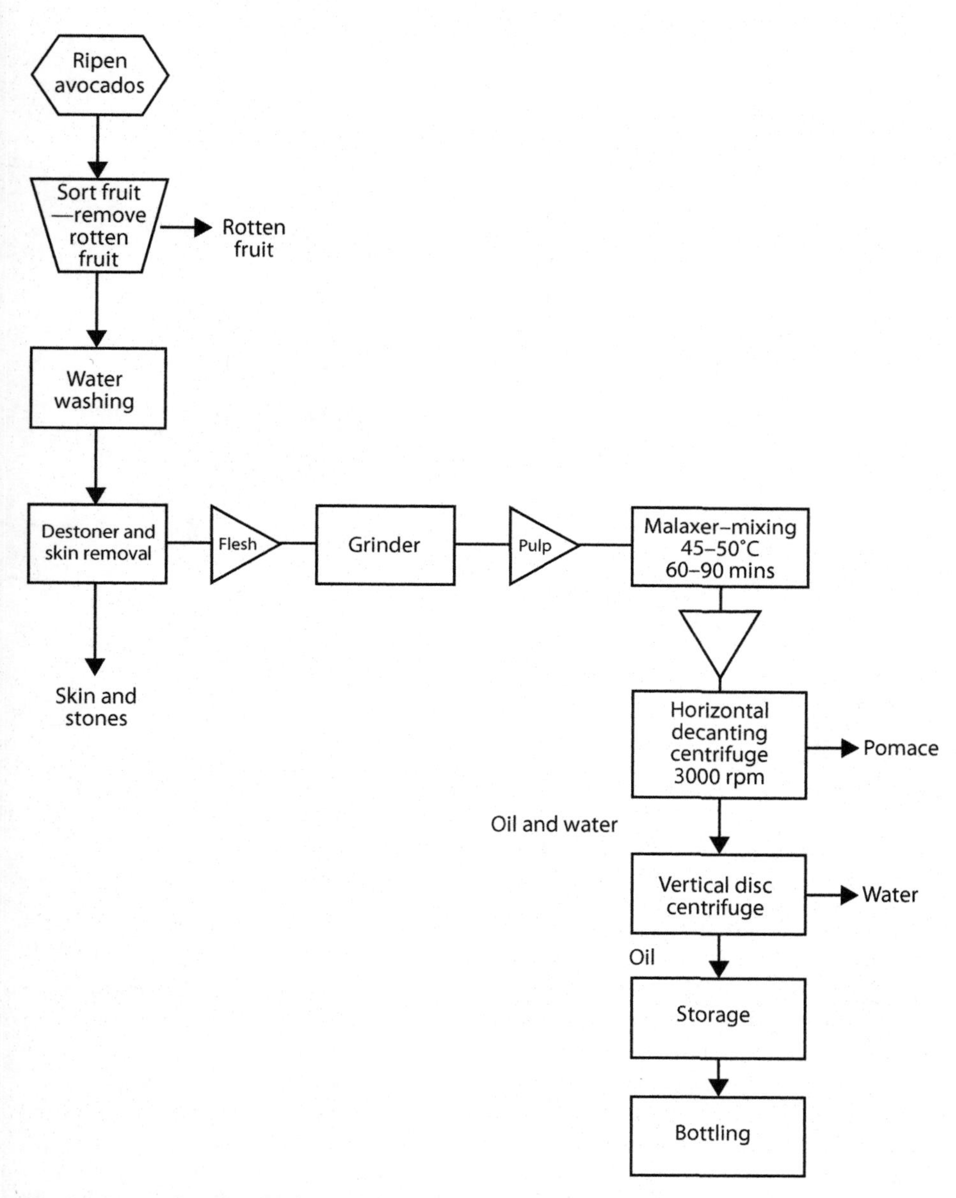

Fig. 2.3. Process flow diagram for cold pressed extraction of avocado oil.

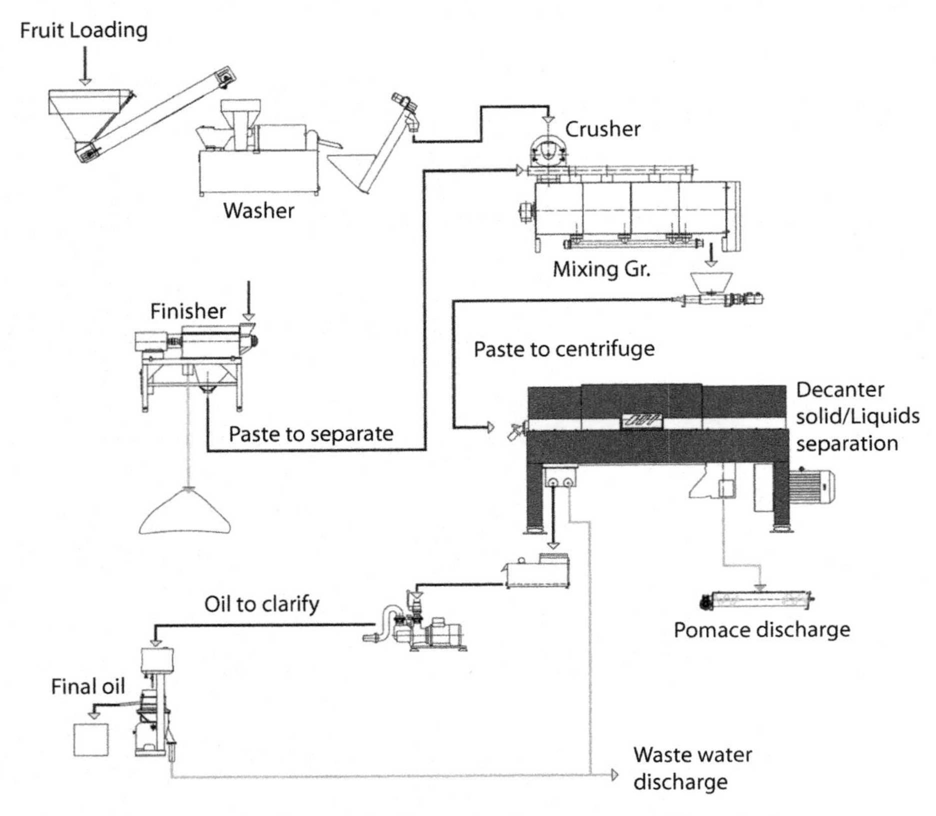

Fig. 2.4. Typical process flow for avocado oil extraction (Image courtesy of Amenduni).

of skin (14%) to flesh (68%) to stone (18%) has been found to be consistent for "Hass" avocados whatever the fruit weight (as a percentage of fresh weight) (Wong et al., 2008).

Avocado fruit are harvested once they are considered to be mature. Maturity is easily determined from the dry matter content of the fruit. If harvested before they are mature, the fruit will not ripen normally. Once physiologically mature, the avocado fruit can remain on the tree for 7 to 18 months before it is harvested, and will still ripen normally. Ideally, the fruit should be harvested within the first seven months to allow the tree to set fruit for the next season. The dry matter content of the avocados remaining on the tree will increase over the season with a corresponding increase in oil accumulation. Dry matter determination is used by the avocado industry to determine maturity and oil content, as strong correlations between these two factors have been found for the "Hass" cultivar from fruit grown in several locations worldwide (Woolf et al., 2009).

Fruit harvested with a dry matter content of <20% are likely to have poor sensory attributes on eating and will not ripen properly. Recommended minimum dry matter percentages to harvest at and when the avocado is considered mature have been decided in each avocado growing country by their grower organizations; this is important for fruit export. The recommended minimum dry matter content for the "Hass" cultivar is 20.8% in Mexico and the United States, 21% in Chile and Australia, 25% in South Africa, and 24% in New Zealand (Wong et al., 2008).

Oil levels in "Hass" avocado flesh can increase 1–2% early in a season to >30% late in a season, depending on the cultivar (Lewis, 1978; Kaiser & Wolstenholme, 1994; Requejo-Tapia, 1999; Woolf et al., 2009). Experiments performed by Requejo-Tapia (1999) in New Zealand showed that oil content in "Hass" avocados also differed between growing regions and that the oil content increased to 16–17% in September and then increased to 25–31% in April at the end of the season. The oil content and oil composition of different avocado cultivars including "Hayes," "Pinkerton," "Santana," and "Fujikawa" were reported by Requejo-Jackman et al. (2005).

Due to market forces or unpredictable weather conditions (e.g., high winds), fruit are sometimes harvested early with little or no oil that can be economically extracted by aqueous methods. These fruit will not ripen properly, and very little oil can be recovered. Extraction of the oil from mature avocado fruit harvested early in the season has been found to be more difficult than extraction from mature late season fruit. The fruit from both ends of the season may contain the same total oil content when determined by solvent extraction, but during cold pressed extraction, the oil is not easily released from the idioblast cells holding the oil. Woolf et al. (2009) showed the variation in oil yield between early and late season fruit (Fig. 2.5).

Fig. 2.5 shows that commercial cold pressed oil yield is only about 12% early in the season even though the dry matter indicates there is the maximum amount of oil present in the fruit. The commercial press yield increases up to 20–25% later in the season; dry matter and concomitant oil content have increased slightly from early season but not by 10%. It is unclear why this occurs, whether it is due to the presence of endogenous enzymes in the fruit or due to cell wall breakdown.

Fruit Receipt and Storage

Once harvested, the avocados are either sorted and graded for fresh fruit export, leaving the reject fruit for oil extraction or sent directly to the oil processor. On receipt of the fruit, the oil processor will need to access the fruit; in most cases, the fruit direct from the fruit packing houses or orchards will not have ripened. To ensure maximum oil yields, the fruit should be ripened to the equivalent of "firm ripe," equal to a hand firmness score of 4 or a Firmometer reading between 65–95 Fv (White et al., 2005). To obtain uniform and controlled ripening, ethylene is used commercially by the industry to ripen avocados to ensure a more uniform ripeness. Treatment with 100 μL^{-1} ethylene at 20–22°C for 24 hours will ripen fruit in approximately four to six days at 20–22°C (Hofman et al., 2002).

It is important to process good quality ripened fruit for virgin avocado oil. Avocados suffer from a number of postharvest disorders, including rots, bruising, and

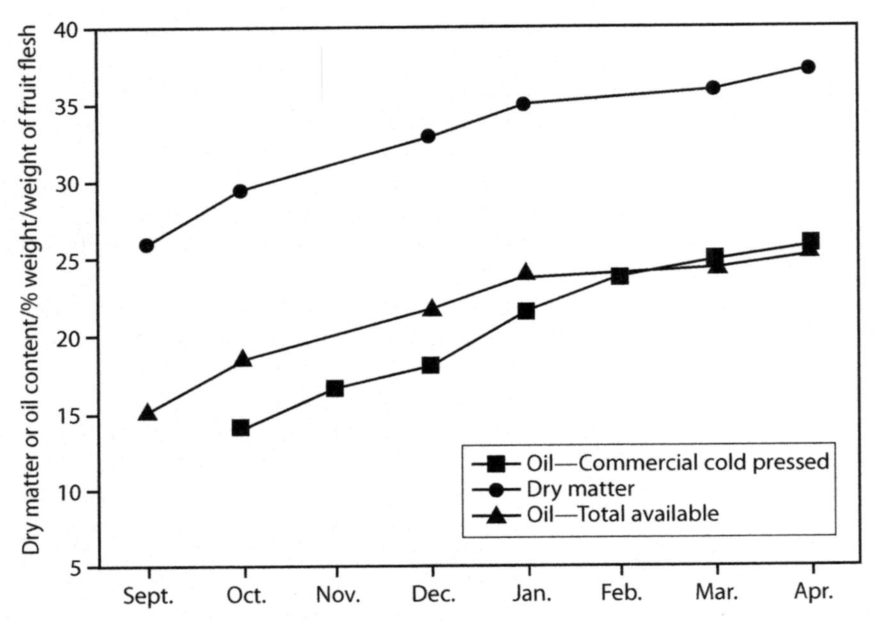

Fig. 2.5. Schematic of typical changes in dry matter and oil content over a commercial harvest and oil processing season in New Zealand "Hass" avocados (Woolf et al., 2009).

greying (White et al., 2005). Fruit should be handled carefully to minimize bruising, and as the fruit ripens, rots become more prevalent especially if harvested wet; hence, any fruit with rots should be removed. Greying is a chilling injury due to long term storage at low temperatures. If there is too much fruit to process, the fruit can be stored in coolstores at 2°C for up to 4 weeks with minimal effect on the fruit quality (unpublished data). Storing the fruit at appropriate temperatures and for a limited time will result in minimal loss of quality in the oil, when the fruit stored is of good quality. This is in contrast to olives, where the fruit ideally must be processed immediately or soon after harvesting as mentioned earlier in this chapter. Woolf et al. (2009) showed that cold storage of fruit at 6°C results in ≅40% unsound fruit with a slight increase in the resulting oil's peroxide value but no significant increase in the percentage of free fatty acids (%FFA). Results from previous research show that increasing the percentage of rotten fruit will lead to increased %FFA in the resulting cold pressed oil (Fig. 2.6). It was observed that the increase in rots did not reduce the oil yield significantly.

Free fatty acids in oil are formed as a result of lipolysis—the cleavage of a fatty acid from tri-, di-, or mono-acylglycerols. When excessive lipolysis occurs in an oil, results are high levels of fatty acids (hydrolytic rancidity), the oil tastes soapy, is more acidic, and is very susceptible to oxidation. In avocado, this reaction is due to internal lipases or from fungal lipases, from fungi that have infected the fruit. Internal

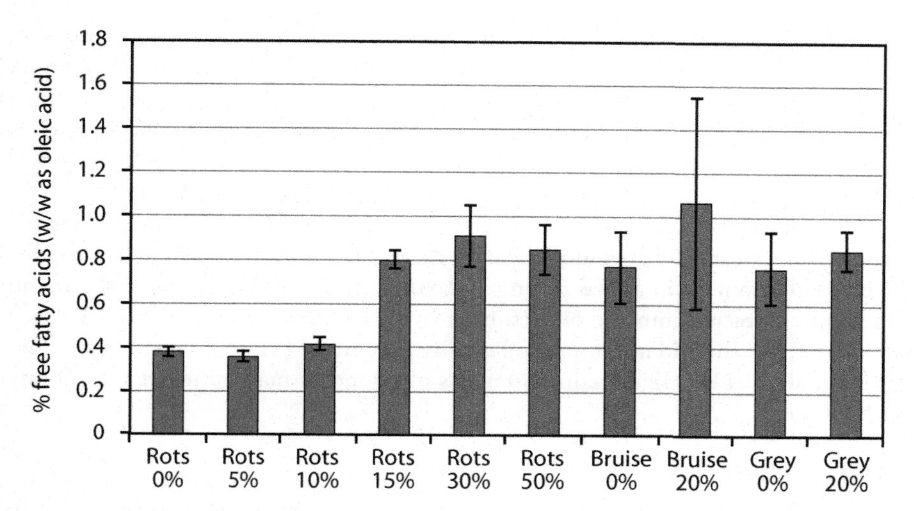

Fig. 2.6. Impact of avocado fruit postharvest disorders on the % free fatty acid content of the resulting cold pressed extracted oil.

lipase is less active in avocado fruit than most other oil containing fruits, giving crude avocado oil a naturally low free fatty acid content (<0.1%). Using older or damaged fruit for avocado oil extraction will result in much higher free fatty acid levels (0.82–2.0%) due to the reaction of avocado lipase to wounding and probable action of fungal lipase in older fruit. It is therefore important to process only avocados of good quality for oil extraction.

Washing, Destoning/Skin Removal

Washing the fruit is important to remove dirt, spray residues, and other undesirable matter. Residual copper spray on the skins of avocados was found to be removed in the water phase during extraction as the copper is not soluble in the oil (unpublished data). The wash water is routinely used to irrigate the orchards.

After washing, the skin and stones are then separated from the flesh (Fig. 2.3). The fruit enters the destoner/finisher which uses paddles that rotate at speed. The fruit is literally smashed against the chamber wall which consists of a stainless steel screen. The aperture size in the screen can be changed to meet different fruit conditions. The resulting pulp is fine enough for effective malaxing. The skin and stone are expelled from the chamber and are removed as waste. Incorrectly ripened or hard fruit is expelled with the stones and is wasted. This can dramatically affect the percentage of oil recovery and the economic viability of an operation. A small percentage of flesh adheres to the skin; this flesh layer contains the highest concentration of chlorophylls and carotenoids of the flesh. The concentration of these pigments decreases as one moves in from the skin toward the stone (Ashton et al., 1996).

Earlier operations included a grinder or hammer mill prior to the malaxer to reduce the particle size of the flesh and to aid in disrupting the plant cell structure containing the oil. Newer systems no longer use grinders. This step is achieved during the destoning phase, thus eliminating potential cleaning problems and paste contamination. It also allows for more flesh close to the skin to be extracted. During skin removal, it is important to remove only skin and minimal avocado flesh with the skin. The riper the avocado, the easier this is to achieve. Some earlier systems to remove the skin and stone from the avocado left some skin in the pulp which then entered the malaxer. Additional skin added during malaxing resulted in higher chlorophyll and carotenoids extracted into the oil, resulting in a dark green oil. In laboratory based extraction trials, this did not impact oil quality in terms of %FFA and peroxide values (Wong et al., 2011). Higher concentrations of pigments may influence oil stability during storage.

Malaxing Systems

The purpose of the malaxing step is to allow time for oil release from the flesh to occur. The pulpy flesh is mixed by slowly turning the paste inside jacketed D-shaped tanks. The malaxers can vary in size from 250 to 1000 kg. The temperature of the pulp in the malaxer is controlled by a hot water jacket around the tanks. During avocado oil extraction, the recommended extraction temperature is 45–50°C. At this temperature, the viscosity of the pulp decreases, hence allowing mixing at lower shear rates. It also aids the release of oil from the cells as the oil viscosity is reduced. The quality of the oil is not affected by these slightly higher temperatures.

Water is often added during malaxing to help reduce pulp viscosity. Freitas et al. (1996) examined the impact of water dilution on the apparent viscosity of avocado pulp. They reported that, at shear rates >129 s^{-1}, the increased addition of water did not reduce the apparent viscosity of the pulp and water emulsion. Hence, pulp–water ratios of 1:3 or 1:5 could be reduced to 1:1 without affecting oil yield. This means less wastewater is produced in the subsequent separation steps, and also allows use of smaller malaxer tanks. Freitas et al. (1996) also reported that the addition of exogenous pectolytic enzymes at 40°C resulted in reduced pulp-water viscosity and an increase in oil yield.

The length of time the pulp is malaxed can range from 40 to 90 minutes, on average 60 minutes. Werman and Neeman (1987) chose a mixing time of 30 minutes to optimize their experimental conditions. The actual time required will depend on the ease at which the oil is released from the pulp. Early season fruit may require longer times, while for late season fruit the oil is released rapidly into the pulp-water-oil paste. The released oil is visible on the top of the pulp-water-oil paste; at this point, the operator can gauge the time required. During malaxing, oxidation and paste deterioration can occur if the paste has too much air contact. Hence, covering the malaxers is recommended to reduce oxidation. New malaxer designs that have a smaller paste-to-air surface are most suitable. Typically these are no larger than 830 kg which allows better batch control due to the high variability of fruit inputs.

Addition of Exogenous Enzymes during Malaxing

In the avocado, oil is mostly found in the idioblast cells which are distinguished by their large size and lignified walls (Platt-Aloia & Thomson, 1981; Werman & Neeman, 1987). These cell walls must be broken in order to release oil droplets before the oil can be separated. These cell walls can either be broken by mechanical force (mechanical grinding) (Werman & Neeman, 1987), enzymatic action on the cell walls of the idioblasts (Buenrostro & Lopez-Munguia, 1986), heat degradation, or any combination thereof. Avocado fruit naturally contains cell wall hydrolyzing enzymes that are present in high levels when the fruit is ripe. These cause the gradual degradation of the primary cell wall structure and middle lamella as part of the senescence process (Platt-Aloia & Thomson, 1981). Therefore soft, mature fruit are desirable for oil separation by centrifugation (Human, 1987; Werman & Neeman, 1987) since natural enzymatic degradation of the cell wall can take place during mixing without adding extra enzymes.

Commercial exogenous enzyme mixes are added during malaxing for two purposes; first (and primarily), to increase oil yield by assisting with the breakdown of cell walls and the release of oils from the idioblast cells; second, to reduce the viscosity of the pulp-water-oil paste resulting in less energy and water addition required. Using additional cell wall hydrolases, Buenrostro and Lopez-Munguia (1986) found that mixing an avocado-water slurry for 1 hour allowed maximum extraction yield following centrifugal extraction. Since added enzymes were used, oil release using the lower concentration of natural enzymes within the avocado fruit could be expected to be longer based on a longer lag phase and a lower maximal enzyme reaction rate (less enzymes present). Bizimana et al. (1993) noted that oil was released after only 5 minutes at 98°C, but oil release at these high temperatures was due to thermal degradation of the cell wall, not due to enzymatic activity, and was likely to result in lower oil quality.

Enzyme trials completed with early and mid-season fruit showed no significant improvement in oil yields. The enzymes added to the avocado pulp were a simple pectinase preparation or a mixed enzyme containing macerating and pectolytic activity, added at 45°C and at concentrations of less than 0.1% w/w. The results indicated that standard pectinase formulations developed for most other fruit extraction processes were not suitable for early or mid-season fruit. The ideal enzyme mix will need to be determined with more experimental work.

Fig. 2.7 shows very low oil yields for early season avocados. Various additives were trialled including citric acid, a cellulase enzyme mix, reducing the particle size of the pulp with a Waring blender, and using a higher malaxing temperature of 75°C compared to 45°C. Fig. 2.8 shows the results from enzyme trials with late season fruit. The addition of pectinase or cellulase-based enzymes led to increased oil yields.

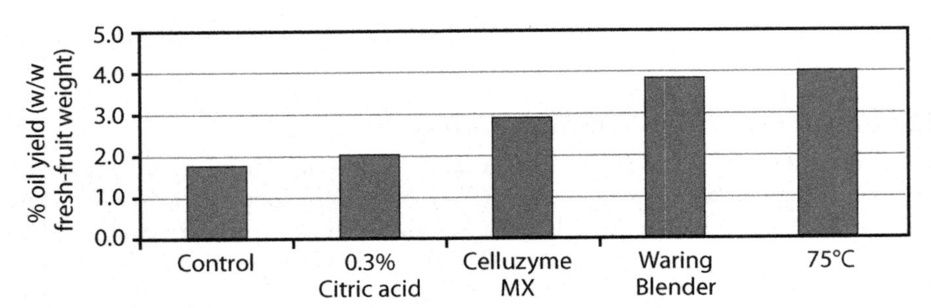

Fig. 2.7. Impact of various process additives on % oil yield (based on fresh fruit weight) from early season "Hass" avocados.

Adding citric acid was also found to increase the oil yield, a logical result as acidifying the emulsions to a pH of approximately 4.0 in other rendering systems breaks the emulsion, allowing better separation.

Oil Separation Systems

The purpose of oil extraction by centrifugation is to extract the maximum amount of oil from the mesocarp without damaging oil quality (Human, 1987). Cold-press oil extraction methods involve the creation of an avocado paste by crushing or grinding the avocado mesocarp followed by the addition of water, mixing at a specific temperature, and separating the oil layer from the resulting slurry using centrifugal force (Werman & Neeman, 1987; Buenrostro & Lopez-Munguia, 1986; Bizimana et al., 1993; Freitas et al., 1996). Centrifugal extraction is based fundamentally on the difference in density and miscibility between the lighter oil and heavier aqueous/solid phase. Thus, when exposed to the high G forces of a centrifuge or gravity for long periods of time, the oil will separate to a layer above the aqueous solution and remaining solids (Bizimana et al., 1993).

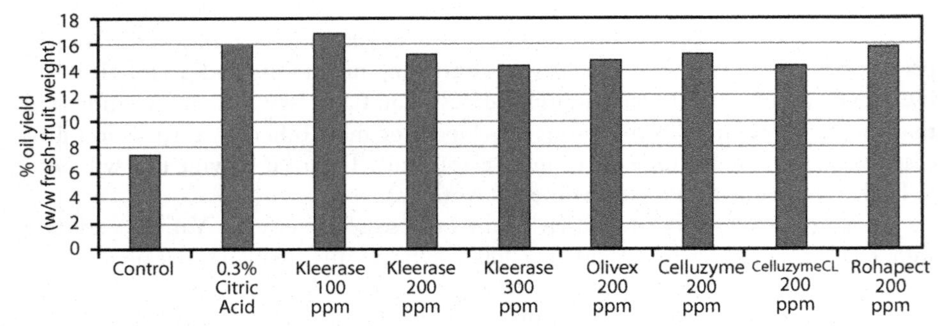

Fig. 2.8. Impact of enzyme addition on % oil yield (based on fresh fruit weight) from late season "Hass" avocados.

It is now standard practice in commercial operations to use a horizontal decanting centrifuge, similar to olive oil as shown in Fig. 2.2. The paste is pumped into the decanter at an adjustable rate with varying amounts of filtered water to gain the optimum oil extraction. This is still determined by the operator who must constantly monitor the oil and water streams. This skill is only achieved by training. A combination of residence time in the decanter and water at optimum levels will produce the best result. Separation of the oil at 50°C is far easier than at 30°C due to the differences in dynamic viscosity as mentioned earlier. Two streams are recovered from the decanter, one stream is the pomace and the other is the oil with some water. The oil and water phase from the decanter is then separated in vertical centrifuges as described for olive oil. As with olive oil, these polishing centrifuges must be scrupulously cleaned. Operation at 45–50°C ensures good separation between the oil and aqueous phases.

Filtration, Decanting, and Storage

After polishing the oil in the polishing centrifuges and ensuring moisture is below 0.2%, the oil is nitrogen sparged to simultaneously remove oxygen and to chill the oil. In-line filtering of the oil through 100 μm filter bags or cartridges is advisable to remove protein and solid residues. To ensure minimal oxidation of the oil during processing and storage, it is vital to minimize oxygen and light and to use stainless steel where possible with no contact to brass or mild steel plating.

Oils processed through pumps and pipelines will pick up dissolved oxygen unless the plant is totally hermetically sealed, a process condition that is complex and expensive to achieve. A standard practice in oil processing is to sparge inert gas into the oil, removing dissolved oxygen and preventing peroxide value rise (Fig. 2.9). The technique has been described for soybean oil processing (Erickson, 1980). When

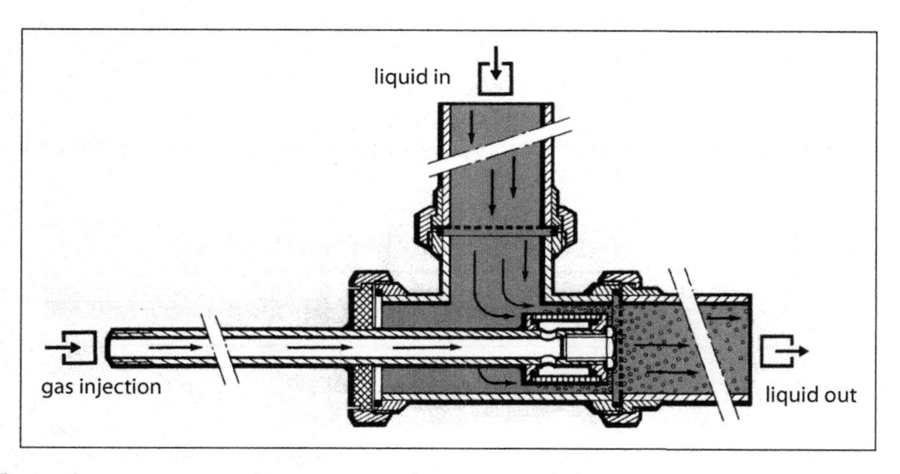

Fig. 2.9. Gas sparging system (Image courtesy of Air Liquide NZ).

producing vegetable oils without nitrogen sparging, one can expect the dissolved oxygen level to be approximately 40–50 ppm at 30–50°C. This can easily lead to a peroxide value of >5 meq/kg on storage (Berger, 1994). In addition to removing dissolved oxygen, the oxygen in the headspace of tanks, drums, and bottles should be reduced from 21% to less than 2%. If the avocado oil is cooled slowly and stored in tanks under nitrogen at 12–20°C, the higher melting glycerides will crystallize out at the bottom of the tank, resulting in a partial winterization of the oil. A typical analysis of avocado oil in comparison to olive oil is shown in Table 2-C.

To control the oxidation kinetics during the storage of bottled oil when considering alternatives to glass, it may be helpful to use well-designed plastic bottles (polyethylene terephthalate, PET), innovative plastic materials containing an oxygen scavenger, and to perform the bottling operations under a nitrogen atmosphere (to reduce the oxygen pressure in the bottle headspace). These practices are often underestimated by the oil industry which defines empirically the period of shelf life (date of recommended consumption) of bottled virgin oil, without carefully considering oil characteristics, packaging properties, and the temperature conditions during product distribution. It is becoming more and more common now for the industry to use polycons (polymer based containers that can hold 1,000 liters of oil) for storage and transportation. It is important to consider the surface area-to-volume ratio of the container to the oil. With a higher ratio, more oxygen will diffuse into the oil. Del Nobile et al. (2003) showed that the shape of the bottle and the surface area have a significant impact on the oxygen partial pressure in the headspace of the bottle when modelling storage of olive oil in different shaped bottles.

Waste and Disposal—Pit and Skins, Decanter Waste

An example of a mass balance over an avocado oil processing plant for extraction of "Hass" avocados processed early and late season is shown in Table 2-D. The major waste materials are the stone and skin, pomace from the decanter, and the waste water from the separators.

The traditional and most cost effective method of handling the waste stones and skin from avocados is by landfill on the areas used to grow the avocado trees. Alternative uses for the skin and stones are to burn them and to use the heat generated for

Table 2-C. Typical Analysis of Extra Virgin Avocado and Olive Oil.

Analytical result	Avocado oil	Olive oil
Viscosity at (25°C) cP	62	63.28
FFA (as oleic) (%)	0.08–0.17	0.15–0.25
Specific gravity (25°C)	0.915–0.916	0.914–0.918
Iodine value (from GLC)	82–84	75–82

Table 2-D. Typical Mass Balance for an Avocado Oil Processing Plant.

	Early season				Late season			
	Total (kg)	Oil (kg)	Water (kg)	Solids (kg)	Total (kg)	Oil (kg)	Water (kg)	Solids (kg)
Process input								
Fruit	796	92	454	251	673	151	354	167
Water (destoner)	98	0	98	0	83	0	83	0
Water (rinse)	244	0	244	0	443	0	443	0
Total	1138	92	796	251	1199	151	880	167
Process output								
Stones & skin	195	<2	0	≅195	113	<2	0	≅113
Pomace (ex-decanter)	120	4	98	19	152	7	123	22
Oil reservoir	73	73	0	0	129	129	0	0
Waste water	750	15	698	37	805	15	757	32
Total	1138	≅92	796	≅251	1199	~151	880	≅167
Fruit oil content (w/w)	11.5 %				22.5%			
Available oil (kg)	92				151			
Oil recovered (kg)	73				129			
Recovery efficiency	79.8%				85.4%			
Yield (kg oil/kg fresh(fruit)	9.2%				19.2%			

water heating, which is now being used by several companies. Waste water is handled in oxidation ponds, and is then sprayed onto the land for irrigation purposes. Using the pomace for stock/animal feed has been investigated by a number of researchers. An analysis of the pomace for stock food is given in Table 2-E.

Table 2-E. Analysis of the Avocado Pomace for Stock Food.

Analytical result	Avocado Pomace
Total Sugar (% w/w)	0.55–0.70
Carbohydrate (% w/w)	18.0–20.0
Sodium (mg/kg)	186–190
Average energy (Joule/kg)	414
Protein (% w/w)	2.2–2.5
Fiber (% w/w)	1.0–1.2
Total Fat (%)	1.5–5.5

Other Fruits/Nuts/Seeds for Which Aqueous Extraction Will Work

Coconut Oil

Virgin coconut oil is defined by Codex alimentarius as oil extracted from fresh coconut meat and processed using physical and natural processes (Codex alimentarius, 1999). There is no prior drying (e.g., use of copra), refining or chemical reactions before the oil is extracted from coconut meal by mechanical means (expelling, pressing, heating, washing, settling, filtration, and centrifugation). Sant'Anna et al. (2003) and Che Man et al. (1996) extracted coconut oil from the coconut milk by breaking the emulsion. Sant'Anna et al. (2003) used enzymes from Novo Nordisk, Viscozyme® and Neutrase® to increase the extraction of fat and coconut milk from the coconut flesh. McGlone et al. (1986) and Tano-Debrah & Ohta (1997) have investigated aqueous extraction of coconut oil with added exogenous enzymes (proteases, amylases, polygalacturonases, cellulases, pectinases or combinations of enzymes) achieving oil yields ranging from 12–80% from fresh or dried coconut meat. More recent research has been carried out to optimize the aqueous extraction process for virgin coconut oil from fresh coconut meat. Fresh coconut meat was ground to particle sizes ranging from 1–4 mm. A greater oil yield was achieved with smaller particle size. The ground meat was then diluted with water to an optimum ratio of 1:4, with no pH adjustment. This mixture was then malaxed with and without Alcalase 2.4L FG (Novozyme) protease enzyme. A temperature range of 40–50°C and a malaxing time of 60 minutes was optimum. After malaxing, the oil fraction was recovered from the malaxing paste with centrifugation. Using the protease enzyme the oil yields were approximately 75% of total oil available in the fresh coconut flesh. The protease enzyme also assisted with breaking any emulsion formed with the coconut proteins and the oil. The coconut oil produced was white with a pleasant odor, with a PV < 1 meq/kg and acid values < 0.6 mg KOH/g oil (McCormick, 2012).

Macadamia Nut Oil

The aqueous extraction process for macadamia nut oil can, in general, be improved by any treatment that enhances the dissolution of the other components, for example, by using enzymes or increasing the temperature (Rosenthal et al., 1996). Process development and optimization of aqueous extraction techniques thus need to take into account factors facilitating dissolution of water-soluble components (Rosenthal et al., 1996; Kashyap et al., 2007). Hence, the aqueous extraction process was investigated for the extraction of macadamia nut oil (Zhang, 2007). Compared to peanuts and soybeans, the oil bodies in macadamia nut kernel were observed to be more abundant than protein bodies and actually surround the protein bodies (Walton & Wallace, 2005). But macadamia nuts contain enough protein to cause problems with emulsion formation during aqueous extraction. To reduce the formation of emulsions, pH adjustment or salt addition can alleviate or minimize this problem.

Aqueous extraction of macadamia nuts was achieved by ensuring the nut to water ratio was greater than 1:1; a ratio of 1:3 was found to be optimal. Adjustment of pH ensured that the emulsion stayed away from the isoelectric point of the proteins; an acidic pH of 3.5 and addition of 1% sodium chloride reduced emulsion formation and provided oil yields 20% higher than a nut to water ratio of 1:1 and no pH adjustment or salt addition. Peroxide values remained low at 0.85 to 1.5 meq/kg and % FFA 0.09 to 0.16%. Various Novozyme enzymes were trialled during aqueous extraction: Celluclast®, Viscozyme®, Neutrase®, and Flavourzyme®. The impact of these enzymes on oil recovery is shown in Fig. 2.10.

Pumpkin Seed Oil

Limited trials have been completed with the aqueous extraction of pumpkin seed oil. Oil was extracted using a seed to water ratio of 1:1 to 1:2. The addition of pectinase and macerating enzymes increased the oil yield, and the adjustment of pH to acidic conditions also resulted in an increased oil yield.

Rice Bran and Peanut Oils

Sharma et al. (2001, 2002) have described the aqueous extraction of rice bran oil and peanut oil. For rice bran oil, a mix of protease, amylase, and cellulose enzymes without pH adjustment at 65°C gave the best oil recovery. Aqueous extraction of peanut oil was achieved with a mix of protease enzymes in the form of Protizyme®. The pH was adjusted to 4.0, and the slurry was incubated at 40°C. There are definite

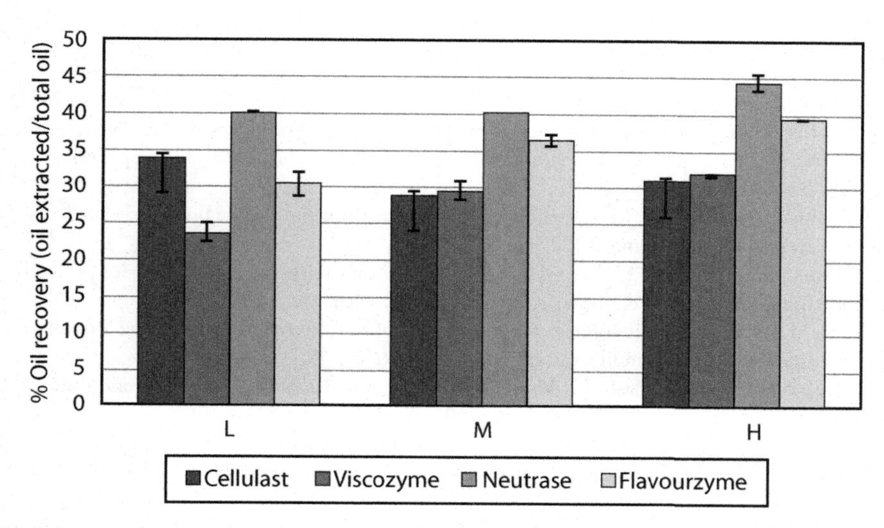

Fig. 2.10. Aqueous extraction of macadamia nut oil with the addition of exogeneous enzymes. Enzyme addition rates % of total nut weight: Celluclast® 0.5% (L), 1.0% (M), 1.5% (H); Viscozyme® 0.5% (L), 1.0% (M), 1.5% (H); Neutrase® 0.04% (L), 0.08% (M), 0.12% (H); Flavourzyme® 0.04% (L), 0.08% (M), 0.12% (H).

environmental benefits with the aqueous extraction of peanut and rice bran oils (Sharma et al., 2001, 2002). For peanut oil extraction, the protein cake from an aqueous extraction can be recovered without the use of solvents. The main factor limiting the implementation of this technology was considered to be the cost of commercial enzymes.

Conclusions

Aqueous oil extraction is a technology with great potential in the oil industry because it offers a number of advantages compared to conventional solvent extraction. As discussed in this chapter, the aqueous extraction process has already been used successfully and commercially for virgin olive oil and avocado oil production. The aqueous extraction process has made recent advances to become a modern technology with the incorporation of modern efficient separation techniques and the application of processing aids such as exogenous enzymes. Each raw material source requires different handling practices prior to the formation of a pulp-water-oil paste which is mixed to aid oil release from the oil containing cells. After this point, very similar technologies can be used to separate the oil from the aqueous phase and solids. Aqueous extraction of oils combined with centrifugation can produce high quality oils with unique and desirable organoleptic characteristics without the use of solvents. With good processing management, the production of waste streams can also be reduced.

References

Alba, J.; Martínez, F.; Moyano, J.; Hidalgo, F.; Cárdenas, R.; Hruschka, S. Elaboration of extra virgin olive oil in Spain. The technological change. In: Culture and Knowledge Inspired by Olive Oil. Humanes, J., Vilar, J., Fialho, M., Higueras, P., Eds.; GEA Westfalia Separator: Ibérica, SA, 2011; 241–259.

Alba, J.; Muñoz, E.; Martinez, M. Obtención del aceite de oliva: Empleo de productos que facilitan su extracción. Alimentaria. 1982, 138, 25–55.

Alba, J.; Ruiz, M.A.; Hidalgo, F. Control de elaboración y características analíticas de los productos obtenidos en una línea continua ecológica. Dossier Olea. 1992, 2, 43–48.

Amouretti, M. Olive oil production: An original history of technology. In World Olive Encyclopaedia. International Olive Oil Council: Madrid, 1996; 26–29.

Angerosa, F.; Mostallino, R.; Basti, C.; Vito, R. Influence of malaxation temperature and time on the quality of virgin olive oils. Food Chem. 2001, 72, 19–28.

Ashton, O.B.O.; Wong, M.; McGhie, T. K.; Vather, R.; Wang, Y.; Requejo-Jackman, C.; Ramankutty, P.; Woolf, A.B. Pigments in avocado tissue and oil. J. Agr. Food Chem. 2006, 54, 10151–10158.

Berger, K.G. Practical measures to minimise rancidity in processing and storage. In Rancidity in Foods; Allen, J.C.; Hamilton, R.J., Eds.; Aspen Publishers, Maryland, 1994; pp. 68–83.

Bizimana, V.; Breene, W. M.; Csallany, A. S. Avocado oil extraction with appropriate technology for developing-countries. J. Am. Oil Chem. Soc. 1993, 70, 821–822.

Boselli, E.; Di Lecce, G.; Strabbioli, R.; Pieralisi, G.; Frega, N. Are virgin olive oils obtained below 27°C better than those produced at higher temperatures? Lebensm. Wiss. Technol.—Food Sci. Tech. 2009, 42, 748–757.

Buenrostro, M.; Lopez-Munguia, A. C. Enzymatic Extraction of Avocado Oil. Biotech. Letters 1986, 8(7), 505–506.

Che Man, Y.B.; Suhardiyono, A.B.; Azudin, M.N.; Wei, L.S. Aqueous enzymatic extraction of coconut oil. J. Am. Oil Chem. Soc. 1996, 73, 683–686.

Chouliaras, V.; Tasioula, M.; Chatzissavvidis, C.; Therios, I.; Tsabolatidou, E. The effects of seaweed extract in addition to nitrogen and boron fertilization on productivity, fruit maturation, leaf nutritional status and oil quality of the olive (Olea europaea L.) cultivar Koroneiki. J. Sci. Food Agr. 2009, 89, 984–988.

Codex-alimentarius. Codex standard for named vegetable oils. 1999.

Cruz, S.; Yousfi, K.; Perez, A.; Mariscal, C.; Garcia, J.M. Salt improves physical extraction of olive oil. Eur. Food Res. Technol. 2007, 225, 359–365.

Del Nobile, M.A.; Bove, S.; La Notte, E.; Sacchi, R. Influence of packaging geometry and material properties on the oxidation kinetic of bottle virgin olive oil. J. Food Eng. 2003, 57, 189–197.

Di Giovacchino, L. Incidenza delle tecniche operative nell'olio dale olive con il sistema continuo. La Rivista delle Sostanze Grasse. 1988, 65, 283–289.

Di Giovacchino, L.; Sestili, S.; Di Vincenzo, D. Influence of olive processing on virgin olive oil quality. Eur. J. Lipid Sci. Technol. 2002, 104, 587–601.

Erickson, D.R. Handbook of Soy Oil Processing and Utilisation. American Soybean Association, Missouri and The American Oil Chemists' Society, Champaign, Illinois, 1980; pp. 310–311.

European Commission. 2002. http://europa.eu/legislation_summaries/other/l11054_en.htm (accessed 12/04/2011)

Eyres, L.; Sherpa, N.; Hendricks, G. Avocado oil: A new edible oil from Australasia. J. Lipid Technol. 2001, July, 84–88.

Fedeli, E. Oil production and storage technology. In World Olive Encyclopaedia. International Olive Oil Council, Madrid 1996, pp. 251–294.

Freitas, S.P.; Da Silva, F.C.; Lago, R.C.A.; Qassim, R.Y. Rheological behaviour of processed avocado pulp emulsions. Int. J. Food Sci. Tech. 1996, 31, 319–325.

Garcia, J.M.; Yousfi, K.; Oliva, J.; Garcia-Diaz, M.T.; Perez-Camino, M.C. Hot water dipping of olives (Olea europaea) for virgin oil debittering. J. Agr. Food Chem. 2005, 53, 8248–8252.

García-González, D.L.; Romero, N.; Aparicio, R. Comparative study of virgin olive oil quality from single varieties cultivated in Chile and Spain. J. Agr. Food Chem. 2010, 58, 12899–12905.

Hofman, P.J.; Fuchs, Y.; Milne, D.I. Harvesting, Packing, Postharvest Technology, Transport and Processing. In The Avocado. Botany, Production and Uses. Whiley, A.W.; Schaffer, B.; Wolstenholme, B.N., Eds.; CABI Publishing, Oxon, 2002; pp. 363–402.

Human, T. P. Oil as a byproduct of avocado. South African Avocado Growers' Assoc. Yearbook, 1987, 10, 159–162.

International Olive Council (IOC). 2011. http://www.internationaloliveoil.org/ (retrieved 12/04/2011).

Kaiser, C., Wolstenholme, B.N. Aspects of delayed harvest of "Hass" avocado (Persea Americana Mill.) fruit in a cool subtropical climate. I Fruit lipid and fatty acid accumulation. J. Hort. Sci. 1994, 69, 437–445.

Kashyap, M.C.; Agrawal, Y.C.; Ghosh, P.K.; Jayas, D.S.; Sarkar, B.C.; Singh, B.P.N. Oil extraction rates of enzymatically hydrolyzed soybeans. J. Food Eng. 2007, 81, 611–617.

Kiritsakis, A.K. Olive oil: From the tree to the table. 2nd ed., Food & Nutrition Press, Trumbull, USA, 1998; 348 pp.

Lewis, C.E. Maturity of avocados—general review. J. Sci. Food Agr. 1978, 29, 857–866.

Linares, H.M.; Mendoza, J.A.; Perez, M.J.M.; Jackisch, B.O.; Influences of using enzymatic complexes in the second centrifugation of olive paste. Grasas y Aceites 2006, 57, 301–307.

Mailer, R.; Ayton, J.; Graham, K. The influence of growing region, cultivar and harvest timing on the diversity of Australian olive oil. J. Am. Oil. Chem. Soc. 2010, 87, 877–884.

McCormick, D. Extraction of extra virgin coconut oil. Massey University, College of Sciences Project Report. 2012.

McGlone, O.C.; Lopez-Munguia, A.; Carter, J.V. Coconut oil extraction by a new enzymatic process. J. Food Sci. 1986, 51, 695–697.

Moya, M.; Espínola, F.; Fernández, D.G.; de Torres, A.; Marcos, J.; Vilar, J.; Josue, J.; Sánchez, T.; Castro, E. Industrial trials on coadjuvants for olive oil extraction. J. Food Eng. 2010, 97, 57–63.

Perez, A.; Luaces, P.; Rios, J.; Garcia, J.M.; Sanz, C. Modification of volatile compound profile of virgin olive oil due to hot water treatment of olive fruit. J. Agr. Food Chem. 2003, 51, 6544–6549.

Petrakis, C. Olive Oil Extraction. In Olive oil: Chemistry and Technology. 2nd Edition. Boskou, D. Ed.; AOCS Press, Champaign, Illinois, 2006; pp. 191–224.

Plant & Food Research. Fresh Facts. New Zealand Horticulture. New Zealand Plant & Food Research Ltd., Auckland, New Zealand, 2008; pp. 12–13.

Platt-Aloia, K.A.; Thomson, W.W. Ultrastructure of the mesocarp of mature avocado fruit and changes associated with ripening. Ann. Bot.—London. 1981, 4, 451–465.

Ranalli, A.; De Mattia, G. Characterization of olive oil produced with a new enzyme processing aid. J. Am. Oil Chem. Soc. 1997, 74, 1105–1113.

Ranalli, A.; Gomes, T.; Delcuratolo, D.; Contento, S.; Lucera, L. Improving virgin olive oil quality by means of innovative extracting biotechnologies. J. Agr. Food Chem. 2003, 51, 2597–2602.

Requejo-Tapia, C. International Trends in Fresh Avocado and Avocado Oil Production and Seasonal. Variation of Fatty Acids in New Zealand-grown cv. Hass. Massey University, Palmerston North, New Zealand, 1999, 212 pp.

Requejo-Jackman, C.; Wong, M.; Wang, Y.; McGhie, T.; Petley, M.; Woolf, A.B. The good oil on avocado cultivars—A preliminary evaluation. The Orchardist. 2005, 78(10), 54–58.

Rosenthal, A.; Pyle, D. L.; Niranjan, K. Aqueous and enzymatic processes for edible oil extraction (review). Enzyme Microb. Tech. 1996, 19, 402–420.

Salvador, M.D.; Aranda, F.; Gómez-Alonso, S.; Fregapane, G. Influence of extraction system, production year and area on Cornicabra virgin olive oil: A study of five crop seasons. Food Chem. 2003, 80, 359–366.

Sant'Anna, B.P.M.; Freitas, S.P.; Coelho, M.A.Z. Enzymatic aqueous technology for simultaneous coconut protein and oil extraction. Grasas y Aceites 2003, 54(1), 77–80.

Servili, M.; Esposto, S.; Lodolini, E.; Selvaggini, R.; Taticchi, A.; Urbani, S.; Montedoro, G.; Serravalle, M.; Gucci, R. Irrigation effects on quality, phenolic composition, and selected volatiles of virgin olive oils cv. Leccino. J. Agr. Food Chem. 2007, 55, 6609–6618.

Sharma, A.; Khare, S.K.; Gupta, M.N. Enzyme-Assisted Aqueous Extraction of Rice Bran Oil. J. Am. Oil Chem. Soc. 2001, 78(9), 949–951.

Sharma, A.; Khare, S.K.; Gupta, M.N. Enzyme-assisted aqueous extraction of peanut oil. J. Am. Oil Chem. Soc. 2002, 79, 215–218.
Solinas, M.; Di Giovacchino, L.; Di Magcola, A. Influencia della temperatura e della durata della gramolatura sul contenuto in polifenoli de oli. La Rivista Italiana delle Sostanse Grasse 1978, 19–23.
Standards Australia. AS5264 Olive oils and olive-pomace oils. Australian Standards. Standards Australia Limited, Sydney, Australia, 2011.
Tano-Debrah, K.; Ohta, Y. Aqueous extraction of coconut oil by an enzyme-assisted process. J Sci. Food Agr. 1997, 74, 497–502.
Uceda, M.; Jiménez, A.; Beltrán, G. Olive oil extraction and quality. Grasas y Aceites. 2006, 57, 25–31.
USDA. 2010. United States Standards for Grades of Olive Oil and Olive-Pomace Oil. http://olivecenter.ucdavis.edu/publications/USDA.pdf. (retrieved 10/10/2011).
Walton, D.A.; Wallace, H.M. Ultrastructure of macadamia (proteaceae) embryos: Implications for their breakage properties. Ann. Bot.—London 2005, 96, 981–988.
Werman, M.J.; Neeman, I. Avocado Oil Production and Chemical Characteristics. J. Am. Oil Chem. Soc. 1987, 64(2), 229–232.
White, A.; Woolf, A.; Hofman, P.; Arpaia, M.L. The International Avocado Quality Manual. The Horticulture & Food Research Institute of New Zealand, 2005, 72 pp.
Wong, M.; Ashton, O.B.O.; Requejo-Jackman, C.; McGhie, T.; White, A.; Eyres, L.; Sherpa, N.; Woolf, A.B. Avocado Oil—The Colour of Quality. In Color Quality of Fresh and Processed Foods. ACS Symposium Series 983. Culver, C.A.; Wrolstad, R.E., Eds.; American Chemical Society, Washington DC, 2008; pp. 328–349.
Wong, M.; Requejo-Jackman, C.; Woolf, A. What is unrefined, extra virgin cold-pressed avocado oil? Inform. 2010, 21(4), 198–201, 259.
Wong, M.; Ashton, O.B.O.; McGhie, T.K.; Requejo-Jackman, C.; Wang, Y.; Woolf, A.B. Influence of Proportion of Skin Present During Malaxing on Pigment Composition of Cold Pressed Avocado Oil. J. Am. Oil Chem. Soc. 2011, 88(9), 1373–1378.
Woolf, A.; Wong, M.; Eyres, L.; McGhie, T.; Lund, C.; Olsson, S.; Wang, Y.; Bulley, C.; Wang, Y.; Friel, E.; Requejo-Jackman, C. Avocado Oil. From Cosmetic to Culinary Oil. In Gourmet and Health-Promoting Specialty Oils. Kamel-Eldin, A; Moreau, R., Eds., The American Oil Chemists' Society, AOCS Press. 2009; pp. 73–125.
Woolf, A.; Requejo-Jackman, C.; Lund, C.; McGhie, T.; Olsson, S.; Eyres, L.; Wang, Y.; Bulley, C.; Wong, M. Avocado oil and other niche culinary oils in New Zealand. In Handbook of Australasian Edible Oils. O'Connor, C.J., Ed., Oils and Fats Specialist Group of NZIC, Auckland, 2007; pp. 12–37.
Zhang, N.Y. Development and optimisation of the aqueous extraction technique for macadamia nut oil. Bachelor of Technology (Food Technology) Final Year Report, Massey University, Auckland, New Zealand, 2007; 67 pp.

3

Aqueous Extraction of Corn Oil After Fermentation in the Dry Grind Ethanol Process

Robert A. Moreau, David B. Johnston, Kevin B. Hicks, and Michael J. Haas
Sustainable Biofuels and Co-products Research Unit, U.S. Department of Agriculture, Wyndmoor, PA, USA

Introduction—A Comparison of Various Processes to Produce Corn Oil

Unlike most edible plant oils that are obtained directly from oil-rich seeds by either pressing or solvent extraction, corn seeds (kernels) have low levels of oil (~4%), and currently almost all commercial corn oil is obtained from pressing or extracting the isolated corn germ (embryo) which is an oil-rich portion of the kernel. Commercial corn oil could actually be called "corn germ oil." The current worldwide production of corn oil is about 2.4 million metric tons, of which about half (2,400 million pounds) is produced in the U.S., followed by China in second place (Moreau, 2011). There are at least 11 different types of corn oil that have been either produced commercially or described in the literature (Table 3-A), and these can have very different compositions and appearance.

Corn Oil from Pressing and/or Solvent Extraction of Wet Milled Corn Germ

Most commercial corn oil is obtained from corn germ that is a by-product of the wet milling industry (Moreau, 2002; 2005). The wet milling process was developed in the late 1800s to optimize the production of corn starch from corn kernels. Wet milling efficiently removes the ~70% starch in corn kernels, and it produces three major by-products that are usually called "co-products" because the considerable revenues from their sale provide additional profit to the corn wet mill. The first co-product is corn germ, which contains 40–50% oil (Table 3-A). The corn germ fraction from the wet milling process represents about 5% of the mass of the kernel (Moreau et al., 1999b). In our experience, when the corn kernel is carefully dissected with a scalpel, the germ (embryo), hull (pericarp and aleurone cells), and endosperm comprise approximately 5, 5, and 90%, respectively, of the mass of the kernel.

Table 3-A. A Comparison of the Feedstock, Process Details, and Yields of Various Corn Oils.

Type of Corn Oil	Feedstock (% oil)	Process	Oil Recovered (wt%)	U.S. Production Million Pounds
Corn oil from pressing and/or solvent extraction of wet milled corn germ	Germ from corn wet mill (40–50%)	Pressing or hexane extraction	90–99%	2200[a]
Corn oil from pressing and/or solvent extraction of dry milled corn germ	Germ from corn dry mill (15–25%)	Pressing or hexane extraction	90–99%	200[a]
Corn oil from pressing and/or solvent extraction of germ dry fractionated before dry grind ethanol production	Germ from dry grind ethanol (15–25%)	Pressing or hexane extraction	90–99%	~20[a]
Corn oil from pressing and/or solvent extraction of germ wet fractionated before dry grind ethanol production	Germ from dry grind ethanol (40–50%)	Pressing or hexane extraction	90–99%	0
Corn oil from aqueous enzymatic extraction of germ from corn wet mills	Germ from wet mill (40–50%)	Water, enzymes, and centrifugation	70–90%	0
Corn oil from aqueous enzymatic extraction of germ from corn dry mills	Germ from corn dry mill (15–25%)	Water, enzymes, and centrifugation	40–60%	0
Corn oil from aqueous enzymatic extraction of germ dry fractionated before dry grind ethanol production	Germ from dry grind ethanol (15–25%)	Water, enzymes, and centrifugation	40–60%	0
Corn oil from aqueous enzymatic extraction of germ wet fractionated before dry grind ethanol production	Germ from dry grind ethanol (40–50%)	Water, enzymes, and centrifugation	80–90%	0
Corn oil extracted from ground corn with ethanol	Ground kernels (~4%)	Ethanol (100%) extraction	90%	0
Corn fiber oil from extraction of wet milled corn fiber with solvents	Corn fiber from corn wet mill (~2%)	Hexane extraction	70–95%	0
Corn oil from dry grind ethanol production obtained after fermentation	Ground corn (~4%)	Typically, evaporation and centrifugation[b]	25–80%	571[b]

[a]Moreau, 2011b.

[b]Estimates for 2012, Jenssen 2013a.

The oil is usually removed by either hexane extraction or mechanical pre-pressing followed by hexane extraction (Moreau, 2002; 2005a). The two additional wet milling co-products are corn gluten feed (an animal feed with high levels of fiber and low levels of protein, ~20%) and corn gluten meal (an animal feed with low levels of fiber and about 60% protein). Corn gluten feed is obtained by mixing corn fiber with condensed steepwater. Corn oil ranks tenth in world production of edible oils, with an annual production of about 2 million tons (Moreau, 2011b).

Corn Oil from Pressing and/or Solvent Extraction of Dry Milled Corn Germ

Approximately 10% of the edible corn oil (corn germ oil) in the United States is obtained from corn germ that is produced by a process called dry milling. Corn dry milling was also developed in the 1800s as a process to remove bran and germ from corn to create a corn meal with a longer shelf life (because of the reduced amount of germ that is susceptible to oxidation) or to produce "flaking grits" for corn flakes and other breakfast cereal products. The corn germ fraction from the dry milling fraction represents about 10% of the mass of the kernel (Moreau et al., 1999b), and it contains about 15–25% oil.

Corn Oil from Pressing and/or Solvent Extraction of Germ Dry Fractionated before Dry Grind Ethanol Production

A third industrial corn milling process is the corn dry grind ethanol process (Fig. 3.1). In its traditional form it does not produce corn oil, but the process has recently been modified by adding two separate ways to produce corn oil. In the traditional corn dry grind ethanol process, corn kernels are milled to a flour or meal, and the entire kernel contents are fermented with yeast to make ethanol. Because only the starch is fermented to ethanol, all of the other components in the kernel (protein, cellulose, hemicelluloses, oil, etc.) remain in the fermentor after the ethanol is removed via distillation. The remaining components (whole stillage, 5–15% solids) are then separated into solids (wet cake, 30–50% solids) and thin stillage (5–10% solids) via centrifugation. Water is removed from the thin stillage via evaporators. The wet cake and concentrated thin stillage are then combined and dried to make an animal feed called "distillers dried grains with solubles" (DDGS). Corn germ can be removed prior to fermentation and the corn germ pressed or extracted with hexane to produce corn oil. It is our understanding that although some corn dry grind ethanol plants contain the fractionation equipment to remove germ before fermentation, very little corn oil is actually produced via this route, mainly due to the current economics.

Corn Oil from Pressing and/or Solvent Extraction of Germ Wet Fractionated before Dry Grind Ethanol Production

In addition to the dry fractionation process described above, it is also possible to remove corn germ in a dry grind ethanol plant before fermentation using "wet" techniques. The first wet fractionation process was called the "quick germ" process (Singh

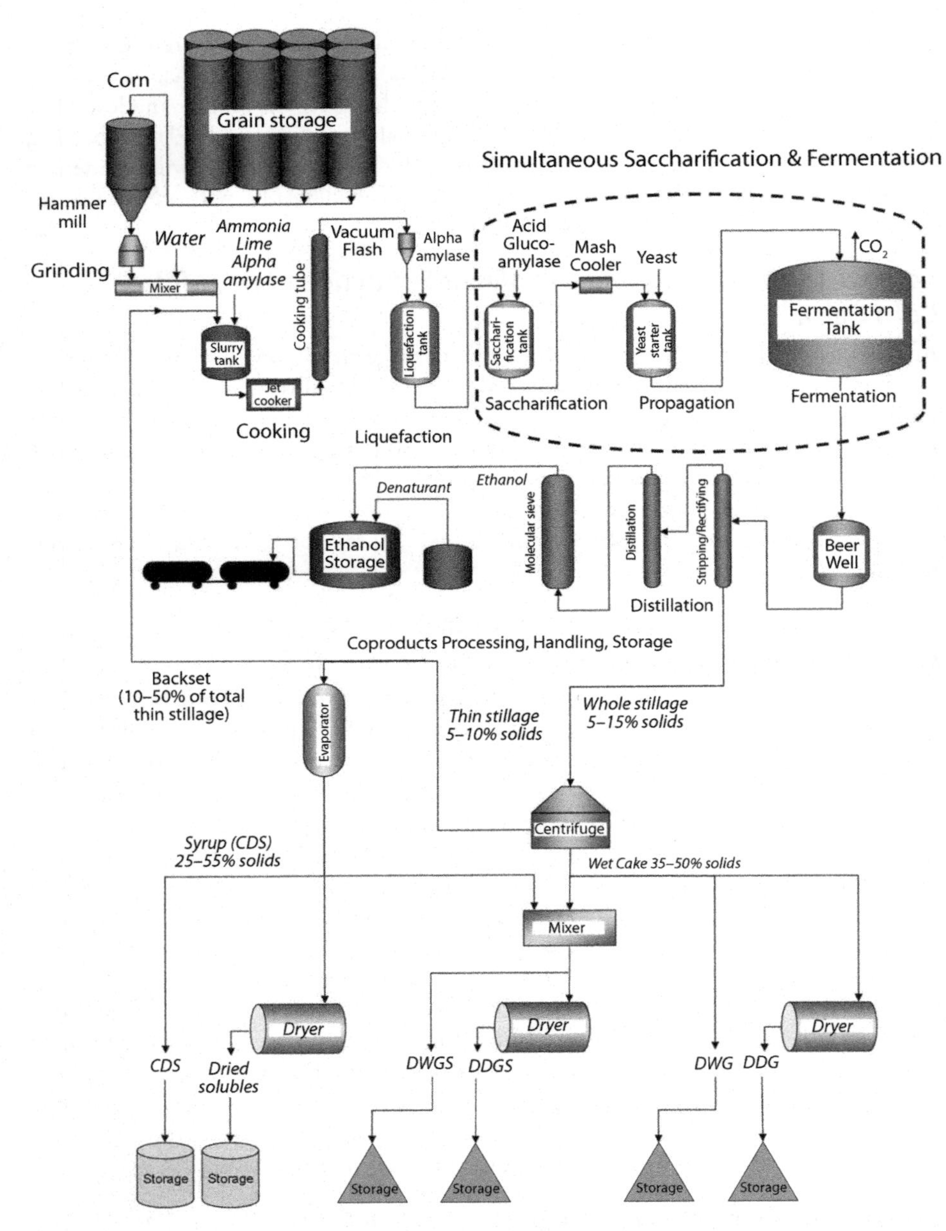

Fig. 3.1. Flow chart of typical corn dry grind ethanol process (Adapted from Rosentrater, K.A., *International Sugar Journal 109:* 1-12, 2007).

& Eckhoff, 1996). It involved soaking kernels in water, gently grinding, and separating the fiber and germ by floatation. Other wet fractionation methods have been developed using enzymes to improve the separation and purity of the germ, fiber, and endosperm fractions (Johnston et al., 2005; Singh et al., 2005).

Corn Oil from Aqueous Enzymatic Extraction of Germ from Wet Milled Corn Germ

An aqueous enzymatic oil extraction (AEOE) process was reported to extract corn oil from wet milled corn germ (Moreau et al., 2004). The process involved milling the germ, incubating it with cellulase enzyme (Multifect GC or GC220), and centrifugation to float the free oil. Maximum yields were about 80% of the total oil in the germ, relative to hexane extraction.

Corn Oil from Aqueous Enzymatic Extraction of Germ from Dry Milled Corn Germ

The above AEOE process was also adapted to extract corn oil from dry milled corn germ (Moreau et al., 2009). Using a combination of an acidic cellulase (GC220) and an alkaline protease (Alcalase), we were able to extract corn oil from dry milled corn germ, but the oil yields were lower (65%, relative to hexane extraction of the germ).

Corn Oil from Aqueous Enzymatic Extraction of Germ Dry Fractionated before Dry Grind Ethanol Production

The above AEOE process was also adapted to extract corn oil from germ that was dry fractionated before dry grind ethanol production (Moreau et al., 2009). Using a combination of an acidic cellulase (GC220) and an acidic protease (GC106), we were able to extract corn oil from dry fractionated corn germ, and the oil yields were 60–70% (unpublished results).

Corn Oil from Aqueous Enzymatic Extraction of Germ Wet Fractionated before Dry Grind Ethanol Production

The above AEOE process was also adapted to extract corn oil from germ that was wet fractionated before dry grind ethanol production (Moreau et al., 2009). Using a combination of an acidic cellulase (GC220) and an alkaline protease (Alcalase) and no heat or microwave treatment, we were able to extract corn oil from wet fractionated germ, "e-germ," prepared using an enzymatic method (Johnston et al., 2005; Singh et al., 2005). The corn oil yield was the highest that has been reported for any type of corn germ, about 90% (Moreau et al., 2009).

Corn Oil Extracted from Ground Corn with Ethanol

Two additional processes have been developed to produce corn oil, but currently no commercial corn oil is being produced in the United States via these methods. The first type of experimental corn oil is called "corn kernel oil," and it is produced by the extraction of ground corn with ethanol. Although it has long been known that oil could be obtained directly by extracting it from ground corn, the low levels of oil in most corn kernels (3–5%) have indicated that obtaining corn oil directly from corn kernels may not be economical (Table 3-A). However, if the corn oil extraction process could be linked to a second profitable process such as extraction of zein protein, then the economics of corn kernel oil production might be favorable. In 1992, Hojilla-Evangelista et al. (1992) at Iowa State University developed a "sequential extraction" process which involved an initial extraction of corn oil from flaked corn with ethanol (100%), followed by an extraction of zein protein with 70% ethanol and finally an extraction of the remaining proteins and starch. Although the economics of the sequential extraction process have been rigorously debated, some recent modifications of the process have improved the economics (Hojilla-Evangelista et al., 1992; Feng et al., 2002). A group at the University of Illinois developed a similar two step COPE (Corn Oil and Protein Extraction) process which also focused on the ethanol extraction of corn oil and zein, but it included the use of membrane filters to process both products (Kwiatkowski & Cheryan, 2002). Recently, our laboratory demonstrated that corn kernel oil should have health-promoting properties because of its very high levels of lutein and zeaxanthin (valuable for preventing macular degeneration) and its moderate levels of tocopherols, tocotrienols, and phytosterols (Moreau et al., 2007).

Corn Fiber Oil from Extraction of Wet Milled Corn Fiber with Solvents

The second type of experimental corn oil is "corn fiber oil." Corn fiber oil was first developed in our laboratory in the mid to late 1990s (Moreau et al., 1996; 1998). The corn fiber coproduct from wet-milling was known to contain oil (2–3%), and most experts assumed that this small amount of oil was probably attributable to contamination of the fiber with a small amount of corn germ (Table 3-A). We reported that the chemical composition of corn fiber oil, obtained by extracting wet milled corn fiber with either hexane or supercritical CO_2, was very different from that of commercial corn oil (Table 3-A). Whereas commercial corn oil contains about 1% phytosterols (plant sterols), corn fiber oil contains 10–15% phytosterols (Moreau et al., 1996). During the 1990s there was much international interest in phytosterols because of several clinical studies that had demonstrated that eating 1–2 grams per day of phytosterols could reduce the levels of low density lipoprotein (LDL) cholesterol in the blood by 10–15%. In 1995 the first phytosterol nutraceutical, a margarine called Benecol, was marketed in Finland by Rasio. In 2000, the U.S. Food and Drug Administration approved the sale of phytosterol-enriched margarines in the United States, followed by approval of other types of phytosterol-enriched functional foods (Moreau et al., 2003).

Corn Oil from Dry Grind Ethanol Production Obtained after Fermentation via Centrifugation

The most widely used method for corn oil recovery by dry grind ethanol producers today involves removal from the thin stillage stream after fermentation using processes developed by GreenShift Corporation. The processes are described in six patents (Cantrell and Winsness, 2009, 2011a,b, 2012; Winsness and Cantrell, 2009 and Winsness, 2012). The GreenShift processes were extensively reviewed by the U.S. Environmental Protection Agency (EPA) during promulgation of the RFS2 (EPA, 2011, EPA, 2012). The EPA found that the economic payback for back-end oil recovery was much more favorable than that for front-end processes, and estimated that more than 60% of dry grind producers will be using this process to recover oil from thin stillage by 2013 (EPA 2011). A recent article estimated that oil recovery is now being used in approximately 80% ~200 corn dry grind ethanol plants currently in operation in the United States (Jenssen, 2013b), and the approximately 571 million pounds (72 million gallons) of post fermentation corn oil were produced in the United States in 2012 (Table 3-A) (Jenssen, 2013a). The 2012 yield of post fermentation oil (571 million pounds) represents a yield of about 23% compared to the 2010 yield of commercial corn oil (2.4 billion pounds, all from germ) (USDA, 2011).

In addition to the Greenshift patents, recent U.S. patents on back-end corn oil extraction have also been issued to ICM (Gallop et al., 2012) and Primafuels (Woods et al., 2012). Several patent applications on back-end oil extraction have also been have also been submitted and published by POET (Bootsma, 2010, 2011, 2012).

Recent publications have described physical, chemical, and enzymatic processes to increase the amount of oil that can be recovered from condensed corn distillers soluble (Majoni et al., 2011a,b). Researchers at Iowa State University reported the effect of corn breaking method (Wang et al., 2008) and low-shear extrusion on corn fermentation and the oil partition in a simulated corn dry ethanol and post-fermentation centrifugation process. Liu (2013) recently described enhanced chemical and physical processes to fractionate oil from condensed distillers solubles. Moss (2012) also suggested that two technologies for corn oil extraction could potentially be used in tandem in a corn dry grind ethanol plant—corn germ could be removed by dry degermination process and corn oil could be produced from it by hexane extraction (with an estimated oil yield of 0.25–0.50 lbs of oil per bushel of corn) and the remaining corn oil could be extracted at the back end by heating and centrifugation (with an estimated oil yield of 1.3–1.5 lbs of oil per bushel of corn).

The Processes for Obtaining Corn Oil After Fermentation in a Dry Grind Ethanol Plant

The Dry Grind Ethanol Process

The conventional corn dry grind process for ethanol production begins with the transfer of corn from storage silos (Fig. 3.1). The corn is cleaned to remove foreign material and ground using a hammer mill with a screen size typically less than 4 mm. The flour produced has a particle size distribution with more than 80% passing through a 2 mm screen (Rausch et al., 2005). The ground corn flour is mixed with water to form a slurry, and a thermostable alpha amylase is added. The temperature of the slurry is increased to gelatinize the starch using a jet cooker and held at an elevated temperature for a short period of time. During the heating and hold periods, the alpha amylase converts the gelatinized starch into dextrins (short alpha linked glucose chains), which reduces the viscosity of the slurry and "liquefies" the slurry.

The slurry is cooled through heat exchangers, and the pH is reduced to about 4.5. A glucoamylase enzyme is added to convert the dextrins into glucose (saccharification). Yeast is also added to biologically convert the glucose into ethanol and carbon dioxide during the fermentation. The glucose production and fermentation will occur simultaneously and is referred to as simultaneous saccharification and fermentation (SSF). Fermentation takes place for about 60 hours.

After fermentation, the beer is passed through a series of distillation columns to remove and concentrate the ethanol from the remaining water and solids. Through distillation, ethanol can only be concentrated to about 95% because ethanol and water form an azeotrope. The remaining water is removed using molecular sieves to produce 100% (200 proof) ethanol.

The liquid and solids remaining after ethanol recovery, called whole stillage, are separated using a decanter centrifuge. The liquid (thin stillage) is concentrated into syrup using evaporators to about 30% solids. The solids (wet cake) are mixed with the syrup, and dried together to produce distillers dried grains with solubles.

Back-end Oil Recovery

Downstream (or sometimes called back-end) corn oil recovery refers to the separation of oil after fermentation in the dry grind process. The majority of information available on how the oil separation is accomplished is disclosed in patent literature described in the previous section.

The downstream oil recovery in dry grind ethanol plants is generally completed by concentrating the thin stillage stream by evaporation, recovering the oil by centrifugation to form two easily separable phases, and then further evaporating the stream. For plants that centrifuge the concentrated thin stillage for oil recovery, much of the oil in the thin stillage stream is not available as free oil and cannot be recovered efficiently by simple centrifugation. GreenShift's patents disclose a number of means to help release the oil from the thin stillage, including thermal treatment and/or chemical treatment prior to centrifugation. The chemical treatments are compounds added to the processing stream, and are generally described as a family of proprietary compounds that are designed to release emulsified oil, thereby increasing recovery during centrifugation. The specific amounts, conditions, times, and temperatures used are proprietary.

The general process for oil recovery is outlined in Fig. 3.2. Two locations are shown where corn oil recovery integrates with the ethanol plant. The first is recovery of corn oil from the thin stillage circuit, and the second treats the whole stillage for subsequent enhanced recovery of oil from the thin stillage.

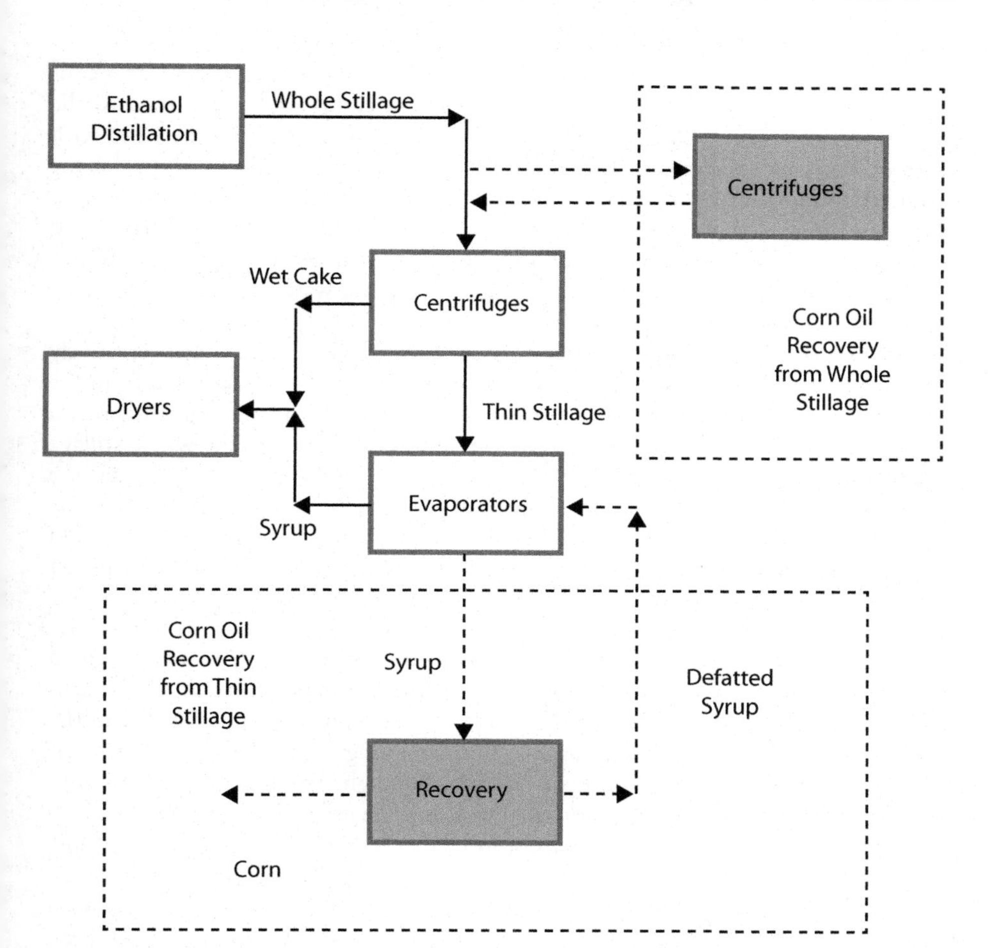

Fig. 3.2. General processing steps used for oil recovery from thin stillage. (Figure was kindly provided by GreenShift, Inc.)

The thin stillage is concentrated in the ethanol plant using multi-effect evaporators. This process removes water and concentrates the solids, including the oil, to about 40%. It also increases the specific gravity of the stream, increasing the difference between the aqueous and lipid components of the stream. During the evaporation process, at the appropriate solids content, the concentrated thin stillage is pretreated by heating to as high as 100°C and holding at the elevated temperature. The optimum conditions for holding times and temperatures are not disclosed; however, they range from a few minutes to several hours and from 65–100°C. It should be noted that if the conditions used are too extreme, darkening of the syrup will occur and result in adverse effects on DDGS quality.

Following pretreatment, the syrup stream is generally processed through a continuous centrifuge to separate the oil. The de-fatted syrup is then returned to the evaporation process or finished syrup tank. The separation can be accomplished with different types of centrifuges. The disc-stack style of centrifuge appears to be the preferred option because its design and application of higher centrifugal forces than other centrifuge designs render disc-stack centrifuges capable of removing bound and emulsified oil; however, reports of using three-phase decanter centrifuges are also available. Both types of centrifuge can handle the oil and water separation as well as the small amount of solid material that is also present in the syrup stream.

When the second recovery system is added, additional centrifuges effectively "wash" the whole stillage, freeing bound corn oil so that it can be recovered from the thin stillage. The entire whole stillage stream is run through the new centrifuges and returned to the plant just downstream of the point of extraction. The whole stillage proceeds to the plant centrifuges for separation of the liquids (thin stillage) and the solids (wet cake). Corn oil recovery can be doubled with the addition of this step.

Valicor, Inc. also developed a unique technology for increasing oil extraction rates and recovering other co-products. According to James Bleyer, program manager at Valicor, "Instead of focusing on thin stillage, Valicor has developed a technology to process whole stillage. This gives us access to a much broader range of extractables including corn oil, proteins, and fiber." This technology called VFRAC was presented at the 2013 FEW conference in St. Louis, Missouri, USA. According to the presentation, the VFRAC process delivers oil yields greater than 1 lb/bushel of corn. The technology for processing whole stillage is outlined in Fig. 3.3. Whole stillage is separated into a heavy fraction and a light fraction. The light fraction, which is high in oil and protein, undergoes a hydrothermal treatment that enables facile separation of the oil from the protein. Valicor uses its proprietary centrifuge technology specifically designed for a three-phase separation: corn oil, a clarified liquid termed stickwater, and a solids fraction termed VFRAC solids. The heavy fraction is milled in a disc mill then rinsed with stickwater from the processing of light fraction. This mixture is then centrifuged using the plants existing decanting centrifuges. This process washes starch and oil from the heavy fraction into the centrate. The decanter centrate is the used as backset in the front end of the process. The oil in the backset can then be recovered as it is cycled through the front end of the plant to the VFRAC process. A portion of the stickwater is evaporated, combined with the wet cake from the decanter and dried to form dried distiller grains with solubles (DDGs).

The Chemical Composition of Corn Oil Extracted from Corn Germ, Ground Corn, and Corn DDGS with Various Organic Solvents and via Centrifugation after Fermentation

The chemical composition and appearance of the various corn oils can be very different based on the feedstock (corn germ, ground corn, corn fiber, etc.) and extraction

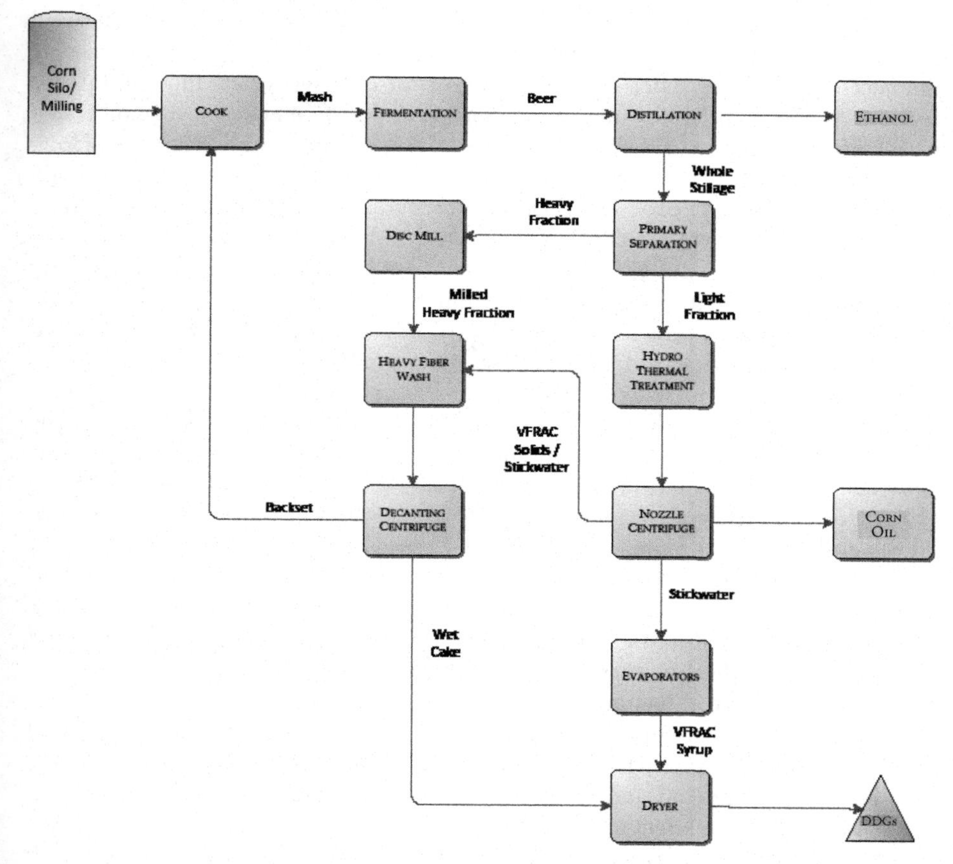

Fig. 3.3. Processing steps for a new process to recover corn oil from whole stillage (Figure was kindly provided by Valicor, Inc.)

process (pressing, hexane extraction, aqueous enzymatic extraction, post fermentation centrifugation, etc.) used to produce the oil (Table 3-B). Also, the levels of various components in corn oils is usually altered when corn oils are processed, customarily using conventional refining, bleaching, and deodorization. Refining, bleaching, and deodorization are performed on most edible oils to improve the flavor and smell and to increase shelf life (Moreau et al., 2010).

All crude (unrefined) edible oils contain free fatty acids, usually in the range of 1–3%. Free fatty acids are very prone to oxidation, and are usually removed by refining (with alkali or via physical refining which utilizes heat and vacuum). The lengthy high-temperature thermal processes during corn fermentation are a probable cause of the higher free fatty acid concentration in post fermentation corn oil. Another possible explanation is that the process used in all current dry grind ethanol plants diverts and recycles about half of the thin stillage, or "backset," for use during starch liquefaction (Fig. 3.1). The corn oil in this recycled backset is re-introduced into the pretreatment and fermentation process, subjecting the oil to factors causing the release of free fatty acids (e.g., heat, lipases, or other enzymes) in accumulative fashion. A unique and unfortunate characteristic of post fermentation corn oil is that most contain exceptionally high levels of free fatty acids, often more than 10%. However, POET's Viola brand corn oil (produced with raw starch hydrolyzing enzymes, eliminating the need to cook and gelatinize the starch) is marketed as a premium oil because it is certified to contain less than 5% free fatty acids (Winkler-Moser and Breyer, 2011; Kotrba, 2013). At one time, removal of these high levels of free fatty acids in post fermentation corn oil via refining was thought to be cost prohibitive, but recently a refined post fermentation corn oil—Corn Oil One™—with reduced levels of waxes, moisture and fatty acids, is being marketed by FEC solutions (www.cornoilone.com).

In our experience, the fatty acid composition of all types of corn oil are nearly identical, with linoleic acid being the most abundant fatty acid, followed by oleic, and then palmitic acids (Moreau, 2009; Moreau et al., 2011). Majoni and Wang (2010) recently reported that the precipitate often found in post fermentation corn oil was enriched in free palmitic acid.

The levels of total phytosterols in post fermentation corn oil were reported to be about 5% (Moreau et al., 2010); however, the accuracy of this measurement needs to be confirmed because the authors noted that the normal phase HPLC method used for this quantification may not be reliable (Moreau et al., 2010).

The levels of tocopherols and tocotrienols in post fermentation corn oil are higher than in corn germ oil, and are comparable to those that we previously reported in ethanol extracted corn kernel oil (Moreau & Hicks, 2005). The presence of these high levels of tocopherols and tocotrienols may contribute to the enhanced oxidative stability of post fermentation corn oil as reported by Winkler-Moser and Breyer (2011) and noted in the last section of this chapter.

Polyamine conjugates (diferuloylputrescine and p-coumaroylferuloylputrescine, DFP and CFP) were previously reported to be present at levels of 0.5–0.8% in ethanol extracted corn fiber oil (Moreau & Hicks, 2005) and not in any other types of corn oil (Table 3-B). Polyamine conjugates (DFP and CFP) were not detectable in post fermentation corn oil (Table 3-B). A possible explanation for their absence in post fermentation corn oil is that because the process to obtain this oil involves centrifugation to separate the oil from a larger aqueous phase, it is possible that the polyamine conjugates partition into the aqueous phase, similar to the "degumming" process to remove phospholipids from crude edible oils (Moreau et al., 2010).

The levels of xanthophylls (lutein and zeaxanthin) in post fermentation corn oil impart the oil with a reddish-orange or amber color, similar in appearance to ethanol extracted corn kernel oil (personal observation). This observation is confirmed by the fact that ethanol extracted corn kernel oil and post fermentation corn oil both contain nearly identical levels of xanthophylls. Lutein and zeaxanthin are important in human health because they are present in the macula region of the retina, and it

Table 3–B. The Major Compositional Characteristics of Various Types of Corn Oil (wt%).

Type of Corn Oil	Free Fatty acids	Phyto–sterols	Toco–pherols	Toco–trienols	DFP & CFP	Xanthophylls	Reference
Corn oil from pressing and/or solvent extraction of wet milled corn germ	1–3	~1	0.12	0.005	0	0.0002	Moreau & Hicks, 2005
Corn oil from pressing and/or solvent extraction of dry milled corn germ	1–3	~1	0.03	0	0	0.00020	Moreau & Hicks, 2005
Corn oil from pressing and/or solvent extraction of germ dry fractionated before dry grind ethanol production	1–3	~1	NR	NR	NR	NR	Moreau et al., 2009
Corn oil from pressing and/or solvent extraction of germ wet fractionated before dry grind ethanol production	1–3	~1	NR	NR	NR	NR	Moreau et al., 2009
Corn oil from aqueous enzymatic extraction of germ from corn wet mills	1–3	~1	NR	NR	NR	NR	Moreau et al., 2004
Corn oil from aqueous enzymatic extraction of germ from corn dry mills	1–3	NR	NR	NR	NR	NR	Moreau et al., 2004
Corn oil from aqueous enzymatic extraction of germ dry fractionated before dry grind ethanol production	1–3	NR	NR	NR	NR	NR	Moreau et al., 2009
Corn oil from aqueous enzymatic extraction of germ wet fractionated before dry grind ethanol production	1–3	NR	NR	NR	NR	NR	Moreau et al., 2009
Corn oil extracted from ground corn with ethanol	1–3	2–3	0.15	0.03–0.05	0.5–0.8	0.02	Moreau & Hicks, 2005
Corn fiber oil from extraction of wet milled corn fiber with solvents	1–3	10–15	0.08	0.01–0.02	0	0.005	Moreau et al., 1996
Corn oil from dry grind ethanol production obtained after fermentation via centrifugation	10–15	5	0.10	0.03–0.04	0	0.02	Moreau et al., 2010; Moser-Winkler & Breyer, 2011

Abbreviation: NR, not reported

was estimated that consumption of about 30 mL of ethanol extracted corn kernel oil per day would supply the ~6 mg of lutein plus zeaxanthin per day thought to be necessary to slow the progression of age-related macular degeneration (Moreau et al., 2007). The high levels of xanthophylls in post fermentation corn oil make it attractive as a feed for chickens because xanthophylls (usually from dried marigold flower petals or from corn gluten meal) are often added to poultry feed to enhance the yellow color of the egg yolks and sometimes to add yellow color to the fat of chicken meat (Moreau et al., 2007).

The Changes in Corn Oil Composition That Occur During the Dry Grind Ethanol Process

A study was conducted to investigate changes in the lipid composition of distillers dried grains with solubles (DDGS) and other intermediate fractions during the dry grind ethanol process (Moreau et al., 2011). Before this study, the high levels of free fatty acids in DDGS and in post fermentation corn oil had been reported, but the levels of other lipids had not been studied in detail. Examination of the fatty acids revealed that during the cooking step (to gelatinize the starch) there was a small but significant decrease in the levels of linoleic (~2%) and oleic acids (~1%) and a small increase in the levels of palmitic acid (~3%). The fatty acid composition remained essentially the same for the remaining steps of the process.

The levels of phytosterols, tocopherols, and tocotrienols remained relatively constant in all fractions during the dry grind ethanol process, although there were some small significant changes in phytosterols in some of the fractions (Moreau et al., 2011). Unlike the previous report of a significant contribution of yeast amino acids to DDGS (Han & Liu, 2010), no evidence of a contribution of yeast lipids was noted in any of the fractions in the dry grind ethanol process, as evidenced by no measurable levels of ergosterol (the primary sterol in yeast) or any other major changes in sterols or tocopherols (Moreau et al., 2011).

Current Applications of Corn Oil Obtained After Fermentation in a Dry Grind Ethanol Plant

As noted in the previous sections, corn oil recovered from dry grind ethanol facilities by centrifugation of condensed syrup or by solvent extraction from DDGS contains a mixture of valuable lipid components, vitamins, nutraceuticals, and xanthophyll pigments along with free fatty acids, waxes, and other impurities which render the product inedible. While refining the oil to produce an edible grade is possible, it is generally not economical to do so. Thus, such oils are currently either used as feedstocks in a wide array of industrial products, such as biodiesel and various specialty chemicals, or as a component of animal feeds.

Of the many applications for such corn oils, biodiesel currently presents the most promise and is the most utilized today. Distillers corn oil (DCO) use for biodiesel was about 40.5 million gallons in 2011, to 76 million gallons in 2012—less than ten percent of the 1 billion-plus gallons of biodiesel produced both years (Biodiesel, 2013). If the EPA's projections are accurate, production of post fermentation corn oil will increase to over 300 million gallons per year by 2013, as biodiesel production increases to meet mandates and other demand drivers (EPA, 2012). Despite the free fatty acid content of post fermentation oil, it remains in significant demand as a feedstock for biodiesel in lieu of higher cost and lesser quality vegetable oils and animal fats. For example, with cloud points ranging between –3°C and –5°C, biodiesel derived from post fermentation corn oil (corn oil methyl ester, or CME) has been shown to achieve superior cold flow properties as compared to biodiesel produced from soybean oil.

The topic of biodiesel production from these oils has been recently and comprehensively discussed (Haas, 2012), and the reader is directed to that source for extensive information. Haas notes that little information has been published on conversion of corn oil to biodiesel, probably because its premium quality for food uses would prevent its use for making biodiesel in normal situations.

Recently, El Boulifi, et al. (2010) determined the optimum process conditions for production of biodiesel from pure, refined corn oil using a simple potassium hydroxide catalyst. Corn oil from DDGS or condensed distillers solubles (CDS or syrup) is not a pure oil, and it often contains more than 10% free fatty acids (Janes et al., 2008; Winkler-Moser & Vaughn, 2009; Moreau et al., 2010; Moreau et al., 2011) plus other impurities, making it an "inedible" oil. To produce biodiesel from this crude oil, a two-stage process is needed (Canacki & Van Gerpen 2003). The first stage, typically an acid-catalyzed process in methanol, converts free fatty acids into methyl esters. Following removal of the water produced in the first stage, a second stage using alkaline-catalyzed transesterification completes the conversion of acylglycerols into fatty acid methyl esters (FAME). Such a process is favorable when compared to refining processes that strip free fatty acids because, on a pound for pound basis, the market value of corn oil methyl esters (CME) is greater than the market value of free fatty acids.

Noureddini et al. (2009) developed a modified acid-base-catalyzed process to produce biodiesel from crude corn oil extracted from CDS. In this process, a strong anion exchange resin was used to neutralize the substrate after the initial acid-catalyzed stage. The process was said to effectively remove the acid catalyst as well as residual free fatty acids prior to the base-catalyzed step, allowing 98% conversion to FAME. By processes such as these, crude corn oil from dry grind ethanol plants, inclusive of its fatty acid content, can be used to make biodiesel. However, some of the impurities in the oil such as waxes, steryl glucosides (Moreau et al., 2008), steradienes and disteryl ethers (Kapicak et al., 2010), and other components as described by Haas (2012) can require further processing such as degumming and other forms of filtration.

Oxidative stability of fatty acids in biodiesel is also important for quality and storage life of biodiesel. El Boulifi et al. (2010) recently studied the oxidative stability of refined corn oil biodiesel, and found that with storage, acid value, peroxide value, and viscosity all increased while the iodine value decreased. Sanford et al. (2009) produced biodiesel from corn oil recovered from DDGS, and found that it met all specifications except for oxidative stability. In contrast, Winkler-Moser and Breyer (2011) studied the composition and oxidative stability of crude corn oil from dry grind ethanol plants. It was found that these crude oils had higher oxidative stability than oil extracted from corn germ (the source of commercial corn oil). It was concluded that the many powerful antioxidants in these crude corn oils, including tocopherols, tocotrienols, phytosterols, steryl ferulates, carotenoids, and xanthophylls, probably are responsible for their oxidative stability. Whether this oxidative stability is conferred to biodiesel made from these antioxidant-containing crude corn oils has not been determined.

With further research and development it is likely that post fermentation corn oil will be used as a feedstock for other types of biofuels, such as "green diesel" and other "drop in" fuels by using hydrotreating and other catalytic processes that are currently being developed. Until other domestic sources of inexpensive oils are developed from algae, jatropha, or other sources now in development, inedible post fermentation corn oil will likely remain one of the most attractive feedstocks for biodiesel and other biofuel producers.

Little has been published regarding the use of crude corn oils from ethanol plants in animal feeds. Corn oil provides energy needed in rations for poultry and other animals. Since this energy can be achieved by any lowest-cost combination of fats and oils, these crude corn oils may not have a premium value, especially if high free fatty acid values are a feed deterrent. One aspect that has not been commercially realized yet and may be in the near future is the use of these crude corn oils for their unique non-fat components. Zeaxanthin, lutein, beta-carotene, phytosterols, tocopherols, and tocotrienols in these oils have great value. It has been suggested that these could be refined from these oils and sold as high value products and then the purified corn oil could be used for food or non-food uses. Although a considerable amount of post fermentation corn oil is being used for feed applications, we are not aware of any companies that are fractionating the oil and marketing products that are enriched in carotenoids or other valuable components. We anticipate that this topic will be pursued in the near future.

Acknowledgments: The authors would like to thank GreenShift Corporation and Valicor, Inc. for providing figures

Note: Mention of trade names or commercial products in this publication is solely for the purpose of providing specific information and does not imply recommendation or endorsement by the U.S. Department of Agriculture. USDA is an equal opportunity employer.

References

Bootsma, J. Oil composition and method of producing the same, U.S. Patent Application Publication, U.S. 2010/0058649, 2010.

Bootsma, J. Oil composition and method of producing the same, U.S. Patent Application Publication, U.S. 2011/0086149, 2011.

Bootsma, J. Oil composition and method of producing the same, U.S. Patent Application Publication, U.S. 2013/0109873, 2013.

Canacki, M.; Van Gerpen, J. A pilot plant to produce biodiesel from high free fatty acid feedstocks. Transactions of the ASAE 2003, 46, 945–954.

Cantrell, D.F.; Winsness, D. Method of processing ethanol byproducts and related subsystems. U.S. Patent 7,601,858, 2009.

Cantrell, D.F.; Winsness, D. Method of processing ethanol byproducts and related subsystems. U.S. Patent 8,008,516, 2011a.

Cantrell, D.F.; Winsness, D. Method of recovering oil from thin stillage. U.S. Patent 8,008,517, 2011b.

Cantrell, D.F.; Winsness, D. Methods of processing ethanol byproducts and related subsystems. U.S. Patent 8,283,484, 2012.

Dickey, L.C.; Kurantz, M.J.; Johnston, D.B.; McAloon, A.J.; Moreau, R.A. Grinding and cooking dry-fractionated corn germ to optimize aqueous enzymatic oil extraction. Ind. Crops Prod. 2010, 32, 36–40.

El Boulifi, N.; Bouaid, A.; Martinez, M.; and Aracil, J. Process optimization for biodiesel production from corn oil and its oxidative stability. Int. J. Chem. Eng. 2010, Article ID 518070, doi:10.1155/2010/518070

EPA, 2011, Regulation of fuels and fuel additives: 2011 Renewable Fuel Standards, Federal Register Vol. 76, No. 127, Friday, July 1, 2011, 38844–38890, 2011. http://www.gpo.gov/fdsys/pkg/FR-2011-07-01/pdf/2011-16018.pdf.

EPA, 2012, Regulation of fuels and fuel additives: 2012 Renewable Fuel Standards, Federal Register Vol. 77, No. 5, Monday, January 9, 2012. 1320–1358, http://www.gpo.gov/fdsys/pkg/FR-2012-01-09/pdf/2011-33451.pdf.

Feng, F.; Myers, D.J.; Hojilla-Evangelista, M.P.; Miller, K.A.; Johnson, L.A.; Singh, S.K. Quality of corn oil obtained by sequential extraction processing. Cereal Chem. 2002, 79, 707–709.

Gallop, C.C.; Cooper, T.; Dieker, A. Bio-oil recovery, U.S. Patent 8,192,627, 2012.

Haas, M.J. Extraction and use of DDGS lipids for biodiesel production, in Distillers Grains: Production, Properties, and Utilization, Liu, K.S.; Rosentrater, K.A., Eds. CRC Press, Boca Raton, 2012; pp. 487–502.

Hojilla-Evangelista, M.P.; Johnson, L.A.; Myers, D.J. Sequential extraction processing of flaked whole corn: Alternative corn fractional technology for ethanol production. Cereal Chem. 1992, 69, 643–647.

Janes, M.; Bruinsma, K.; Cooper, T.; Endres, D. 2008. Solvent extraction of oil from distillers dried grains and methods of using extraction products. 2008, International Patent WO2008039859.

Jenssen, H., Corn oil makes the grade, Ethanol Produce Magazine, April 16, 2013a http://ethanolproducer.com/articles/9755/corn-oil-makes-the-grade.

Jenssen, H., Harnessing corn oil power, Ethanol Producer Magazine, August 12, 2013b, http://ethanolproducer.com/articles/10144/harnessing-corn-oil-power.

Johnston, D.B.; McAloon, A.J.; Moreau, R.A.; Hicks, K.B.; Singh, V. Composition and economic comparison of germ fractions derived from modified corn processing technologies. JAOCS 2005, 82, 603–608.

Kapicak, L.A. Biodiesel from crude corn oil (CCOE). Proceedings of the Annual Meeting of the Center for Process Analysis and Control. http://depts.washington.edu/cpac/Activities/Meetings/Satellite/2010/Thursday/Kapicak%20Final%20CPAC%20Presentation.pdf 2010, Accessed September 30, 2013.

Kotrba, R. The distillers corn oil craze, Biodiesel Magazine May 1, 2013 http://www.biodiesel-magazine.com/articles/9084/the-distillers-corn-oil-craze.

Kwiatkowski, J.R.; Cheryan, M. Extraction of oil from ground corn using ethanol. J. Am. Oil Chem. Soc. 2002, 79, 825–830.

Liu, K.S. Chemical composition of distillers grains, a review, JAOCS 2001, 59, 1508–1526.

Liu, K.S., Barrows, F.T., Methods to recover value-added coproducts from dry grind processing of grains to fuel ethanol, J. Agric Food Chem. 2013, 61, 7325–7332.

Majoni, S.; Wang, T.; Johnson, L.A. Physical and chemical processes to enhance oil recovery from condensed corn distillers solubles, JAOCS 2001, 88, 425–434.

Majoni, S.; Wang, T. Characterization of oil precipitate and oil extracted from condensed corn distillers solubles, JAOCS 2010, 87, 205–213.

Majoni, S.; Wang, T.; Johnson, L.A. Enzyme treatments to enhance oil recovery from condensed corn distillers solubles, JAOCS 2011, 88, 523–532.

Moreau, R.A. Corn oil. In Bailey's Industrial Oil & Fat Products, Shahidi, F., Ed.: 6th ed., Vol. 2, Edible Oil & Fat Products: Edible Oils. Wiley-Interscience: Hoboken, 2005, pp. 149–172.

Moreau, R.A. Aqueous enzymatic oil extraction from seeds, fruits, and other oil-rich plant materials, in RSC Green Chemistry No. 10: Alternatives to Conventional Food Processing, A. Proctor, Ed.; RSC Publishing, Cambridge, 2011a, pp. 341–366.

Moreau, R.A. Corn Oil, in Vegetable Oils in Food Technology—Composition, Properties and Uses, Gunstone, F.D., Ed.; Blackwell Publishing, Sussex, UK, 2011b, pp. 273–289.

Moreau, R.A.; Hicks, K.B. The composition of corn oil obtained by the alcohol extraction of ground corn. JAOCS 2005, 82, 809–815.

Moreau, R.A.; Hicks, K.B. A reinvestigation of the effect of heat pretreatment of corn fiber on the levels of extractable tocopherols and tocotrienols. J. Agric. Food Chem. 2006, 54, 8093–8102.

Moreau, R.A.; Johnston, D.B.; Hicks, K.B. A comparison of the levels of lutein and zeaxanthin in corn germ oil, corn fiber oil, and corn kernel oil. JAOCS 2007, 84, 1039–1044.

Moreau, R.A.; Lampi, A.M.; Hicks, K.B. Fatty acid, phytosterol, and polyamine conjugate profiles of edible oils extracted from corn germ, corn fiber, and corn kernels. JAOCS 2009, 86, 1209–1214.

Moreau, R.A.; Nuñez, A; Singh, V. Diferuloylputrescine and p-coumaroyl feruloylputrescine, abundant polyamine conjugates in lipid extracts of maize kernels. Lipids 2001, 36, 839-844.

Moreau, R.A.; Powell, M.J.; Hicks, K.B. Extraction and quantitative analysis of oil from commercial corn fiber. J. Agric. Food Chem. 1996, 44, 2149–2154.

Moreau, R.A.; Powell, M.J.; Singh, V. Pressurized liquid extraction of polar and nonpolar lipids in corn and oats with hexane, methylene chloride, isopropanol, and ethanol. JAOCS 2003, 80, 1063–1067.

Moreau, R.A.; Scott, K.M.; Haas, M.J. The identification and quantification of steryl glucosides in precipitates from commercial biodiesel. JAOCS 2008, 85, 761–770.

Moreau, R.A.; Singh, V.; Hicks, K.B. A comparison of oil and phytosterols in the seeds of germplasm accessions of corn, teosinte, and Job's tears. J. Agric. Food Chem. 2001, 49, 3793–3795.

Moreau, R.A.; Whitaker, B.D.; Hicks, K.B. Phytosterols, phytostanols, and their conjugates in foods: structural diversity, quantitative analysis, and health-promoting uses. Prog. Lipid Research 2002, 41, 457–500.

Moreau, R.A.; Hicks, K.B.; Johnston, D.B.; Laun, N.P. The composition of crude corn oil recovered after fermentation via centrifugation from a commercial dry grind ethanol process. JAOCS 2010, 87, 895–902.

Moreau, R.A.; Hicks, K.B.; Nicolosi, R.J.; Norton, R.A. Corn fiber oil—its preparation, composition, and use. U.S. Patent 5,843,499, 1998.

Moreau, R.A.; Johnston, D.B.; Powell, M.J.; Hicks, K.B. A comparison of commercial enzymes for the aqueous enzymatic extraction of corn oil from corn germ. JAOCS 2004, 81, 1071–1075.

Moreau, R.A.; Liu, K.; Winkler-Moser, J.K.; Singh, V. Changes in lipid composition during dry grind ethanol processing of corn. JAOCS 2011, 88, 435–442.

Moreau, R.A.; Singh, V.; Nunez, A.; Hicks, K.B. Phytosterols in the aleurone layer of corn kernels. Biochem. Soc. Trans. 2000, 28, 803–806.

Moreau, R.A.; Singh, V.; Powell, M.J.; Hicks, K.B. Corn kernel oil and corn fiber oil, In Gourmet and Health-Promoting Specialty Oils, Moreau, R.A.; Kamal-Eldin, A., Eds.; AOCS Press, Urbana, Illinois, 2009; pp. 409–432.

Moss, P. Combining two proven technologies, Biofuels International Volume 6, Issue 9, November 2012.

Noureddini, H.; Bandlamudi, S.R.P.; Guthrie, E.A. A novel method for the production of biodiesel from the whole stillage-extracted corn oil. JAOCS 2009, 86, 83–91.

Rausch, K.D.; Belyea, R. L.; Ellersieck, M.R.; Singh, V.; Johnston, D.B.; Tumbleson, M.E. Particle size distributions of ground corn and DDGS from dry grind processing. Transactions of the ASAE, 2005, 48, 273–277.

Sanford, S.D.; White, J.M.; Shah, P.S.; Wee, C.; Valverde, M.A.; and Meier, G.R. Feedstock and biodiesel characteristics report. Renewable Energy Group, Inc. Available online: http://www.regfuel.com/sites/default/files/pdf/Feedstock%20and%20Biodiesel%20Characteristics%20Report.pdf, 2009, Accessed October 6, 2011.

Schroeder, J. Corn oil extraction. Biofuels Journal 2011, 3rd Qtr, 16–18.

Singh, N.; Cheryan, M. Extraction of oil from corn distillers dried grains and solubles, Transactions of the A.S.A.E. 1998, 41, 1775–1777.

Singh, V.; Eckhoff, S.R. Effect of soak time, soak temperature and lactic acid on germ recovery for dry grind ethanol. Cereal Chem. 1996, 73, 716–720.

Singh, V.; Moreau, R.A. Methods of preparing corn fiber oil and of recovering corn aleurone cells from corn fiber. U.S. Patent 7,115,295, 2006.

Singh, V.; Moreau, R.A.; Cooke, P.H. Effect of corn milling practices on the fate of aleurone layer cells and their unique phytosterols. Cereal Chem. 2001, 78, 436–441.

Singh, V; Johnston, D.B.; Naidu, K.; Rausch, K.D.; Belyea, R.L.; Tumbleson, M.E. Comparison of modified dry-grind processes for fermentation characteristics and DDGS composition. Cereal Chem. 2005, 82, 187–190.

Srinivasan, R.; Singh, V.; Belyea; R.L, Rausch, K.D.; Moreau, R.A.; Tumbleson, M.E. Economics of fiber separation from distillers dried grains with solubles (DDGS) using sieving and elutriation. Cereal Chem. 2006, 83, 324–330.

Srinivasan, R.; Singh, V.; Moreau, R.A.; Rausch, K.D. Phytosterol composition and yield of oil from fractions obtained by sieving and elutriation of distillers dried grains with solubles (DDGS). Cereal Chem. 2007, 84, 626–630.

USDA, Economic Research Service, 2011 Oil Crops Yearbook, http://usda.mannlib.cornell.edu/MannUsda/viewDocumentInfo.do?documentID=1290 Accessed October, 2011.

Wang, H.; Wang, T.; Johnson, L.A. Effect of low-shear extrusion on corn fermentation and oil partition. J. Agric. Food Chem. 2009, 57, 2302–2307.

Wang, H.; Wang, T.; Johnson, L.A.; Pometto, A.L. II. Effect of corn breaking method on oil distribution between stillage phases of dry-grind corn ethanol production. J. Agric. Food Chem. 2008, 56, 9975–9980.

Winkler, J.K.; Rennick, K.A.; Eller, F.J.; Vaughn, S. Phytosterol and tocopherol components in extracts of corn distiller's dried grain. J. Agric. Food Chem. 2007, 55, 6482–6486.

Winkler-Moser, J.K. Lipids in DDGS, In Distillers Grains: Production, Properties, and Utilization, Liu, K.S.; Rosentrater, K.A., Eds., CRC Press, Boca Raton, 2012; pp. 179–191.

Winkler-Moser, J.K.; Breyer, L. Composition and oxidative stability of crude oil extracts of corn germ and distillers grains. Ind. Crops Prod. 2011, 33, 572–578.

Winkler-Moser, J.K.; Vaughn, S.F. Antioxidant activity of photochemical from distillers dried grain oil. J. Agric. Food Chem. 2009, 86, 1073–1082.

Winsness, D.J., Method and systems for enhancing oil recovery from ethanol production by products. U.S. Patent Application 11/856,150, publication number 20080110577, allowed December 21, 2011.

Winsness, D.; Cantrell, D.F. Method of freeing the bound oil present in whole stillage and thin stillage. U.S. Patent 7,608,729, 2009.

Woods, R.R.; Krikorian, V.; Smithers, J. Recovery of desired products from fermentation stillage streams. U.S. Patent 8,236,977, 2012.

4

Drying and Cooling Collets from Expanders with Major Energy Saving

Walter E. Farr[1] and Farah Skold, P.E.[2]

[1]The Farr Group of Companies, Memphis, Tennessee, USA, and
[2]Solex Thermal Science, Inc. Calgary AB, Canada

Introduction

Soybean processing (as well as most other oilseeds) has been performed basically the same way for the last 50 years. Since soybeans are low oil seeds (18% oil, canola 40% oil), pre-pressing is not needed, and generally not performed.

The process flow can include drying of the seed upon receipt, particularly if the seed was harvested too early and can be high moisture. Upon entering the preparation department (preparation for solvent extraction), the seed is dehulled, hulls removed by aspiration, cracked, and flaked in flaking rolls. The flakes look much like corn flakes, but they must be of the proper thickness to rupture the oil cells to improve solvent extraction. Typically these flakes go directly to the solvent extraction plant with very little residence time to minimize the formation of non-hydratable phosphotides.

Within the last 20 years, the industry has seen the introduction of expanders (first used on cottonseed) in the preparation department in soybean mills. With the use of expanders, the throughput rate through the extractor can be dramatically enhanced due to improved drainage in the extractor due to the very porous collets, as opposed to flakes, while also offering lower residual oil in the soybean meal. Increasing throughput rate in the extractor is not the only advantage of expanders. The first author, Walter E. Farr, was the first to discover the advantage of expanders from the improved quality of the solvent extracted crude oil. When he joined Anderson Clayton & Co., in Houston, Texas, he soon noticed Anderson Clayton was using expanders on soybeans at their plants in Jackson and Vicksburg, Mississippi (plants that were closed some time ago), and that the soybean oil was of superior quality, meaning that the oil could be refined with much improved refining yield and better quality. It was a few years later that he discovered this crude oil was of superior quality because it had much lower non-hydratable phosphotides. It was then concluded that this was the case in that the high heat/high pressure/short time, and quick cooling of the collets reduced the formation of non-hydratable phosphotides.

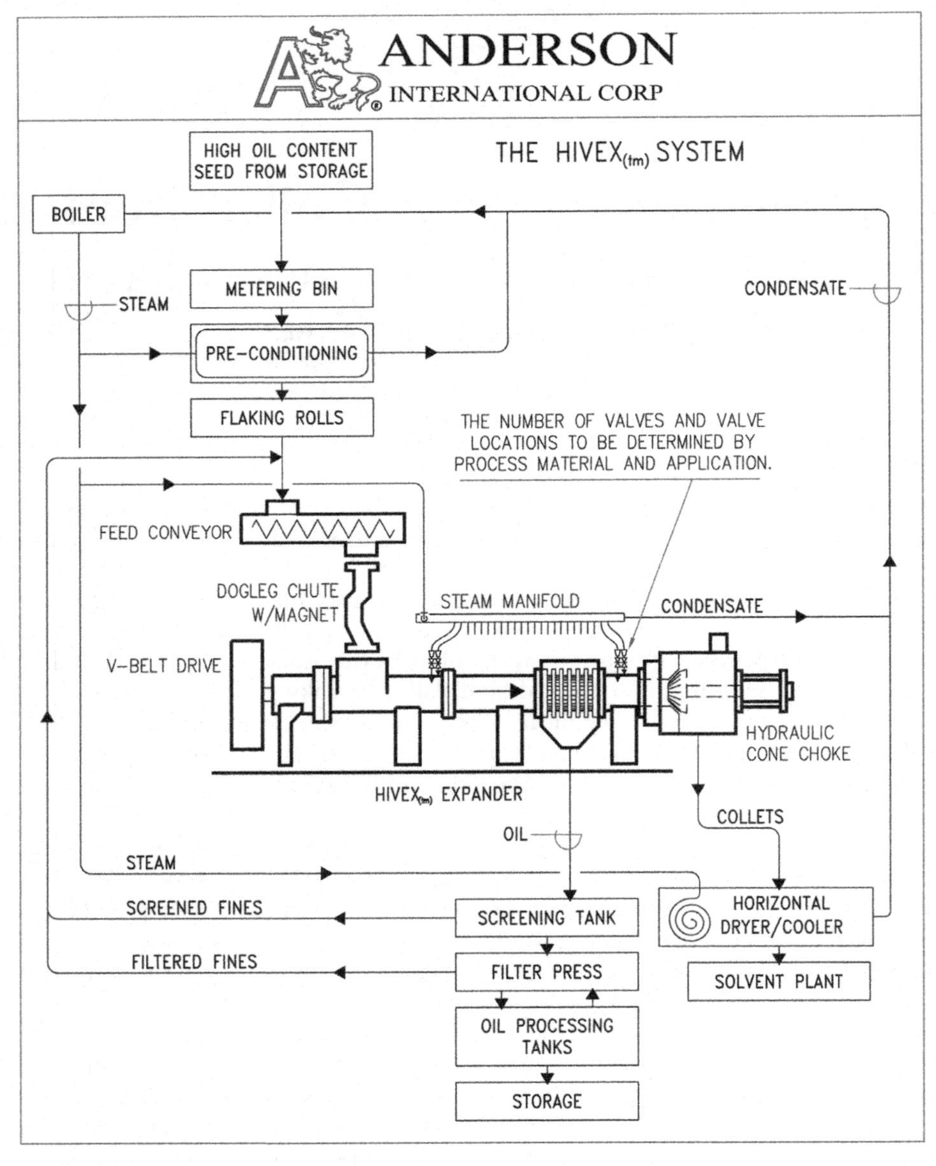

Fig. 4.1. A typical seed conditioning and expander operation with collets drying and cooling (Courtesy Anderson International Corporation, Cleveland, Ohio).

Collets leaving the expanders are 180–200°F, and must be cooled to <140°F before entering the extractor. This is typically done by a large horizontal belt cooler with a high volume of ambient air as the cooling medium. For a large plant in the lower Midwest that uses full expanders in preparation, the HP on the fan is 450 HP!

This paper offers alternatives for major energy savings in cooling of the collets.

Alternative Techniques for Cooling Collets (Offering Energy Savings)

Cooling Collets from Expanders with Carbon Dioxide Snow

Carbon dioxide snow is produced when liquid carbon dioxide at very high pressure is released to atmospheric pressure through a special spray nozzle. The liquid carbon dioxide does not pass to the gaseous phase, but creates a frozen vapor that looks like snow.

Carbon dioxide is an inert gas, much like nitrogen. However, carbon dioxide when released in the presence of moisture can produce a small amount of carbonic acid. This small amount of carbonic acid creates no harm, and in some applications can be an advantage. A Wesson patent, which expired many years ago, described the use of carbon dioxide blanketing of the bleacher vessel in the bleaching of soybean oil (and other vegetable oils). The patent did not mention that the carbonic acid produced was a very active bleaching aid for red color. Carbon dioxide snow has been used for quick chilling of blanched vegetables, and other applications in the food industry. About 15 years ago, the first author envisioned the possibility of using carbon dioxide snow to cool collets from expanders on soybeans. His interest at that time was to determine if there was a quality advantage in the solvent extracted crude soybean oil when the collets were cooled with carbon dioxide snow versus cooling with a high volume of ambient air. An experiment at a plant in the lower Midwest, one that operates full expanders in preparation, did not prove any quality advantage. (Note: This plant trial was supported by Praxair, a producer of liquid carbon dioxide.) Since no quality advantage was achieved, the project was dropped; unfortunately the energy savings that could have resulted for using carbon dioxide snow was not considered at the time.

When this concept was revisited some 15 years later, evaluation began as to possible energy savings. There are two problems in using carbon dioxide snow for collet cooling. The collets need to be dried (to reduce some of the moisture) before the quick cooling. Actually, that would not be a problem, as the collets could go through a heating section before passing through the quick cooling section.

The major problem with cooling collets with carbon dioxide snow is economic. To use carbon dioxide at an oil mill, there would have to be a cryogenic holding tank for the liquid carbon dioxide on site. Plus, if the liquid carbon dioxide had to be transported relatively long distances, this could negate any cost savings by using carbon dioxide snow.

It is thus concluded, that carbon dioxide snow cooling of collets will only be practical for a producer of corn sweeteners/ethanol that produces liquid carbon dioxide as a by-product, or an oil mill that has a Praxair or Air L'iquid liquid carbon dioxide plant in the same community. That is not likely to happen.

Drying and Cooling Collets with Solex Thermal Science, Inc. Calgary, Canada, Vertical Drying/Cooling Units

The most common approach to bulk solids cooling is to use air in direct contact with the bulk solid material. With direct cooling, ambient air is taken in by using large fans and blown across the product using large horsepower fans. The air is then discharged through an emissions stack. The circulating fans have high energy requirements.

The alternative is to cool bulk solids indirectly by using water. The Solex Thermal Heat Exchanger is a unique piece of equipment that consists of one or more banks of vertical, closely spaced, hollow, stainless steel plates. The product (meal collets) flows slowly by gravity between the plates in mass flow. Cooling water flows through the plates, counter-current to the meal collets, resulting in high thermal efficiency. The cooling occurs by heat transfer through the collet particles and is solely based on conduction (Fig. 4.2). At the bottom of the heat exchanger, an extraction screw creates mass flow and regulates the throughput (Fig. 4.3). Often a plant's existing cooling tower is sufficient to provide the necessary water and a low horsepower pump is the only source of energy consumption.

Total Energy Consumption

A detailed comparison of the energy consumed by different types of cooling units demonstrates the significant degree to which indirect cooling of bulk solids outperforms direct cooling in terms of energy efficiency.

Uses Up to 90% Less Energy

The Solex bulk solids cooler is ultra-efficient, using up to 90% less energy than other technologies. The indirect plate cooling design does not require the use of air in the cooling process, which eliminates both the need for large horsepower electric drives and the associated heat losses. This fact, combined with the large heat transfer surface area of the Solex cooling plates, makes this the most efficient bulk solids cooler technology available (Skold, 2011).

Emissions

With indirect cooling, the water is re-used repeatedly by being recycled in a closed loop system. The cooling media does not come into direct contact with the product, so no dust or emissions are created. This eliminates the need for pollution control equipment and makes tight emission limits easier to meet. In addition, indirect bulk solids cooling minimizes product abrasion and degradation.

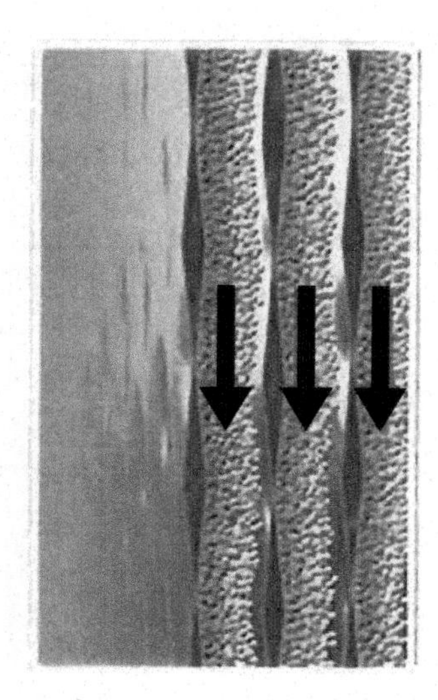

Fig. 4.2. The cooling occurs by heat transfer through the collet particles and is solely based on conduction.

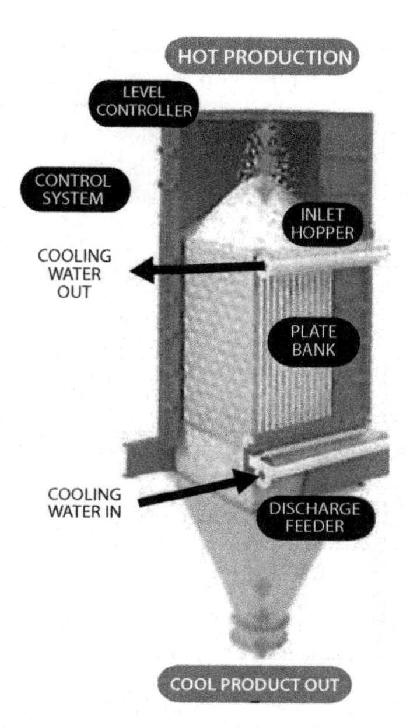

Fig. 4.3. At the bottom of the heat exchanger, an extraction screw creates mass flow and regulates the throughput.

Capacity

Cooling water is often available from a variety of sources, whether from a nearby river or an existing cooling water circuit with cooling towers. The Solex bulk solids cooling unit itself is compact with simple vertical construction and high modularity that make it readily adaptable to new or existing facilities. The only ancillary equipment is the piping running from the water source to the cooling unit. Power consumption is limited to the drive for the small horsepower water pump. With the use of a closed loop system, the water can be recycled repeatedly so that only a limited quantity is required.

Compact and Modular: Adding Future Capacity Is Easy

The vertical configuration makes this design both compact and modular. The clever modular design means that additional heat exchanger plate banks can be stacked if increased cooling capacity is required in the future. The compact installation footprint makes this design easy to integrate in existing plants and is ideal for debottlenecking, revamps, and capacity increases. (See Fig. 4.4.).

Fig. 4.4. The Solex bulk solids cooling unit itself is compact with simple vertical construction and high modularity that make it readily adaptable to new or existing facilities.

Operating Costs

Using indirect bulk solids cooling technology circumvents the problem of saturated air. Cooling water is often available within the plant system. The water-cooled plates allow for low product temperatures even in hot summer months. Furthermore, in situations when it is necessary to chill the water, the plant water cooling system can usually accommodate the extra load, and even if a dedicated water chiller is required, the energy required to cool water is much less than the energy needed to chill air, or using very high volumes of ambient air in hot weather.

Uniform Final Product Temperatures

Mass flow design means the product moves with uniform velocity through the heat exchanger (Skold, 2013). This feature, combined with long residence times (typically 5–10 minutes), enables even temperature distribution as the product passes through the heat exchanger, producing remarkably stable and uniform final product temperatures.

The indirect plate cooling design means air is not used in the cooling process. Installed capital costs are reduced due to the elimination of costly and unnecessary air handling equipment such as large diameter air ducting, motors, fans, scrubbers, chillers, and emissions controls.

Produces a Superior Final Product

The slow and controlled movement of the bulk materials produces a superior final product. It prevents product abrasion and degradation so that there is no change in

Fig. 4.5. It prevents product abrasion and degradation so that there is no change in particle characteristics or shape.

particle characteristics or shape. A superior final product is produced. Since air is not in contact with the product, risk of bacterial contamination, odor contamination, and product moisture content changes are eliminated (Fig. 4.5).

Summary

As with Chapter 1, "Extrusion/Expeller® Pressing as a Means of Processing Green Oils & Meals," this chapter deals with energy savings and savings in equipment in the preparation department, and in solvent extraction.

Both of these chapters contribute significantly to "Green Vegetable Oil Processing and Sustainability."

References

Skold, F., Optimal Processing in Preparation Plant with Accurate Thermal Modelling, paper presented at the 15th Latin American Congress and Exhibition, Santiago, Chile, August 20–23, 2013.

Skold, F., Saving on Steam and Energy Costs in Oilseed Crushing—Efficient Heat Utilization, paper presented at the World Conference on Oilseed Processing, Fats & Oils Processing, Biofuels & Applications, Turkey, 2011.

5

Algae Drying and Extraction

Richard W. Ozer, P.E.
Crown Iron Works, Minneapolis, Minnesota, USA

Solvent Extraction

Algae can be a tremendous source of oils for the production of omega 3s or industrial oil as well as a potential renewable source of oil for energy. This paper will discuss solid/liquid extraction, which is a highly effective and inexpensive means of recovering these oils when starting with a dry feedstock.

Solid/liquid extraction can be defined as "the use of a solvent to dissolve and remove a soluble fraction from an insoluble, permeable solid" (Gertenbach, 1999).

Solvent extraction is widely used in the oilseed industries as a means to recover vegetable oil from soft seeds such as soybeans, canola, sunflower, corn germ, and others. Obvious products are cooking oils, soaps, margarines, biofuels, and other applications. The oil free solids are used as cattle and poultry feeds or as a source for vegetable protein concentrate and other applications.

Solvent extraction has also been used in the nutraceutical and pharmaceutical industries for well over 20 years, extracting omega 3 oils from heterotrophic algae for baby food, or plants such as yucca for pharmaceuticals.

Solvent extraction plants (SEP) are a reliable and well established technology where solidly designed and maintained equipment lasts significantly more than 20 years. Capacities of single plants for traditional oilseeds are range from 1 to 9,000 MTD of whole seed where equipment is expected to operate 24/7 for 350 days per year. Heat recovery is so ingrained into plant design that actual heat input is incredibly low.

To provide a quick example of the economics of a larger scale plant, we start with 9,000 MTD of flaked soybeans at 18.3% oil and 10.5% H_20. Utilities for a SEP using hexane will have the following energy requirements:

Steam	185 kg/ton of feed material, based on whole seed
Power	8.5 kWh/ton of feed material, based on whole seed
Cooling Water	15.4 m^3/hr of feed material, based on whole seed.

Based on fairly recent utility costs, operating expenses (OPEX) are estimated as follows:

Steam	9,000 tpd × 185 kg/ton × $30.87/ MT steam/1000	= $51,400/day
Electricity	9,000 tpd × 8.5 kWh × $0.1/kWh	= $7,650/day
Labor 2 + 1 men/shift	3 × $30/hr × 24 hr/day	= $2,160/day
		Daily OPEX = $61,210/day

Assuming soybeans @ 18.3% oil and a residual oil of 0.6% oil yield is ~1,596.1 MTD

Utility cost per MT of oil = $38.35

There are certainly more costs that need to be added to the above, such as preparation or dehulling, solvent costs, sewage disposal, but when considered as a single unit operation, solvent extraction is an extremely cost effective means to recover oil from a solid material. Solvent loss must be kept to a regulated and minimal level for both environmental and economic reasons.

In the specialty oil industries solvent/liquid extraction is used to recover higher value oils such as omega 3s or 6s and other nutraceuticals (Oils rich in omega 3 and omega 6 fatty acids, DHA/EPA oils, specialty oils with unique fatty acid content for improved health and/or functionality). Capacities are generally lower and can range as low as a few MTD of feed to hundreds of MTD. At lower capacities OPEX are generally higher, but the value of the oil far exceeds the cost to extract.

As noted earlier, our economic model for the algae oil for energy was based upon receiving a feed material at 10.5% H_2O and therein lies the rub as the water content in algae is extremely high. And while wet solvent extraction methods are being developed, they currently use much higher solvent feed ratios than conventional solvent extraction. To take advantage of the economies described earlier, the algae must be dry.

Drying

For the purposes of this discussion, drying is a step required in the preparation of the solids prior to solvent extraction. Prior to drying, it is almost always beneficial to dewater the material to the maximum solids concentration before drying. The general rule of thumb is that mechanical dewatering costs less than thermal drying. The benefits of improved dewatering techniques is shown in Table 5-A.

Assuming that the heat input required to evaporate a kilo of water is the same for all drying techniques (and this is a big assumption), it is easy to see the benefit of increasing solids concentration on the overall cost of producing oil.

The traditional approach to dewatering was dissolved air flotation followed by a centrifuge. Typical solids concentrations were as low as 12 to 20% on a good day with results varying between algae species and growing conditions. Recent advances by centrifuge company Evodos suggest that solids concentrations ranging from +20 to 30% are possible and will eventually become the norm. As methods of lysing the

Table 5-A Effect of Feed Solids on Dryer Evaporation Rate.

Dryer feed solids	10%	15%	20%	25%	30%
Feed rate to dryer T/HR	1000	666.7	500	400	333.3
Product @ 10% H_2O T/HR	111.1	111.1	111.1	111.1	111.1
Evaporation of water T/HR	888.9	555.6	388.9	288.9	222.2

cell wall without totally releasing the oil become available, we expect that further increases in solids concentration can be expected. (For the purposes of this chapter, lysing of cell walls will mean the disruption of the cell wall creating avenues for the water to leave or solvent to enter the cell.)

Drying Techniques Available Today

Spray Drying

A spray dryer may be defined as a co-current pneumatic conveying dryer that directly contacts the product with the drying medium which could be air, inert gas, steam, or some combination thereof. The spray dryer is capable of atomizing and drying pumpable liquids to a discrete and dry solid in a single pass.

The spray dryer generally employs pressure nozzles or rotary atomizers to break-up pumpable liquids (slurry or solution) into a series of fine droplets which are then mixed with the hot air stream. The spray dryer chamber is designed so that the product is at least non-sticky if not fully dry before it reaches a wall. As a result, spray dryers will require a lot of floor space as capacity increases. One significant advantage of the spray dryer is that the material is fully treated in a single pass so heat input to the solids is extremely low. The spray dryer can often control finished particle size, eliminating dust although the powder may still be too small for solvent extraction.

The downside of the spray dryer is that it requires an atomizable or pumpable liquid so it cannot take advantage of increases in solids concentrations currently possible in the dewatering step. Capital cost is usually very high compared with other direct contact dryers such as flash dryers.

Flash Drying

A flash dryer is also a co-current pneumatic conveying dryer where the product comes into direct contact with the drying medium. Material residence times in the hot air stream are extremely short so that heat input to the solids is kept very low, maintaining product quality. Air can be once through or for heat recovery can recirculate up to 70% of the exhaust gases (Svonja, 2007). Flash dryers are capable of drying friable and non-sticky filter or centrifuge cakes directly on a once through basis. When drying pumpable slurries or thixotropic filter cakes, dry product can be recirculated and mixed with the incoming feed material in a unit operation commonly called back-mixing. The objective of backmixing is to create a crumbly and free flowing material that will disperse readily in the heated air without sticking to internal drying surfaces.

Recycle of dried algae solids is generally required, given the current solids concentration achieved to date. Further increases in solids concentration will result in a reduction in the amount of algae solids that must be recycled to make an acceptable feed for flash dryers. The flowsheet below shows a system that incorporates recycle of dry solids to condition the feed material and recycle of air to recover heat. See Fig. 5.1.

Advantages of flash drying are that residence times in the hot air are extremely short thus maintaining product quality. Flash drying can take advantage of higher solids concentrations that reduce the evaporative load and improve overall economics. It is also easy to set up recycle of air to further improve overall drying economics. To date, flash drying maintains the integrity of the cell walls, limiting production of fines although some fracturing could be a positive as we will discuss later.

Downsides are that currently, recycle of solids is required to condition the feed material. When the dewatering step can achieve high solids concentrations, then minimal solids need to be recycled and product quality is generally maintained. If low solids are achieved in the dewatering step, than high solids recycle is required making overheating of the solids more likely. Another downside of the recirculating flash dryer is that the bleed off air will likely require some sort of odor control.

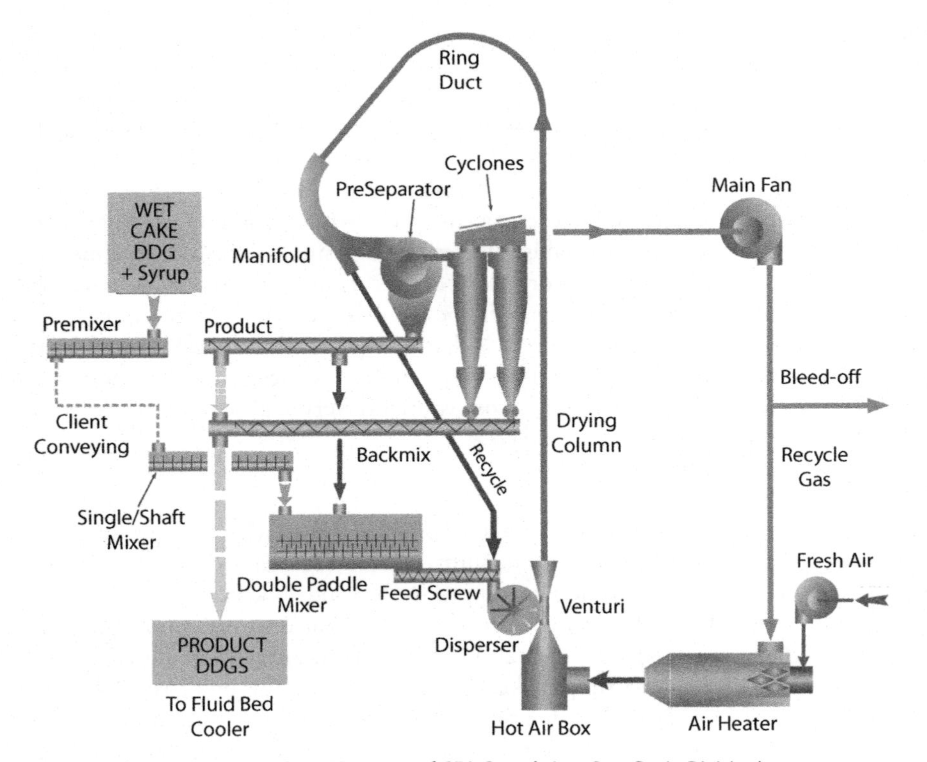

Fig. 5.1. Partial gas recycle ring dryer (Courtesy of GEA Canada Inc., Barr-Rosin Division).

Flash Drying in Superheated Steam

Another co-current drying process uses superheated steam as the drying medium. One hundred percent of the vapors are recycled where the evaporated water is returned to the system as a low quality source of steam. When steam can be used elsewhere in the process, overall utility cost for the drying/solvent extraction plant can be reduced significantly (see Fig. 5.2).

Advantages of drying in superheated steam vapors are that the evaporated water can be used in the extraction process as low pressure (up to 60 psig) steam. If water is in short supply, the excess steam not used in the process can be condensed in atmospheric condensers and returned to the process. Material residence times are extremely short so thermal degradation is limited.

Potential downsides are the same as with flash drying in that solids must be recycled. However, results to date indicate that minimal product degradation occurs when the system is operated near atmospheric pressure. Improvements in dewatering will reduce the recycle, and with improvements in product quality expected.

Solar drying is the tried and true technique used in drying sludge from sewage treatment plants. Large areas and favorable weather conditions as well as high manpower requirements are required. Positives are that energy costs are very low. A downside is that the quality of the oil can be negatively affected by extended drying times, as algae resist diffusion through the cell wall.

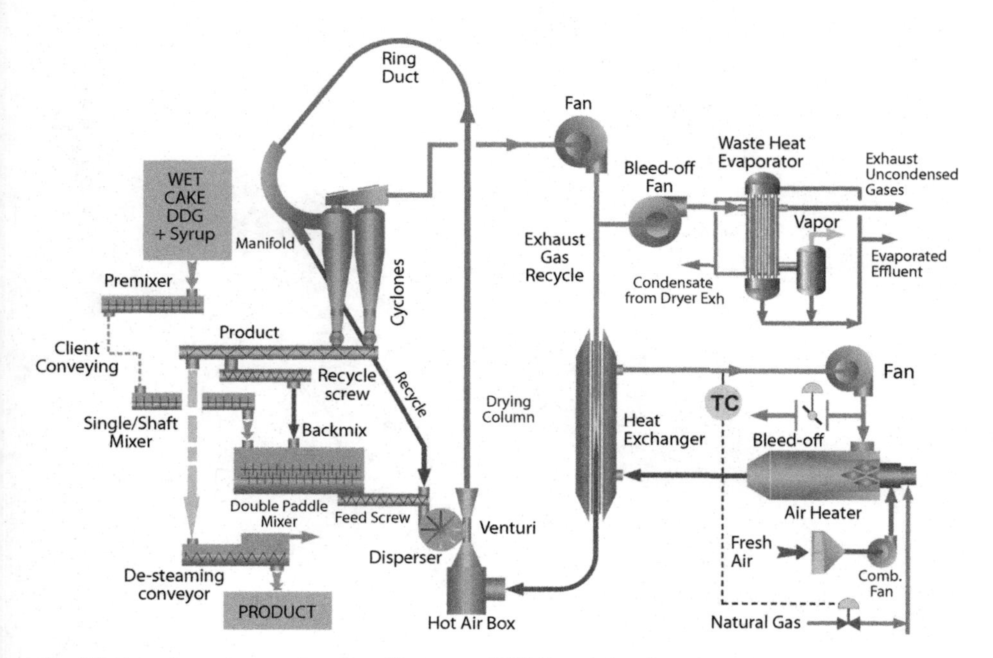

Fig. 5.2. Superheated steam ring dryer (Courtesy of GEA Canada Inc., Barr-Rosin Division).

A brief synopsis of heat required for the various direct contact drying techniques is as follows:

Spray or Flash Drying	Once through air basis	18–2,200 Btu/lb water evaporated
Spray or Flash	Partial recycle	1,250–1,400 Btu/lb water evaporated
Spray or Flash	Superheated steam	~1,400 Btu/lb water evaporated*
Indirect	Steam	~1,160 Btu/lb water evaporated

*When evaporated water can be used in the process as low pressure steam, overall Btu/lb input can be as low as 250 Btu/lb.

Indirect Drying

Indirect dryers separate the heat source from the material being dried with a metal heat transfer surface. Heat sources can be steam, hot oil, or water from the material being dried. The material is kept moving over the metal heat transfer surface by an agitator. There are several types of indirect dryers to select from. See Fig. 5.3 for examples of screw (Thermascrew), thin film paddle (Torus Disc), and thin film

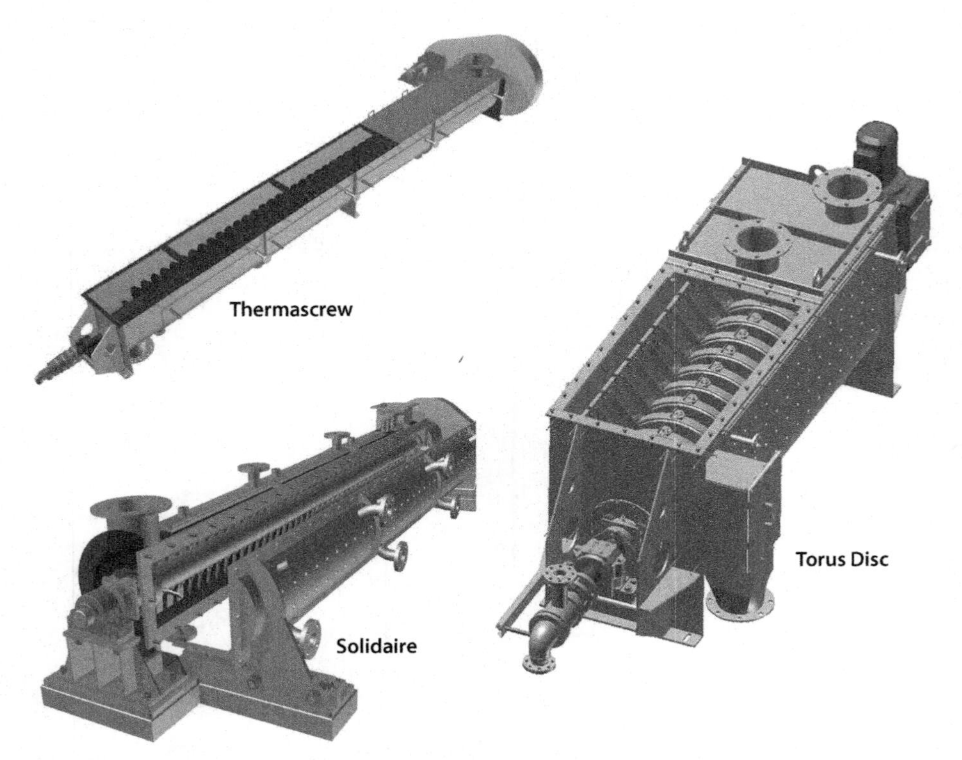

Fig. 5.3. Indirect dryers with screw (Thermascrew), thin film paddle (Torus Disc), and thin film (Solidaire). (Courtesy Bepex International of Minneapolis.)

(Solidaire) dryers. The preferred dryer depends on the nature of the material being dried and how long the product takes to dry.

Indirect dryers are extremely energy efficient because the majority of the energy goes directly into the solids or the liquid to be evaporated, requiring just enough air to carry away the evaporated vapor. Air pollution equipment is extremely small. Evaporative cooling with indirect dryers, however, is not as great as direct dryers, due to the limited amount of air present. Indirect dryers can also be expensive, on a basis of kilogram/water evaporation, compared to direct dryers.

Preparation

While preparation techniques will be discussed in more detail in other chapters, simply put, "90% of extraction is preparation"—that is, a well-prepared feed material will extract quickly and in the most economical manner.

For instance, a spray or flash dried algae will often look like the fine powder shown in Fig. 5.4. While there is a great deal of surface area to enable extraction, the fine material is not easily separated from the solvent, creating issues within the extractor and in distillation or solvent recovery. Preparation is the conversion of the fine algae that looks like the powder in Fig. 5.4, to a feed material that looks like Fig. 5.5. The fines are tied up and there are many avenues for the solvent to get to the center of the collet or pellet to extract the oil.

Extraction

Definitions

Miscella: Combination of extract and solvent. "Full miscella" is the highest concentration of extract in the solvent just prior to discharge from the extractor.

Fig. 5.4. Flash dried algae.

Fig. 5.5. Pelleted algae and colleted algae.

Driving force: Simply put, this means that the lowest concentration oil-depleted solids come in contact with pure solvent. At the feed end of the extractor the feed material is contacted by the richest miscella. We will discuss what happens inside the algae particle later when we describe the extraction process in detail.

Counter current extraction: Solvent/miscella flows counter current to the solids. Extract-rich solids enter the extractor and come into contact with the most concentrated

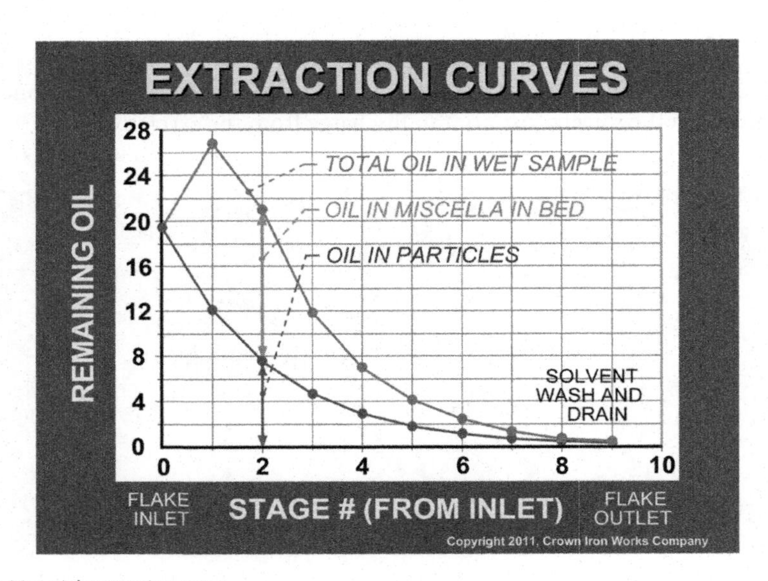

Fig. 5.6. Material extraction curve.

miscella just before the "full miscella" discharges. At the solids discharge end of the extractor, the extract-depleted solids are washed by pure solvent (see Fig. 5.6).

Material extraction curve: Shows the amount of oil in the solvent and material at various stages in the extractor. It does not necessarily reflect the idea of driving force mentioned above, but it does show the benefit of counter-current extraction and its ability to achieve low residual oil by washing the low-extract solids with pure solvent.

Solvent-feed ratio: For the purposes of this chapter, solvent-feed ratio is the mass or kilos of solvent added per kilo of feed material.

Solvent holdup: Amount of solvent held by the solids as it exits the extractor.

Residual oil: Amount of oil left in the solids after extraction on a solids and moisture basis.

As a process, extraction can be broken into four steps:

1. *Diffusion of solvent into the solid particle through its pore structure* (Anderson, 2011). This involves fully wetting the solid material with solvent.
2. *Dissolution of the solute (oil) from the solid particle into the solvent within the solid particle.* The extract will dissolve into the solvent only when the concentration of the miscella within the particle is below the equilibrium concentration.

   First, let's look at any soluble material and assume that the soluble material will continue to dissolve into the solvent as long as the solvent is not saturated.

   Each soluble component will dissolve into the solvent according to the relationship:

   $$K = C_e/C_{dm}$$

   - Where K is the equilibrium constant for each soluble component
   - C_e is the concentration of the extract in the miscella
   - C_{dm} is the concentration of the extract in the dry solid

   The higher the equilibrium coefficient, the more extract will dissolve into the solvent. When equilibrium is reached, the process will become static unless additional solvent is provided.
3. *Diffusion of the dissolved solute from within the particle to the particle surface through the internal pores.*
4. *Washing the solute rich solvent from the boundary layer on the particle surface into the miscella.* This washing is responsible for maintaining the driving force throughout extraction, and counter current extraction ensures that driving force is maximized throughout the extractor itself.

Factors Affecting Extraction

Once thoroughly wetted, there is limited flow of liquid through the particle. Instead, we rely on diffusion of the dissolved extract in the particle from the high concentration area inside to a lower concentration area outside of the particle for continued extraction to occur. Let this boundary layer come to equilibrium and extraction stops.

Diffusion rate is often called the rate limiting step in extraction and is discussed in Fick's second law as follows:

$$\partial C/\partial t = D^*(\partial^2 C/\partial x^2)$$

where

C is the concentration of the extract in solution
T is time
D is the diffusion coefficient
X is the particle diameter

Concentration difference remains the main driving force, but now we see the influence of particle size (particle diameter), giving us another tool to reduce extraction time.

Particle Size

Clearly, the finer the particle, the less distance the extract must travel to the surface of the particle. Since extraction is really mass transfer and thus a function of surface area, one would think that very fine individual solids will extract very quickly. This is indeed the case; however, when solids are too fine, the solids can pack tightly together so that there is no miscella flow past all the material. With no flow around the surface of the particle, the surrounding miscella quickly becomes saturated, halting extraction.

Very small pieces of solids "fines" leave the extractor suspended in the full miscella and should be filtered out before entering the heated surfaces in the evaporators. If they are too large, it will take longer for solvent to penetrate to the center. Large particles will have plenty of room for solvent flow around them, but the time required to bring the extract from the center of the particle to the surface extends required residence time, thus increasing the size of the extractor.

Material Preparation (Requirements)

- Need to rupture as many of the oil bearing cells as possible to allow the solvent to enter the cell and dissolve the oil. If possible, it is preferable to rupture the cell wall without creating fines for the reasons mentioned earlier.
- Provide surface area for rapid diffusion of solvent into the solids.

Solvent Type/Composition

What is the affinity or specificity of the solvent to the extractor? Clearly, the higher the allowable concentration of extract in the solvent, the less solvent required, and the overall cost of operation is reduced. The perfect case would be the rifle shot where the solvent only removes the component of interest. This is in contrast to the shotgun approach where many components (some undesirable) are extracted.

When dealing with some algae or other organics or nutraceuticals, a single solvent will not remove all of the components of interest.

On occasion binary solvents are used to accomplish more than one task. For instance, in nutraceuticals, alcohol/water combination solvents are often used. The water is added to "swell" the cell, creating fissures in the cell wall. These fissures become pathways for the solvent to enter the cell while maintaining the overall integrity of the cell reducing fines.

Another example of binary solvents would be combinations of polar and non-polar solvents where the majority of the extractables are soluble in the nonpolar solvent while several components may be soluble in the polar solvent. Combining the two solvents results in a higher extraction efficiency than could be expected with a single solvent.

Solvent Temperature

Generally speaking, the solubility of the extract in a solvent increases with temperature. Therefore, the simple addition of heat reduces the amount of solvent required to do the extraction. Lower viscosities also accompany higher solvent temperatures with noticeable improvements in extractability.

Operation at elevated temperature also improves operational safety. Keep in mind that every cubic meter of feed material brings with it a certain amount of air. Higher solvent temperatures result in quickly saturated air in the headspace above the solvent in the extractor generally well above the higher explosive limit of hazardous solvent. Safer operation is a result.

Thermal values for the solvent and effect on cost of operation:

Thermal properties for typical solvents

Solvent	Boiling Point F/C	Latent Heat Btu/lb, kcal/kg
Hexane	156°F 69°C	144 Btu/lb 80 kcal/kg
Ethanol	173°F/78°C	367 Btu/lb 203.8 kcal/kg
Methanol	149°F/65°C	473 Btu/lb 262.7 kcal/kg
Water	212°F/100°C	970.1 Btu/lb 538.8 kcal/kg

Keeping in mind that each kilogram of solvent needs to be evaporated and condensed, even if solvent feed ratios were similar (and they are not), it is not hard to see that thermal values can have a big effect on cost of operation.

Miscibility with Water

Later on in this discussion, we will talk about the overall solvent extraction plant (SEP) process. For now, let us just point out that many non-polar solvents are immiscible with water, making mechanical separation of the solvent from water very easy.

In contrast, a polar solvent such as ethanol or methanol is fully miscible in water, requiring thermal separation at additional cost. This will be a huge consideration later on in our discussion.

Environmental and Safety Concerns

Environmental issues can play a large role in how easily permits are obtained and solvent loss monitored. Volatile solvents such as hexane are generally very closely regulated with solvent loss strictly monitored by state and federal environmental authorities. Solvent losses for new products such as algae have yet to be determined, but you can be sure, as soon as plants come on line, standards will quickly approach those similar to soybeans where allowable losses are 0.2 gallon per ton of beans processed.

Safety

Many of the solvents are extremely hazardous, requiring very strict guidelines on construction, resulting in significantly higher construction costs compared with non-hazardous plants. NFPA 36 is the standard solvent for hexane and solvents immiscible with water, and NFPA 30 is for solvents miscible with water. A quick review of the two standards will show that NFPA 36 is much more helpful regarding plant and equipment design.

When all things are considered, hexane is the current solvent of choice for most lipids extractions. Its ability as a solvent at low solvent feed ratios is not duplicated elsewhere and by itself would make hexane a good choice. But when operating costs are added to the mix, then hexane runs away from the pack.

Factors Affecting Operating Cost

Once again, keeping in mind that every kilogram of solvent must be evaporated and condensed, we are looking for a combination of preparation and extraction techniques that will:

- Maximize the oil concentration in the full miscella
- Reduce solvent holdup at the solids discharge of the extractor
- Resulting in the lowest effective solvent feed ratio that will give an acceptable residual oil

Current Extraction Techniques

There are basically two types of continuous extractors in common use today. Both are continuous and counter current flow, solvent to solids. These techniques are percolation or immersion extraction.

Percolation Extraction

Gravity solvent flows over a bed of material similar to old style coffee percolators. A screen of some sort holds the bed in place and allows drainage of miscella past the solids. A hopper located under the screen collects the miscella and directs it toward a pump inlet. The pump, in turn, recycles the miscella over the bed of material, continuing the process.

Typical applications are for materials that have good drainage but provide some resistance to the gravity flow of the solvent such as flakes, collets, press cakes. See Fig. 5.7.

Immersion Extraction

Here the material is drawn through a bath of solvent and is totally immersed in the solvent throughout the extraction process until final drainage. The approach shown in Fig. 5.8 can be described as a mixer-settler type motion where the solids are drawn either up or down a tray in a moving bed. When the solids reach the end of the tray, they fall off the tray, gently dispersing into the miscella that is going in the opposite direction. The solids quickly settle and are drawn up or down the next tray repeating the process at the end of the tray. Material depths are generally less than 250 mm.

Typical applications are for granular solids that sink quickly in the solvent. Another application may be for products that drain too quickly for percolation extractors, making it difficult to keep the product wetted.

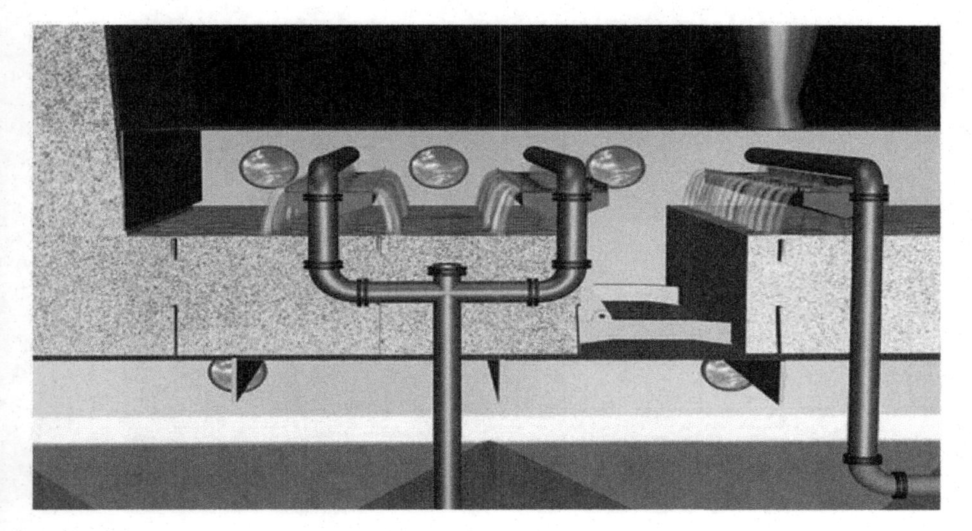

Fig. 5.7. Percolation extractor, overflow weirs.

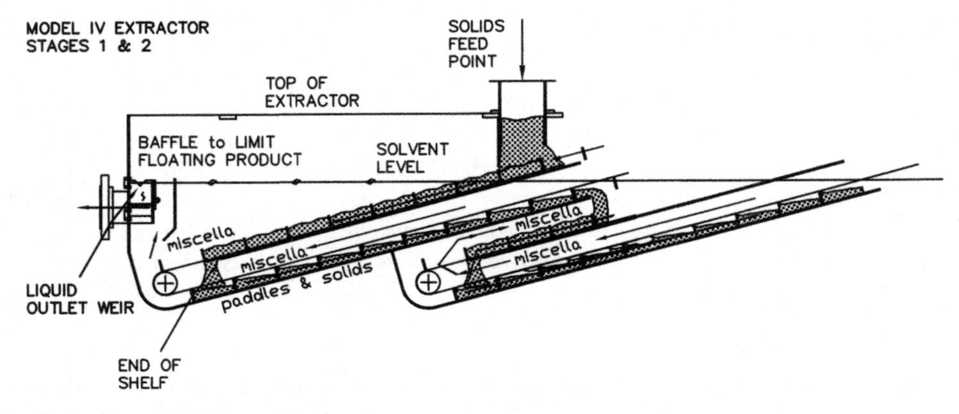

Fig. 5.8. Internals of immersion extractor.

Extraction Selection Considerations

Extractors have many design considerations. Prominent among these are Bed Length (L) and Bed Depth (D). L:D ratio is a common description for many extractors (see Fig. 5.9).

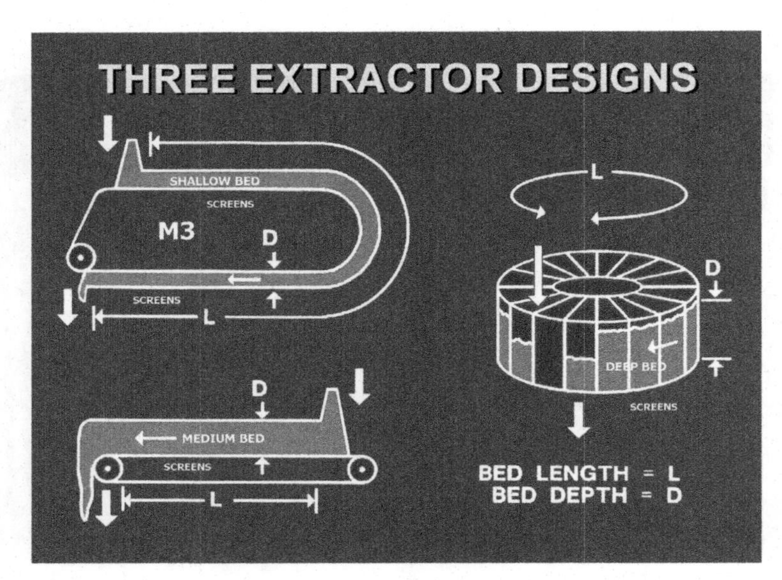

Fig. 5.9. Different approaches to percolation extraction (clockwise, from top left: High L:D, shallow bed <1 m; High L:D, deep bed 1–3 m; Low L:D, deep bed 1–3 m).

Common Points for all Percolation Extractors Solvent Distribution

- Uniform liquid distribution across full bed width (see Fig. 5.10)
- V-notch weirs give even distribution whatever the flow rate
- Continuous loading and discharge

Considerations of Deep and Shallow Bed Extractors, High & Low L:D

- Less tendency for solvent channeling	+ for shallow bed
- Less weight per unit area bottom of screen	+ for shallow bed
- Liquid may migrate to side walls or path of least resistance	+ for shallow bed
- Miscella strength control	+ Hi L:D
- Bed turnover during extraction	+ shallow bed
- Even distribution of feed material (unless wet loaded)	+ shallow bed
- Less space required	+ Low L:D
- Larger surges at extractor discharge (affect desolventization)	+ shallow bed
- Higher capacity per unit area	+ deep bed
- Use collets to tie up fines	+ deep bed

Examples of extractors with various L:D ratios are shown in Figs. 5.11–5.14.

The Solvent Extraction Plant (SEP)

There are two basic feeds that enter the solvent extractor: fresh solvent and solids. The solids are made up of the solid feed material with oil, which is accompanied

Fig. 5.10. Photo of miscella over weirs.

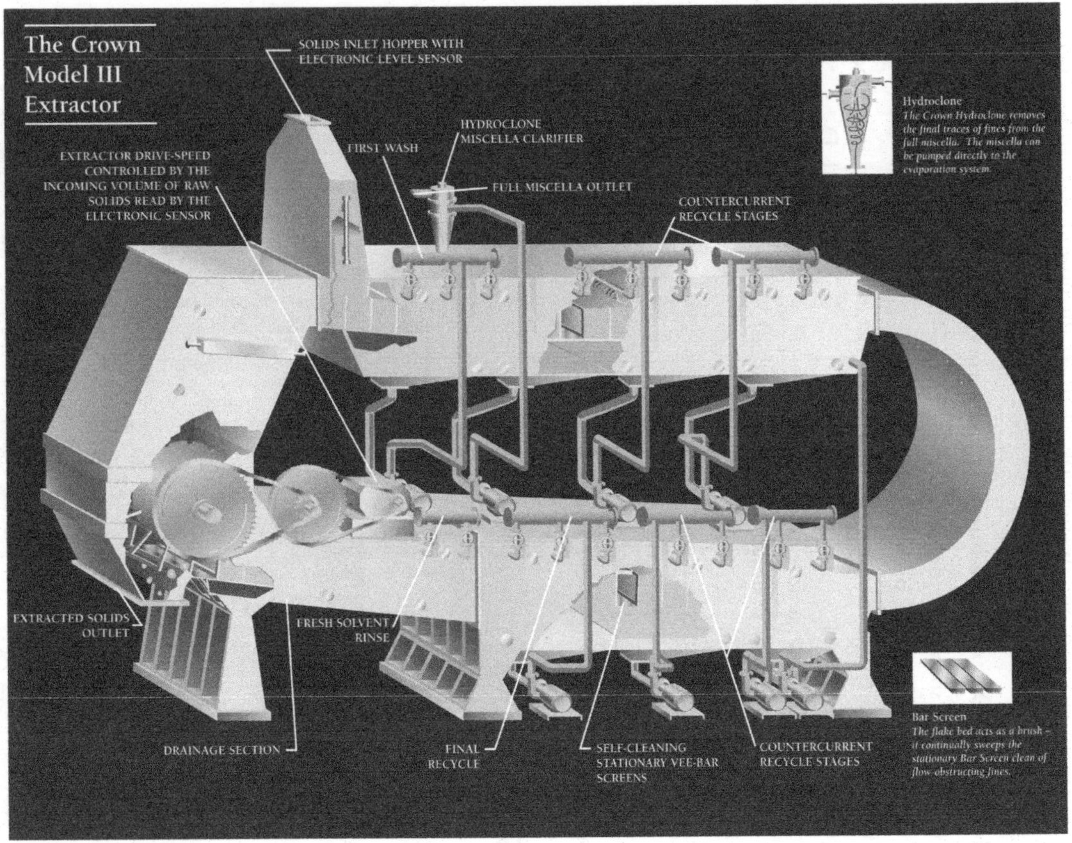

Fig. 5.11. Crown Iron Works Model III extractor.

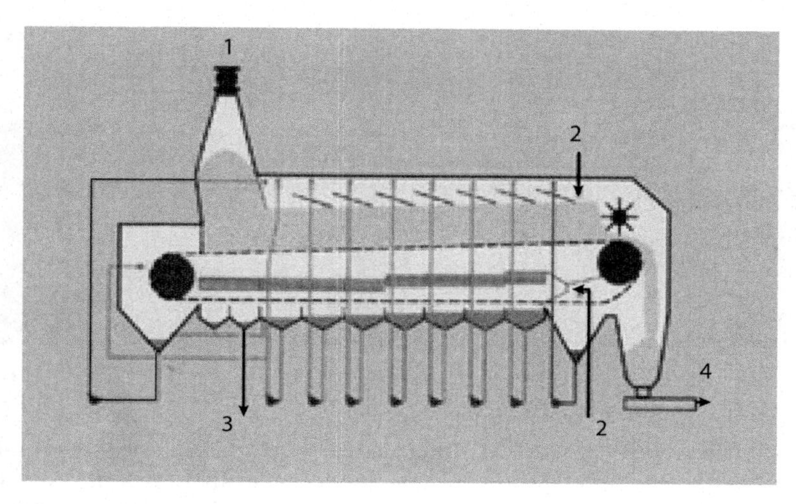

Fig. 5.12. Desmet extractor.

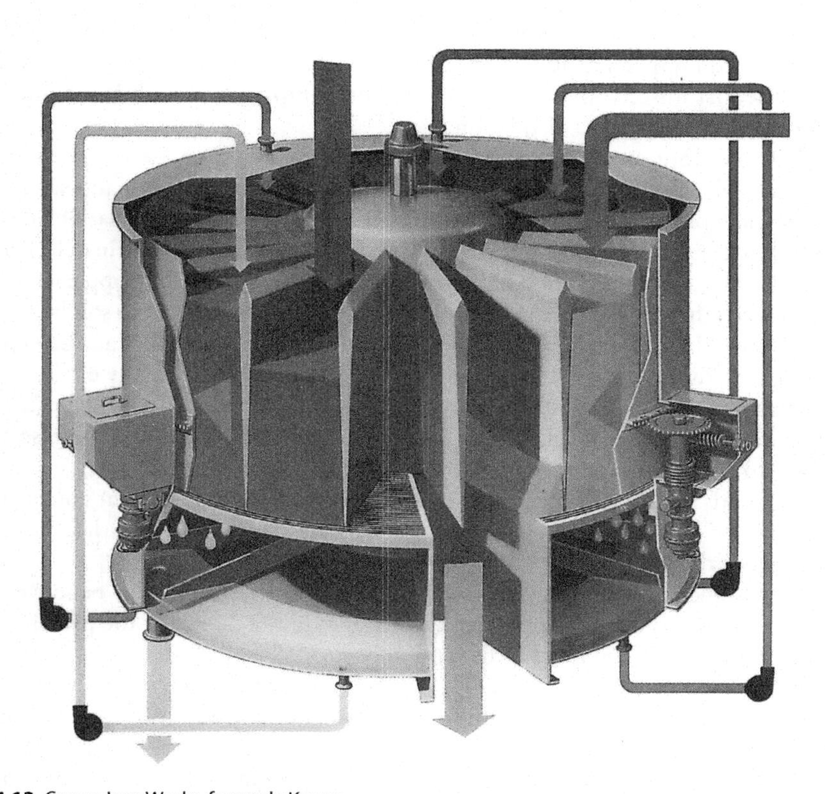

Fig. 5.13. Crown Iron Works, formerly Krupp.

Fig. 5.14. Crown Iron Works Model IV immersion style extractor (High L:D).

by a certain amount of air. Two product streams exit the extractor: spent solids and full miscella. This air is in essence a non-condensable, and leaves the extractor fully loaded with solvent and water via the extractor vent (see Fig. 5.15).

Feed material enters the extractor (Fig. 5.16) via the feed hopper which is kept fairly full, thereby providing a vapor seal for the extractor. A level sensor senses the level in the feed hopper and adjusts the extractor speed to maintain a level in the feed hopper. The first miscella that the feed solids come into contact with is the full miscella just before discharge to distillation.

The meal continues atop the stationary screens passing under a series of weirs from which solvent is always flowing. Miscella drains through the bed and screens to a hopper that collects the solvent and directs it to the miscella recirculation pumps. The pumps return the miscella to the weirs with piping installed so the miscella can be directed to the hopper above, or one back or front. Liquid level in the hoppers is maintained by weirs that make sure that the miscella flows counter current to solids above.

The material travels along the length of the top of the extractor, and then enters the tail where the solids are completely turned over. The realignment of the solids encourages liquid flow once again. The solids continue to the outlet of the extractor which is located under the solids inlet. The spent solids are washed by pure solvent perhaps 5 to 10 minutes prior to final drainage and discharge.

The solids stream is conveyed by means of a vapor tight drag flight conveyor to the top of the desolventizer or DT. Selection of the desolventization method depends on the projected use of the solids after desolventization.

For instance, should the algae solids have protein value, then it behooves the supplier to desolventize under mild conditions so as to not overcook the protein. But let's not forget that we still must recover the solvent.

In most oilseed plants, the DT (Fig. 5.17) is generally a vertically oriented cylindrical dryer with a series of steam heated trays mounted atop each other. Central to the DT is a shaft with one to two sweep arms per tray that keep the material moving. Each tray is provided with a gate and sail to maintain a constant level thereby

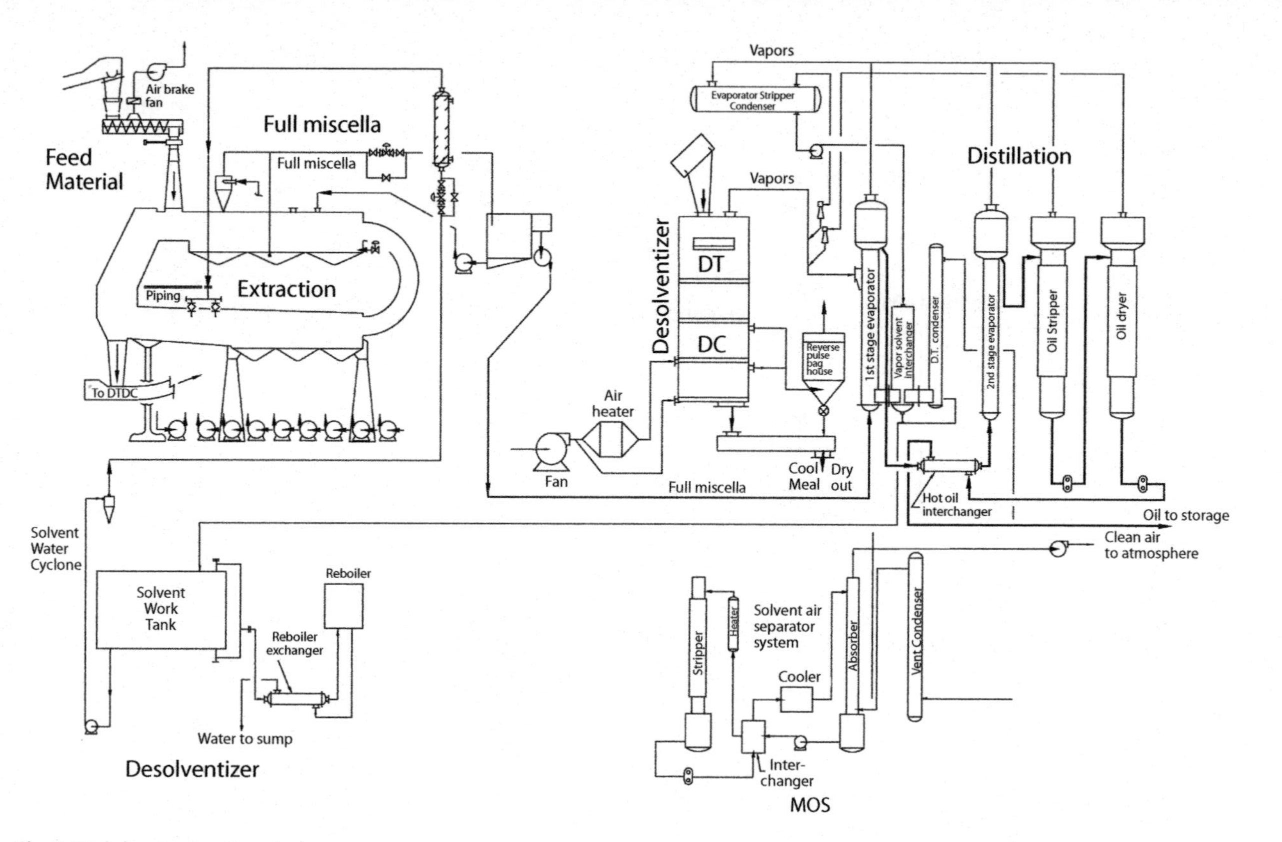

Fig. 5.15. Solvent extraction plant.

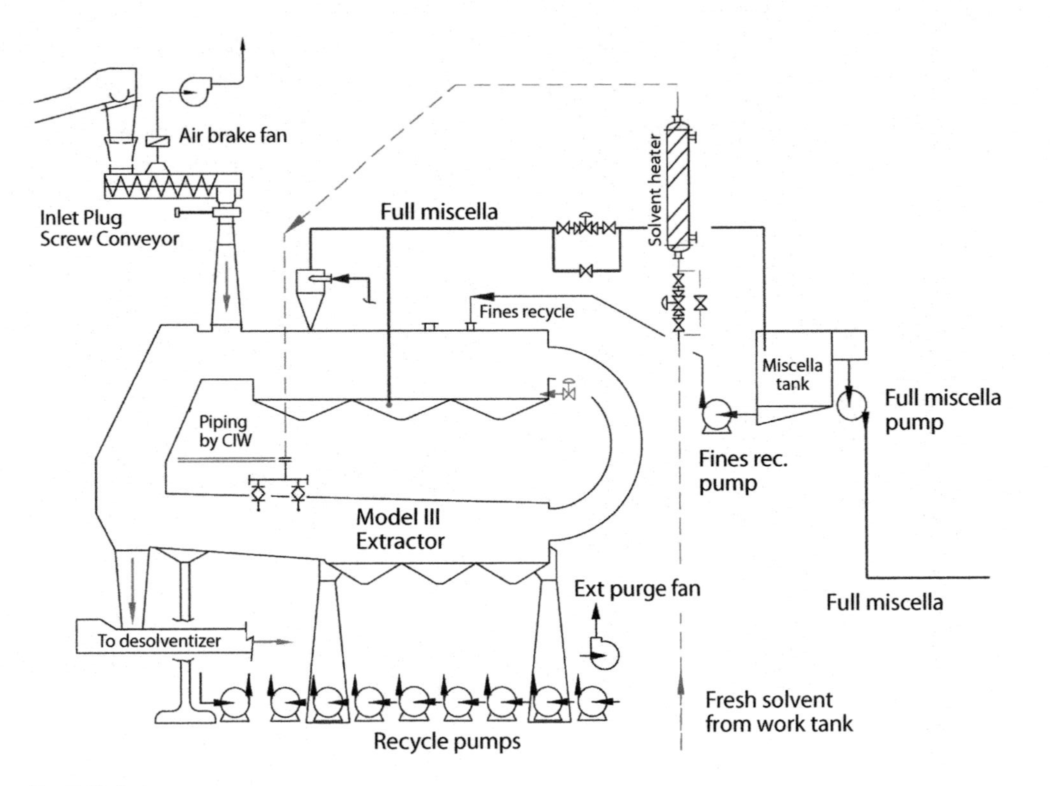

Fig. 5.16. Extractor.

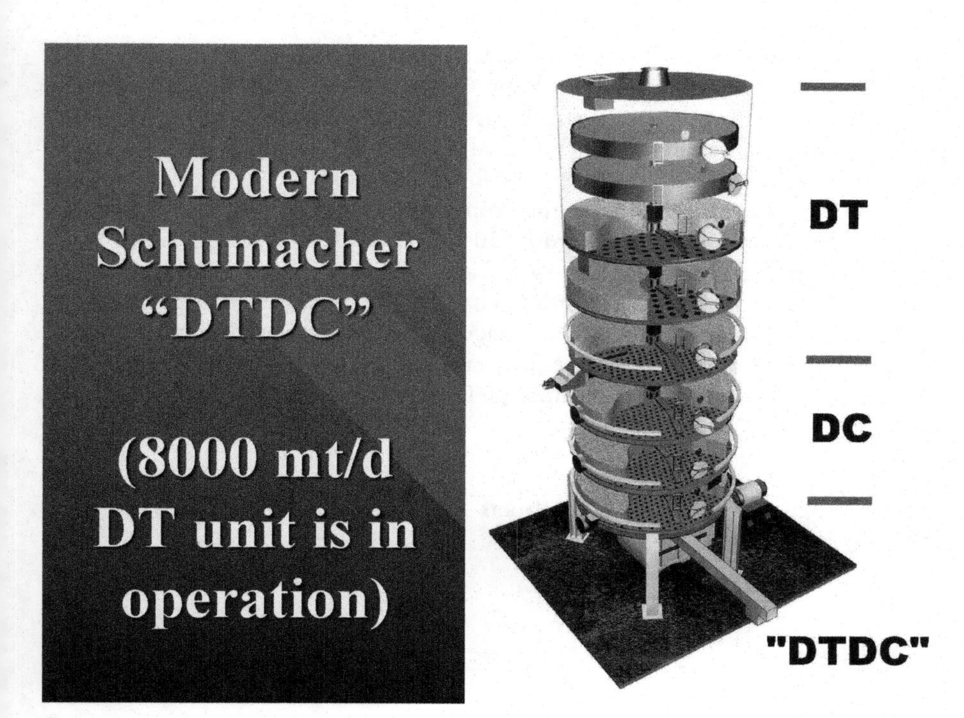

Fig. 5.17. Toaster/dryer-cooler.

providing sufficient residence time for desolventization. The bottom tray has a specially designed rotary valve with level sensor giving more precise control of level on the bottom tray.

With oilseeds, the top few trays are solid trays providing only indirect heat transfer into the material. These trays are generally not the full diameter of the cylinder, allowing vapors to flow upward in the annulus. In oilseed plants, two or three trays down from the top, the trays go to full diameter and are provided with a series of venting holes in the tray deck. The bottom tray is generally a perforated tray that sparges low pressure steam directly into the material on the bottom tray. The combination of solvent vapors and non-condensed steam flows vertically and upward through each tray DT. In oilseed plants, direct sparge steam will provide a large portion (+60%) of the heat required for evaporation of the solvent. The amount of steam added is that which is required to maintain the temperature of the discharged solids at ~100–105°C while the dome atop the DT is maintained at a few degrees C above the boiling point of the solvent.

It should be noted that at this time there is some concern that the direct sparge steam will turn the algae solids into a sticky mass, plugging the machine. If this is indeed the case, then only solid trays will be used with minimal sparge steam.

The vapors from the DT represent quite a heat source, so the vapors are directed to the shell side of the first stage evaporator. On the miscella side, the first stage is a rising film evaporator with a centrifugal vapor separator mounted at the top. The miscella side of the evaporator is maintained under a vacuum by steam ejectors to provide the temperature difference required for evaporation.

The first stage evaporator is the first stop for the miscella after miscella cleaning where oil concentration is increased from ~20% to as much as 75% (Fig. 5.18).

Non-condensed vapors from the shell side of the first stage are used to pre-heat solvent either in a vapor contactor or interchanger. A DT condenser is provided for any other vapors and is sized as if the first stage evaporator were not in place.

Concentrated miscella from the first stage is directed to the steam heated second stage evaporator where oil concentration is increased to the high 90% in a typical

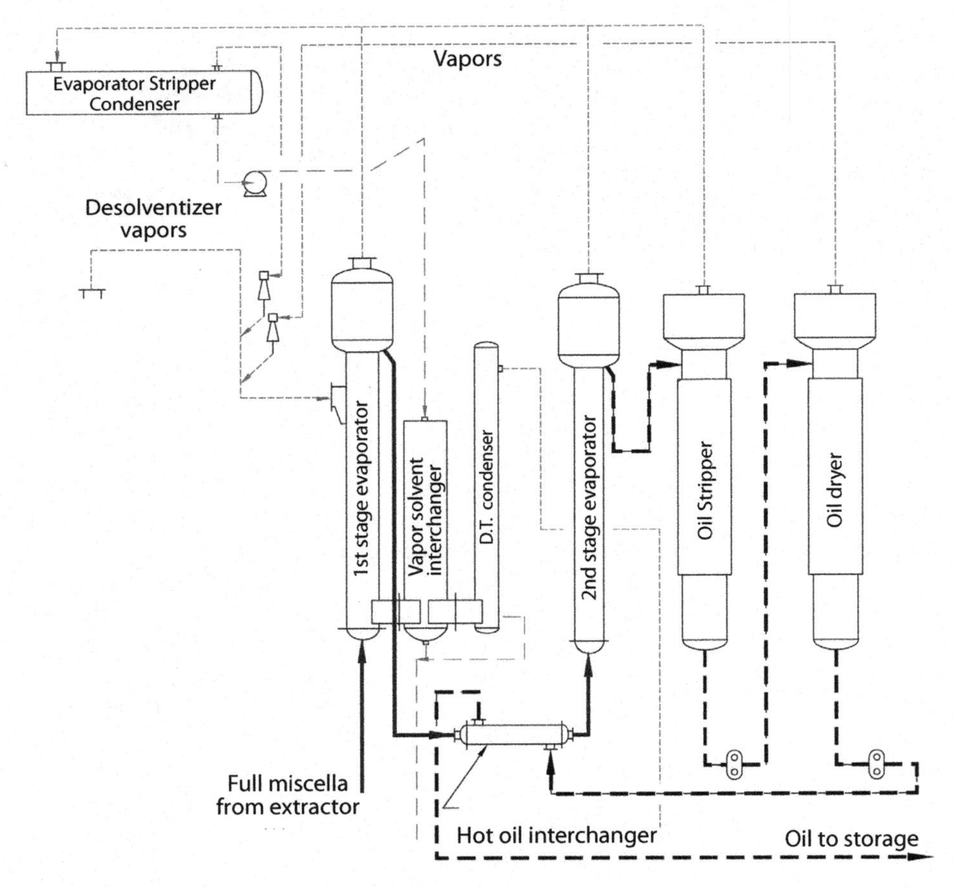

Fig. 5.18. Distillation.

oilseed plant. Vapors from the first and second stage evaporator are condensed in the evaporator condenser where they are returned to the work tank. The evaporator condenser is maintained under a vacuum by steam ejectors with the exhaust of the ejector directed to the shell side of the first stage for heat recovery.

Concentrated oil is preheated to the bubble point and then enters the oil stripper where a thin film of oil is created either by packing or a series of discs and donuts. Steam is added at the bottom of the stripper so that a combination of vacuum and steam strips away hexane to ppm levels. Given the presence of fine solids and other products in the oil, it is anticipated that wiped film evaporators may be required at this point to keep the oil moving and to achieve low ppm of hexane in the finished oil. If the oil is fluid enough, a high vacuum oil dryer is recommended to reduce the ppm of hexane in the finished oil to low levels. Good desolventization of the finished oil is necessary to be sure that hexane is not carried to downstream equipment at high enough levels to present a safety issue.

We had mentioned earlier that a small amount of air enters the extractor along with the solids. This headspace above the solvent is maintained at a temperature 5 to 10°C below the boiling point of the solvent to ensure that evaporated solvent dominates the atmosphere at this point. The objective is to maintain the vapor space well above the explosive limit of the solvent in the air.

All non-condensables are directed to a vent condenser followed by a mineral oil scrubber (see Fig. 5.19) to wash the vapors with mineral oil. Hexane prefers the mineral oil to the air making it possible to reduce hexane levels in the discharge air to less

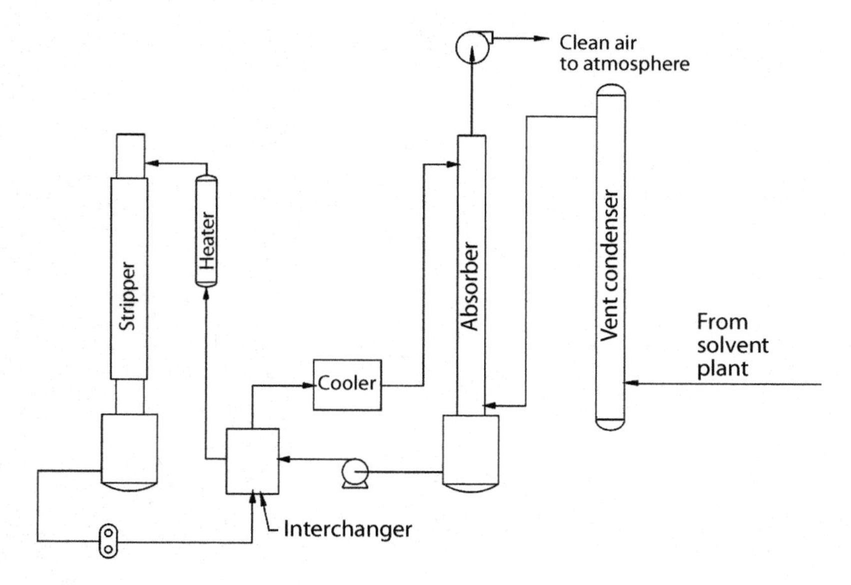

Fig. 5.19. Mineral Oil System.

than 30% of the Lower Exposure Limit (LEL). If alcohol were used as a solvent, then water would be used instead of mineral oil, although recovery of the alcohol from water would be expensive.

Condensed solvent is returned from all condensers and other locations, and contains mostly solvent with some water. As discussed earlier, water is immiscible with hexane making the separation an easy one in the gravity separator commonly called the work tank (see Fig. 5.20). Underflow from the work tank passes through a deboiler with heat recovery to reduce solvent in the wastewater to PPM levels.

Life Cycle Costs

As mentioned earlier, the experience with solvent extraction plants is that when purchased from reputable suppliers and properly maintained, they have an expected life well beyond 30 years. The only caveat here is that many of the algae strains are grown in salt water, making selection of materials of construction more of a concern.

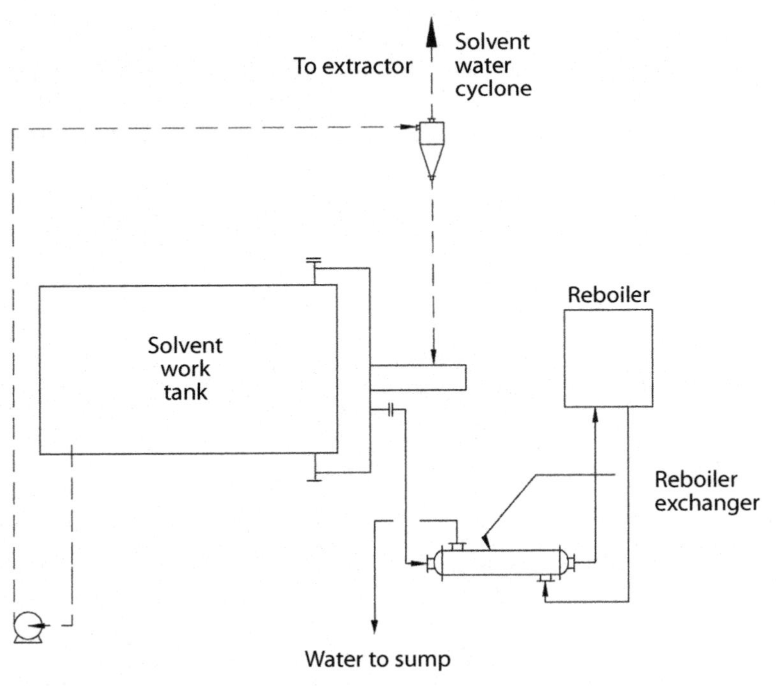

Fig. 5.20. Work tank.

Solvent Losses

Using hexane and soybeans as an example, new plants are installed assuming a regulated solvent loss of less than 1 liter per ton of material processed. This limit was arrived at recently and should be considered to be a moving target as improved solvent recovery techniques become available.

We are not suggesting that the limit for algae should be this low right now and fully expect that higher figures will be allowable for the present. Equally, we expect that experience from the field will be monitored continuously and that best available technologies will be employed to find new lows in the relatively near future.

Future

Several companies are growing algae at ever-increasing rates. These same companies are working on wet means to extract the oil from algae by using hexane as a solvent. These technologies will require improvements to make them cost-effective on a large scale.

It is our feeling that improvements in dewatering are coming and that these improvements along with integrated systems will demonstrate that dry extraction of algae solids will be a viable alternative.

By integrated systems, we mean that harvesting/dewatering/drying and extraction will be integrated to take advantage of each area's strengths. For instance, low pressure steam from a superheated steam flash dryer may be used to power a solvent extraction plant and perhaps pre-dry the algae solids. Equally, the algae growers will look to ways to gain added value from the extracted solids requiring a dry product in the end. We feel that some combination of added value will make drying and extraction a viable alternative.

References

Anderson, G.E. Solvent Extraction 2011.

Gertenbach, D. Solid-Liquid Extraction Technologies for Manufacturing Nutraceuticals from Botanicals, 1999.

Ozer, R.W. Steps Toward the Design and Construction of a Nutraceutical Extraction Facility, 2001.

Svonja, G. The Efficient Drying of Grain Residues & Sludges Using Biomass Fuels, 2007.

Watkins, L.R., Anderson Clayton & Co. and Williams, M., Anderson International. Expanders and Preparation for the Extraction of Oil Seeds.

6

Enzymatic Degumming

Christopher Loren Gene Dayton and Flavio Galhardo
Bunge Global Innovation, White Plains, New York, USA

Introduction

The purification of fats and oils in order to produce a light-colored, transparent, and bland-tasting product or to derivatize them into important chemical intermediates or biofuels is paramount for manufacturers of these products. The single most important step in the purification process is the efficient removal of phospholipids or gums, whether the oil source contains a very small amount of phospholipids, like palm, or oils that contain significant amounts of gums, such as soybean or algal sources. Traditionally, phospholipids have been removed by water or acid degumming followed by a caustic refining process in order to reduce the levels to meet specifications for food or industrial applications. Advances in biotechnology have allowed the development of enzymatic degumming as a sustainable purification method able to achieve the same, if not better, results than the traditional chemical methods while greatly increasing yields and reducing chemicals, wastes, energy, and overall costs.

Crude Oils

The composition of crude oils varies dramatically based on plant, animal or microbial source. These oils are predominately triacylglycerols, but contain many compounds produced in the lifecycle of the source material that may be considered as contaminants in food or industrial applications. The purification process focuses on the removal of naturally occurring contaminates from triacylglycerols. Crude oils typically contain two different types of impurities: oil insoluble and oil soluble. Oil insoluble impurities may include proteins, fibers, dirt and excess moisture. Oil soluble impurities may include free fatty acids, phospholipids, glycolipids, waxes, sterols, tocopherols, color bodies (beta carotene, chlorophylls, gossypols), hydrocarbons (aldehydes and ketones) and a number of trace metals. Table 6-A contains the crude oil compositions of selected oil sources.

Table 6-A. Crude Fats and Oils.

	Range of Composition			
	Soybean	Canola	Palm[a]	Rice Bran
Triacylglycerols (%)	95–97	94–97	90–95	81–84
Diacylglycerols (%)	1.0	0.5–3	4–8	7–6
Monoacylglycerols (%)	<0.5	<DL	1–2	5–6
Free Fatty Acids (%)	0.3–2.0	0.4–2.0	3–5	2–3
Phospholipids (%)	1.5–3.0	1.0–2.5	0.1–0.3	1.6
Glycolipids (%)	trace	trace	0.1–0.3	0.8
Unsaponifiables (%)	1.6	0.5–1.2	—	4
Sterols	0.33	0.7–1.0	0.03	
Tocopherols	0.15–0.21	0.07–0.17	0.05–0.1	
Hydrocarbons	0.014		0.02–0.05	
Trace Metals (ppm)				—
Calcium	50–300	100–300	trace	
Magnesium	50–300	100–300	trace	
Iron	1–3	1–4	4–8	

[a](Sambamthamurthi, et al., 2000)

Gums

The terms "gums" and "lecithins" have been used interchangeably over the years to describe the "sludge" found in crude vegetable oil. Other names used in the literature are: "slim," "mucilaginous material (sticky)," "foots," "gums," "lecithin," and "phosphatides" or "phospholipids." Deslimming was the original process for removing this sticky substance. The first "technical" term utilized was "foots." Foots were defined as solid material that passed through the crude separation system in the oil extraction process. This "solid" material often found itself on the bottom of storage and transportation tanks. Foots consist of phospholipids, gums, proteins, and oil-soluble materials (Kantor, 1950). In this chapter, the following terms are defined:

Gums—the complex mixture of phospholipids, glycolipids, sugars, proteins, and other water soluble compounds that may be present in crude oil. Gums removed from oil will contain at least 33% neutral oil.

Lecithin—the commercial product produced from water degumming of crude filtered soybean, rapeseed, and sunflower oils. The material will consist of roughly 46% phospholipids, 5% sugars, 11% glycolipids, 37% oil, and 1% water (ALC, unknown).

Structure

During the physical or solvent extraction of fats and oils from source material, major components recovered in the lipid phase are triacylglycerols (TAGs), phospholipids, unsaponifiable material (sterols, tocopherols, and hydrocarbons), free fatty acids (FFAs), and trace metals. The structure of triacylglycerols and phospholipids is very similar to one another since they both are built on a glycerol backbone and contain fatty acids of various chain lengths (see Fig. 6.1). Triacylglycerols and phospholipids are prochiral compounds and use the Stereospecific Numbering (*sn*) system to identify the positions of the phosphate functional group (see Fig. 6.2) and the placement of the fatty acids in TAGs, diacylglycerols (DAGs), and monoacylglycerols (MAGs). The *sn*3 position in naturally occurring phospholipids always contains the phosphate group. A phospholipid may also contain an additional phosphate group in the *sn*1 position, commonly called a diphosphatidylglycerol (DPG).

In most cases, phospholipids have a saturated fatty acid in the *sn*1 position and an unsaturated fatty acid in the *sn*2 position. Thus, 1-palmitoyl-2-oleoyl-phosphatidylcholine (Fig. 6.3) is a common constituent in natural membranes, but 1-linoleoyl-2-stearoyl-phosphatidylcholine is not. Cherry et al. (Cherry & Kramer,

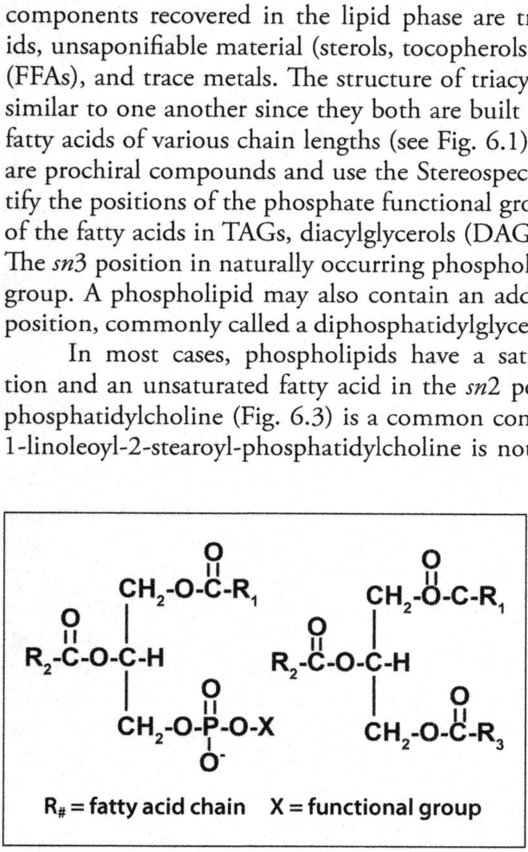

$R_{\#}$ = fatty acid chain X = functional group

Fig. 6.1. Structure of a phospholipid and triacylglycerol.

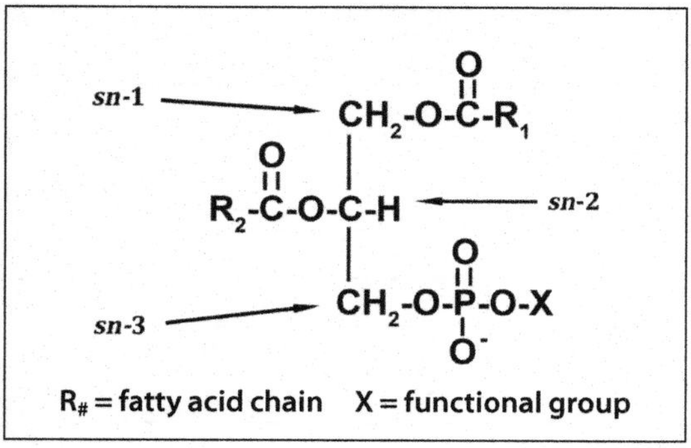

$R_{\#}$ = fatty acid chain X = functional group

Fig. 6.2. Stereospecific numbering of a phospholipid molecule.

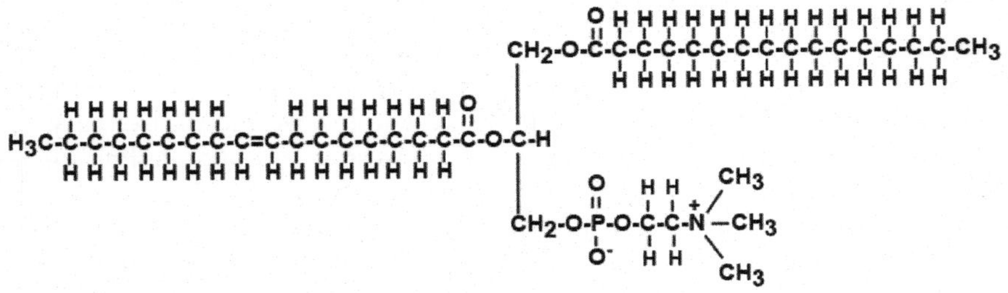

Fig. 6.3. 1-palmitoyl-2-oleoyl-phosphatidylcholine.

1989) reported the fatty-acid composition of soybean lecithin as may be seen in Table 6-B.

Chemistry

Triacylglycerols are nonpolar molecules, meaning the electrical charges are balanced due to the symmetry of the molecule, and thus have no net positive or negative charges. Phospholipids have both a non-polar region and a polar region, making them amphilphiles (possessing both hydrophilic and lipophilic properties). The fatty acid region of the phospholipid's molecule is non-polar, while the phosphate functional group is polar. Phosphatidylcholine (PC) (see Fig. 6.4) is polar due to strong quaternary base (ethylammonium group) attached to the phosphate. The polar functional groups attached to the phosphate group in most gums are choline, ethanolamine, inositol, and serine. A phospholipid without a functional group is known as Phosphatidic Acid (PA). The negatively charged phosphate group in PA will react

Table 6-B. Fatty Acids (%) of Soybean Lecithin (Cherry & Kramer, 1989).

	Range of Composition		
Fatty Acid	**Low**	**Intermediate**	**High**
Myristic (C14:0)	0.3–1.9	—	—
Palmitic (C16:0)	11.7–18.9	21.5–26.7	42.7
Palmitoleic (C16:1)	7.0–8.6	—	—
Stearic (C18:0)	3.7–4.3	9.3–11.7	—
Oleic (C18:1)	6.8–9.8	17.0–25.1	39.4
Linoleic (C18:2)	17.1–20.0	37.0–40.0	55.0–60.8
Linolenic (C18:3)	1.6	4.0–6.2	9.2
Arachidic (C20:0)	1.4–2.3	—	—

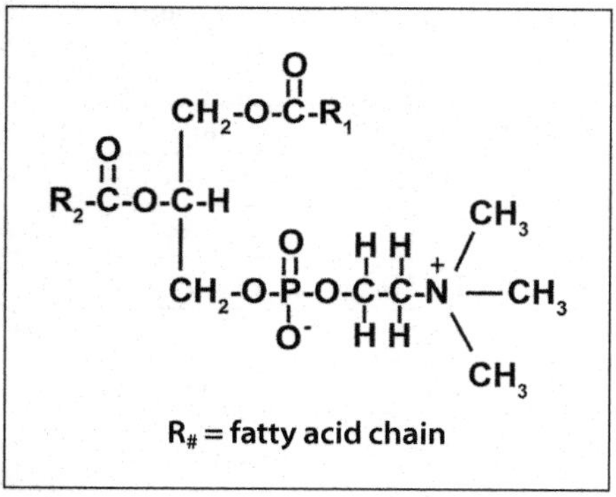

with a multivalent metal to neutralize its charge, forming a metal salt when PA is removed from the cell or when it is formed due to Phospholipase D activity in the preparatory and extraction processes. Calcium, magnesium, and iron salts of phosphatidic acid are present in gums, not phosphatidic acid (due to the neutral pH of the system). The calcium salt of phosphatidic acid may be seen in Fig. 6.5. Metal salts of phosphatidic acid are not polar compounds, but nonpolar as are triacylglycerols.

$R_{\#}$ = fatty acid chain

Fig. 6.4. Phosphatidylcholine.

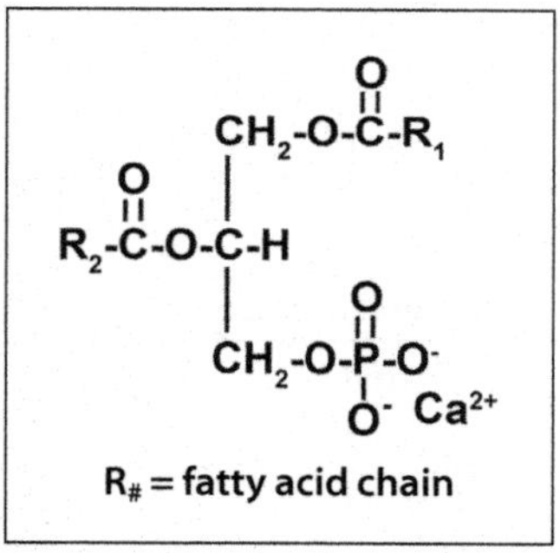

$R_{\#}$ = fatty acid chain

Fig. 6.5. Calcium salt of phosphatidic acid.

The composition of soybean lecithin may be found in Table 6-C. Phospholipids vary in their amphiphilic nature due to their functional groups. Another way to express the polarity of the molecules is by their hydratability. The more polar the compound, the easier it will be to hydrate the compound in the degumming process. The more nonpolar the functional group, the less the phospholipid will hydrate. The relative rates of hydration of phospholipids were first disclosed by Sen Gupta (Sen Gupta, 1986), and later presented by Segers & van de Sande (1990) in Table 6-D. According to them,

> "Sen Gupta used a turbine stirred cell of about 5 liters for the hydration experiments. The contents of the cell were pumped around via a cross flow microfilter. By analyzing the phospholipid composition of the microfiltered material he could follow the rate of hydration of the different phospholipids. The hydrated phospholipids would of course not pass the filter."

Emulsifiers or amphiphiles are described using a system known as hydrophilic lipophilic balance (HLB) in order to express their ability to form different types of emulsions: oil in water (O/W) or water in oil (W/O). The HLB values range from 1 to 20 with a few exceptions like sodium lauryl sulfate which are greater than 20. Low HLB emulsifiers are soluble in oil (lipophilic or non-polar substances) while high HLB emulsifiers are soluble in water (hydrophilic or polar substances). Bancroft's Rule states that the phase in which the emulsifier is most soluble will form the continuous phase of an emulsion (Davis, 1994). Soy lecithin is soluble in oil and has an

Table 6-C. Phospholipids Found in Soybean Lecithin (Cherry & Kramer, 1989).

	Range of Composition		
Phospholipid	**Low**	**Intermediate**	**High**
Phosphatidylcholine	12.0–21.0	29.0–39.0	41.0–46.0
Phosphatidylethanolamine	8.0–9.5	20.0–26.3	31.0–34.0
Phosphatidylinositol	1.7–7.0	13.0–17.5	19.0–21.0
Phosphatidylserine	0.2	5.9–6.3	—
Phosphatidic acid *	0.2–1.5	5.0–9.0	14.0
lyso-PC	1.5	8.5	—
lyso-PI	0.4–1.8	—	—
lyso-PS	1.0	—	—
lyso-PA	1.0	—	—

* Metal salts of phosphatidic acid

Table 6-D. Relative Rates of Hydration of Different Phospholipids (Segers & van de Sande, 1990).

Phospholipid	Rate of Hydration
Phosphatidylcholine	100
Phosphatidylinositol	44
Calcium salt of phosphatidylinositol	24
Phosphatidylethanolamine	16
Phosphatidic acid	8.5
Calcium salt of phosphatidylethanolamine	0.9
Calcium salt of phosphatidic acid	0.6

HLB of 3-4 indicating that it is an excellent emulsifier for water in oil emulsions. The oil is the continuous phase, while the gums and water are the dispersed phase. Lecithins are not soluble in water, but they are able to trap water in oil. Table 6-E contains an abbreviated listing of compounds of interest.

The amphiphilic nature of phospholipids enables it to form several different emulsions with water such as monolayers, bilayer sheets, micelles, and liposomes.

Phospholipids

Phospholipids serve at least four major functions in a plant cell: membrane structure, transduction, fatty acid building, and energy storage. A cell's membrane separates

Table 6-E. Approximate HLB Numbers of Surfactants.

Surfactant	HLB Number
Sodium dodecyl sulfate	40
Potassium oleate	20
Sodium oleate	18
Polyoxyethylene (20) sorbitan monooleate	15
Polyoxyethylene (20) sorbitan trioleate	10.5
Hydroxylated soy lecithin*	10–12
DATEM	8
Acetylated lecithin	9–10
Hydrolyzed soy lecithin	8–9
Deoiled lecithin with ~50 % phosphatidylcholine	6–7
Deoiled lecithin with ~20 % phosphatidylcholine	4–5
Standard soy lecithin	3–4
Glycerol monooleate (monoacylglycerol)	3.4
Glycerol dioleate (diacylglycerol)	1.8
Oleic Acid	1.0

*(Schneider, 1997; Sinram, 1991; McClements, 2005)

the interior of the cell from the outside environment. This membrane is a bilayer (see Fig. 6.6) consisting of two different phospholipid layers that keep ions, proteins, and other molecules where they are needed and prevent others from moving into areas where they should not be (Branton, 1969), while allowing specific molecules to pass through. Phospholipids aid in the biological signal transduction mechanisms across the cell membrane communicating to the cell when various chemical reactions for cellular growth, division, and differentiation should be performed. Phospholipids play a key role in plants' mechanism to transform very small carbon sources

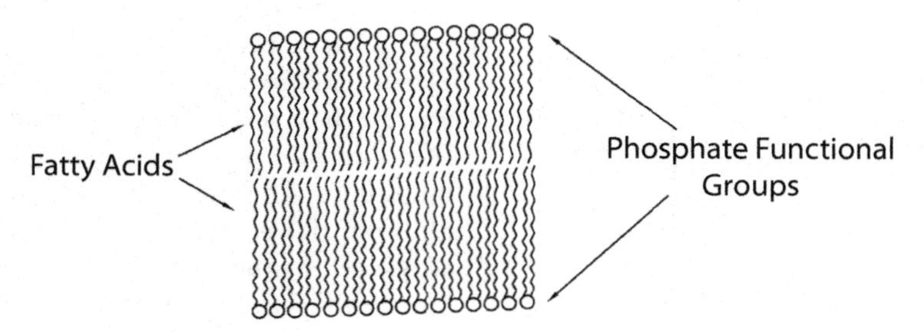

Fig. 6.6. Phospholipid Bilayer.

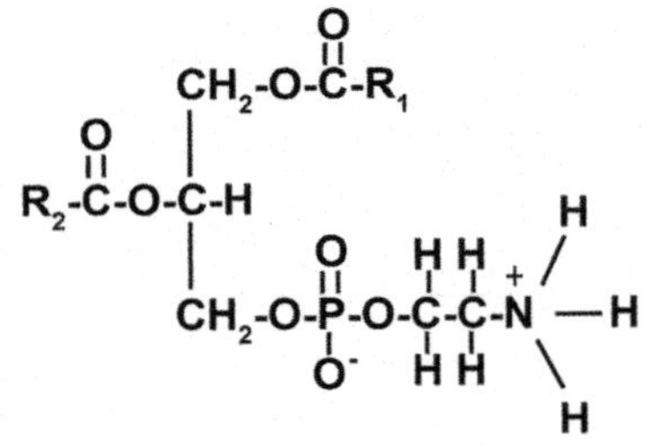

Fig. 6.7. Phosphatidyl ethanolamine.

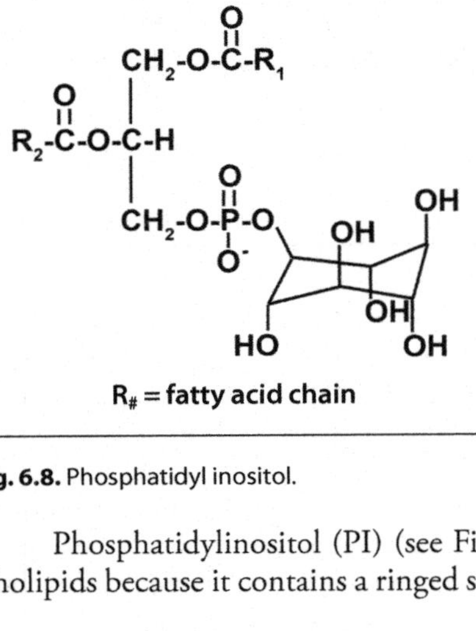

Fig. 6.8. Phosphatidyl inositol.

like carbon dioxide (CO_2) into very long chains of saturated fatty acids. Once these long chain fatty acids have been formed via elongation, they may be converted into monounsaturates and polyunsaturates enzymatically via desaturase enzymes (Harwood, 2005), where phospholipids also play an essential role.

Phosphatidylcholine (PC) is the most prevalent phospholipid in plant sources. Phosphatidylcholine has been shown to play a vital role in many important areas including maintaining cell structure, fat metabolism, memory, and nerve signaling, as a precursor to important neurotransmitters and liver health (Szuhaj & van Nieuwenhyyzen, 2003).

Fig. 6.7 shows the structure for phosphatidylethanolamine (PE) or cephalin. PE is very similar to PC, but does not contain the methyl groups. Cephalin is found in all living cells, although in human physiology it is found particularly in nervous tissue such as the white matter of brain, nerves, neural tissue, and in the spinal cord. Whereas PC is the principal phospholipid in animals, cephalin is the principal one in bacteria (Murzyn, et al., 2005).

Phosphatidylinositol (PI) (see Fig. 6.8), is very different from the other phospholipids because it contains a ringed structure. The phospholipid plays an important role in a cell's ability to signal.

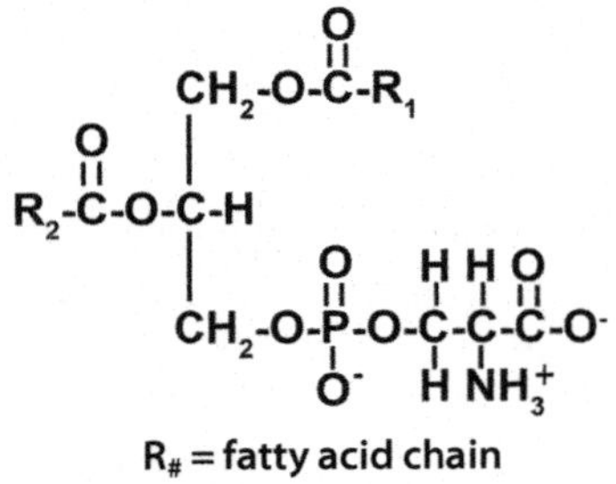

Fig. 6.9. Phosphatidyl serine.

Phosphatidic Acid (PA) is the most misunderstood phospholipid in oil processing. It appears to have no function except making it difficult in the refinery process. It is not a usable emulsifier like PC, and does not appear to have a significant health benefit like PC and PS. However, PA is an essential substrate for enzymatic synthesis of all other phospholipids and triacylglycerols present in cells (Arthenstaedt et al., 1999).

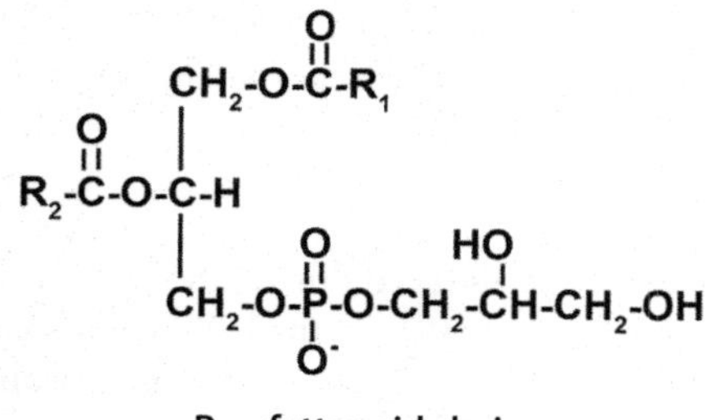

Phosphatidylserine (PS) was first named by Dr. Folch at the Hospital of The Rockefeller Institute for Medical Research. He described in a letter to the editor that a previous paper he had co-authored did not meet the previously accepted rule that all of the phospholipids were thought to be ethanolamine in the brain, but his investigation proved that part of the nitrogen measured was not from an amino acid, but was actually from serine (see Fig. 6.9). He went on to isolate a 92% pure phosphatidylserine from the brain (Folch, 1941, 1948).

$R_{\#}$ = fatty acid chain

Fig. 6.10. Phosphatidylglycerol.

$R_{\#}$ = fatty acid chain

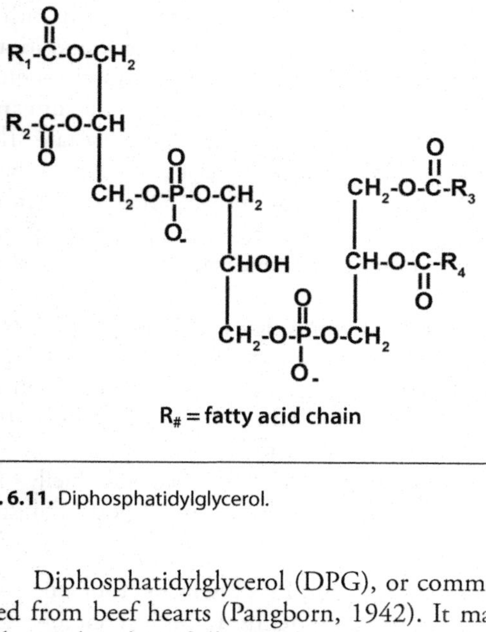

Fig. 6.11. Diphosphatidylglycerol.

Phosphatidylglycerol (PG) is the main component of bacterial membranes (see Fig. 6.10). It was first isolated by Benson and Maruo from an alga source (Benson & Maruo, 1958). This lipid occurs widely, but at varying levels in animals, plants, and bacterial membranes (Hsu & Turk, 2001). Rooney et al. (1972) reported on PG's important role in the function the distribution of surfactants in the lungs.

Diphosphatidylglycerol (DPG), or commonly called cardiolipin, was first isolated from beef hearts (Pangborn, 1942). It may be found in bacterial membranes and mitochondria of all animal tissues. It is unique because it contains four fatty acids and two phosphate groups (see Fig. 6.11). These fatty acids are almost always polyunsaturated in mammalian species (Schlame et al., 1993).

The compositional analysis of phospholipids present in vegetable oils had been historically completed on the recovered gums from oil, not on the phospholipids actually present in the oil. Additionally, most of the identification of the phospholipids was performed using thin layer chromatography (TLC) and/or High Pressure Liquid Chromatography (HPLC) with mass detectors (Racicot & Handel, 1983). Advances in analytical instrumentation have enabled the quantitative analysis of phospholipids in oil or gums. Table 6-F provides the phospholipid analysis of both crude soybean oil and gums by Phosphorus 31 Nuclear Magnetic Resonance (P-31 NMR). The P-31 NMR analytical method was refined and reported by Verenium (Domaille et al., 2007).

Non-hydratable Phospholipids

Non-hydratable phospholipids (NHPs) are formed in the seed during the harvesting, storage, handling, crushing, and extraction of the oilseeds (Pardun, 1988). Phospholipase D (PLD), present in the seeds, reacts with phospholipids, forming phosphatidic acid (PA). The PA then complexes with available divalent metals [(calcium (Ca), magnesium (Mg), and iron (Fe)], present in the seeds, forming salts (Kővári et al., 2008; GEA, unknown). The NHPs are present in crude and water-degummed oils and may be measured by the presence of Ca, Mg, and Fe content (List et al., 1992; Tosi et al., 2002; Kanamoto et al., 1981; Young, 1990). See Fig. 6.5.

Hvolby degummed an oil that had previously been water degummed with a buffered solution at pHs from "–1" to 14 (see Table 6-G). In analyzing the relationship of phosphorus to calcium and magnesium, Hvolby found the ratio was roughly 1:0.9 (P:Ca + Mg). The experiments demonstrate the dissociation of the calcium and magnesium salts of phosphatidic acid, allowing the phosphatidic acid to hydrate. The removal of the phosphorus for all of the degumming experiments using buffered solutions maintained this relationship. Hvolby also water degummed different oils from 1968 to 1970 and then water degummed the same oils a second time and compared the phosphorus, calcium, and magnesium in the two degummed oils. The results are listed below in Table 6-H (on page 118).

The percentage of NHP that is complexed with calcium and magnesium was 77.6% from water degummed oil ("first" water degumming experiments) and 88.8% in the "twice" water degummed oil, confirming the majority of "nonhydratable phospholipids" were salts of calcium and magnesium. The other NHPs present were the slowly hydratable phosphatidyl ethanolamine (PE) (Pan et al., 2000).

Nelson reported, ". . . nonhydratable soybean phospholipids consist chiefly of a mixture of phosphatidic acids and *lyso*-phosphatidic acids." (Nielson,1960). Dijkstra further claimed,

> It should, however, be noted that not all PA and PE present in crude oil are nonhydratable since the gums obtained by water degumming contain

Table 6-F. Phospholipid Compositions of Crude Oil and Gums by P-31 NMR.

	PC (%)	PE (%)	PI (%)	PA (%)	*lyso*-PC (%)	*lyso*-PE (%)	lyso-PI (%)	*lyso*-PA (%)	P (ppm)
Crude	0.54	0.50	0.29	0.23	0.02	0.01	bd	bd	600
Crude	1.18	0.78	0.65	0.29	0.07	0.04	0.04	bd	1600
Gums	9.36	9.32	5.50	4.86	bd	bd	bd	bd	—

Table 6-G. Degumming at Different pHs (Hvolby, 1971).

	Aqueous Phase	Phosphorus (ppm)[a]	Calcium (ppm)[b]	Magnesium (ppm)[c]
Crude Oil	None	523.39	105.81	69.03
Industrial Water Degummed Oil	Water	170.64	92.58	42.29
Starting Water Degummed Oil for pH Study		135.03	95.39	37.67
	Buffered Solution[d]			
pH "-1"	Concentrated HCl	58.22	0	0
pH 0	1 N HCl	9.29	0	0.24
pH 1	1 N HCl, KCl	13.01	4.41	2.19
pH 2	1 N HCl, NaCl, Citric Acid	31.59	17.23	6.56
pH 3	Citric Acid, NaOH, NaCl	36.85	19.24	7.05
pH 4	Citric Acid, NaOH, NaCl	89.50	56.51	20.66
pH 5	Citric Acid, NaOH	119.85	73.35	28.44
pH 6	Citric Acid, NaOH	128.53	78.56	30.62
pH 7	Na_2HPO_4, KH_2PO_4	128.84	82.56	30.62
pH 8	$Na_2B_4O_7$, 1 N HCl	129.45	83.37	30.38
pH 9	$Na_2B_4O_7$, 1 N HCl	118.62	86.57	32.33
pH 10	$Na_2B_4O_7$, 1 N HCl	78.97	61.72	24.79
pH 11	1 N NaOH, NaCl, NH_2CH_2COOH	61.94	54.91	19.20
pH 12	1 N NaOH, NaCl, NH_2CH_2COOH	56.86	42.89	17.26
pH 13	1 N NaOH, NaCl, NH_2CH_2COOH	34.38	26.45	10.21
pH 14	1 N NaOH	0	0	0

[a]molecular weight of phosphorus used for mmole conversion 30.97 to parts per million.
[b]molecular weight of calcium used for mmole conversion 40.08 to parts per million.
[c]molecular weight of magnesium used of mmole conversion 24.305 to parts per million.
[d]phosphorus, calcium, and magnesium were determined after reacting with the buffered solution for 2 hours at 20°C after centrifugation.

appreciable amounts of both PA and PE. Moreover, these gums also contain calcium and magnesium, which means that these ions can also be bonded to hydratable phosphatides. (Gunstone et al., 2007).

Extending the same argument, one may state that not all vegetable oils are non-hydratable since one-third of lecithin is made up of neutral oil. This, of course, is not a valid statement. The presence of calcium and magnesium salts of PA and PE in lecithin does not mean they are hydratable, but means they were removed in the

Table 6-H. Water degumming experiments (Hvolby, 1971).

	First Water Degumming			Second Water Degumming		
Production	P (ppm)[a]	Ca (ppm)[b]	Mg (ppm)[c]	P (ppm)[a]	Ca (ppm)[b]	Mg (ppm)[c]
Oct. 1968	—	—	—	138.21	95.39	37.67
Dec. 1968	165.47	103.81	45.94	138.53	95.79	39.13
Feb. 1969	182.59	94.19	44.96	130.29	80.16	38.16
Feb. 1969	229.51	134.27	54.69	187.66	123.05	50.55
Mar. 1969	173.72	95.97	42.53	128.70	82.97	34.76
Apr. 1969	166.43	83.37	40.10	120.46	71.74	33.78
Apr. 1969	206.68	109.82	48.85	158.18	93.79	43.99
Jun. 1969	165.16	90.18	40.35	145.82	90.18	38.89
Jul. 1969	197.17	103.81	42.78	173.08	104.21	42.05
Aug. 1969	213.66	113.43	46.67	176.25	105.01	43.02
Sep. 1969	186.40	93.39	42.05	147.72	88.58	39.37
Dec. 1969	180.06	93.39	42.05	150.89	88.58	40.59
Mar. 1970	225.39	122.24	50.80	205.10	123.45	51.77
Ave.	191.02	103.16	45.15	153.91	95.61	41.06

[a]molecular weight of phosphorus used for mmole conversion 30.97 to parts per million.
[b]molecular weight of calcium used for mmole conversion 40.08 to parts per million.
[c]molecular weight of magnesium used of mmole conversion 24.305 to parts per million.

emulsion along with other non-hydratable material. Phospholipids that may be present in water degummed oil are dependent not on being complexed with other metals, but on the rate at which PC, PI, and PE are hydrated (Segers & van de Sande, 1990) and the efficiency of the separator.

Hydrated forms of PA, PE, and PI, obtained from the removal of Ca, Mg, and Fe metals also form emulsions when they are removed, resulting in additional oil loss in the gums. The gums removed in the acid degumming process have 40 to 50% neutral oil (Logan, 2004) compared to only 25 to 37% for gums produced in the water degumming process.

Why Remove Phospholipids?

Phospholipids are removed from commodity vegetable oils for three primary reasons: (1) transportation and storage of crude oils, (2) production of lecithins, and (3) purification of oils for food or industrial applications. Crude oils containing high levels of phospholipids will self degum over time due to the differences between the denser gums and the lighter oil. Any water present after the extraction process or obtained from the environment will hydrate the gums, allowing the phospholipid/water complex to fall-out of the oil solution. Tank bottoms will reduce the usable volume of the storage tank, eventually requiring the cleaning of the tank and the physical removal of the sludge. Oil pumped from a tank where the oil has self-degummed will not be uniform and cause further difficulties in downstream oil processing. Finally, any sludge that may have dropped out of the oil during the shipping process will require additional cleaning of the truck, railcar, or ship to prevent any contamination between cargos.

Lecithin is a "clear and brilliant" product used as a natural emulsifier in a large number of food and industrial applications. Crude oil is filtered with the aid of diatomaceous earth to remove any meal fines (protein) and dirt remaining in the oil after the extraction process. It is not industrially feasible to filter lecithin due to its high viscosity. Lecithin produced from soybean oil is always from the water degumming process. It has been reported that some rapeseed and sunflower lecithins are produced by citric acid degumming in Europe. The amount of crude soybean, rapeseed, and sunflower oil degummed to produce lecithin is approximately 40% of oils, as estimated by the authors of this chapter. The remaining oils are degummed for purification.

The removal of phospholipids from crude oils decreases the amount of subsequent chemicals required in the caustic refining process (acid and sodium hydroxide), bleaching process (filter aid, bleaching clays, and silica adsorbents), the hydrogenation process (metal catalysts), and finally the interesterification and biodiesel process (sodium methoxide or enzymes). Phospholipids present in the oil prior to bleaching will be adsorbed on the bleaching clay, decreasing its effectiveness in its ability to remove chlorophyll and its derivatives. Phospholipids present in the oil to be hydrogenated will poison the catalyst, reducing its ability to catalyze the reaction, and modifies the reaction mechanism, thus forming more saturated fatty acids, and increasing the reaction time in the partial hydrogenation process (Higgins, 2010). In the chemical interesterification (CIE) process, phospholipids present in the oil increase catalyst usage, emulsion losses, and blind filters. In the enzymatic interesterification (EIE) process, phospholipids inhibit the active sites of the enzyme and foul the packed bed, causing higher enzyme usage and increasing costs. The presence of phospholipids and galactolipids in oil heated above the "caramelization" temperature of sugar will cause a maillard reaction to occur. The unwanted reaction will result in the darkening of the oil that cannot be completely removed in any subsequent processing. If the phospholipids are not removed prior to hydrogenation or deodorization, the finished oil may become black and have a burnt flavor.

Traditional Techniques for Removal of Phospholipids

The traditional methods for removing phospholipids from triacylglycerols involve creating an emulsion with water in order to take advantage of the natural HLB and the differences in specific gravity for removal from the oil. Water, a polar solvent, is added to the oil to allow the hydratable portion of the phospholipids to form liposomes, trapping neutral oil within an emulsion. The emulsion is then removed from the bulk of the oil via high speed centrifugation in a continuous process. The emulsion will typically contain two molecules of hydrated phospholipids, one molecule of TAG or "neutral oil," and any water soluble carbohydrates and galactolipids. The addition of either acid or caustic does not change the mechanism for the removal of phospholipids, but changes the rate of hydration in the case of acid addition and the increased emulsion for caustic refining due to the formation of soap created in the neutralization process, allowing for lower residual phosphorus in the oil after treatment at the expense of higher neutral oil losses. The inability to successfully acid degum soybean and rapeseed oils with various levels of non-hydratable phospholipids lead to the early failed attempts at designing a rigorous and robust physical refining process for these oils (Forster & Harper, 1983).

Separators

Industrial centrifuges have been developed by the major manufacturers that are very efficient at removing hydratable phospholipids from oil in a continuous process. The light phase (oil) and heavy phase (gums and water) to be separated enter the centrifuge and are quickly separated with a gravitational force greater than 5,000 (Logan, 2004) in less than a second. Centrifuges utilized in the laboratory are typically batch centrifuges, where the material to be separated is under extreme gravitational forces for extended periods of time, enabling excellent separations, but are physically unattainable in an industrial application.

The Gyro-Tester, a laboratory-scale continuous centrifuge equipped with a bowl containing a disc stack for separation of both the oil phase and heavy phase, was developed and marketed by De Laval. The Gyro-Tester is an excellent tool for process development requiring centrifugal separation for accurate and scalable results in oil processing.

Water Degumming

The objective of water degumming is to remove the readily hydratable gums from the oil in order to produce lecithin or reduce the overall phosphorus content in crude oil. The objective is not the complete removal of phospholipids. If a clear and brilliant lecithin is the goal, the crude oil is filtered after the extraction process to remove any impurities that may cause the lecithin to be cloudy or opaque. Table 6-I contains literature references compiled by List et al. for water degumming (1981). Commercially, oil is heated to 70 to 80°C where 1 to 4% "soft" water or steam is added with agitation. The crude oil and water are mixed from 10 to 60 minutes in order to allow the gums to "flocculate" or grow larger. The gums are then separated from the oil via a continuous high-speed centrifuge. The collected gums will contain between 25 to 50% water and may be dried using a thin-film vacuum dryer.

Two critical parameters must be met in order to maximize gum removal and lecithin yield, agitation and time. The water must be vigorously mixed into the oil, allowing for the dispersion of the water and then a slow agitation without shear to allow the gums to flocculate. The phospholipids must have time to hydrate and grow

Table 6-I. Degumming Conditions from Literature (List et al., 1981).

Parameter	Quantity	Reference
Water	75% weight of gums	Crauer (1972)
	1–2.5%	Brain (1976)
	2–3%	Van Niewenhuyzen (1976)
	1%	Norris (1994)
	2%	Carr (1976)
	Equal to weight of gums	Braae (1976)
	2–5%	Anderson (1953)
Temperature	32–49°C	Norris (1994)
	50–70°C	Van Niewenhyyzen (1976)
	60–70°C	Bernardini (1973)
	70°C	Carr (1976)
	95°C	Anderson (1953)
Agitation	Vigorous	Bernardini (1973)
	Mechanical	Carr (1976)
Time	30–60 minutes	Carr (1976)
	10–15 minutes	Braae (1976

into large flake-like material. If time is limited, then the residual phosphorus in the oil will be high and the lecithin yield will be low.

The water-degummed oil will contain approximately 50 to 300 ppm of phosphorus due to the level of non-hydratable phospholipids present in the crude oil, efficiency of the water degumming process, and the allowable amount of phosphorus set by international trading rules. These non-hydratable phospholipids are calcium, magnesium, and iron salts of phosphatidic acid and phosphatidylethanolamine. The degummed oil to be shipped must be dried with a vacuum drier to prevent any gums remaining in oil from settling in the tank and to meet moisture specifications. If the water degumming process was not operated for lecithin production, the gums may be returned to the meal prior to the meal dryer, or dried and sold as a feed ingredient.

Acid Degumming

A number of processes (super-degumming, TOP degumming, ORP degumming, UF™ degumming, and others) have been developed for the removal of non-hydratable phospholipids using acids. These acid degumming processes were designed as a pretreatment for physically refining of "soft" seed oils. It was hoped that the acid degummed oil after the bleaching process would achieve a phosphorus of less than 3 ppm, allowing the oil to be physically refined (Anderson, 2005). Gums produced from an acid degumming process will usually be very dark in color and unsuitable for edible purposes (Anderson, 2005). The gums will also be very viscous and difficult to separate if obtained from crude non-degummed oil. Thus, most acid degumming processes use water degummed oil or pressed oils as their starting material.

Super-degumming

The super-degumming process was developed and patented by Ringers and Segers, working for Unilever for previously water degummed oil (Ringers & Segers, 1977). Crude-water degummed oil was treated with a concentrated acid (typically 50% citric acid) at 65 to 90°C; 2.5% water was added, and the mixture was cooled to below 40°C. The oil was slowly mixed, and the phospholipids were allowed to form a "semi-crystalline" structure according to the inventors. The oil was held at this temperature from 30 minutes to 1 hour. Segers reported that at low temperatures these "crystals" are insoluble in oil and adsorb other minor components (Segers, 1982). The oil could be centrifuged cold or may be heated rapidly (less than one minute) to maintain the "crystal" structure and centrifuged at 70°C. The process resulted in a residual phosphorus of 30 ppm or less (Ringers & Segers, 1977). If a washing step was added, the residual phosphorus was reported to be less than 10 ppm (Segers, 1982). A total of three centrifuges is required for the super-degumming process: water degumming, acid degumming and water washing.

TOP Degumming

The TOP degumming process was developed and patented by Dijkstra and Van Opstal for previously water degummed oil (Dijkstra & Van Opstal, 1987). The process has also been known as the "total" degumming process. TOP is the Dutch acronym for "Totaal Ontslijmings Process" (Gunstone et al., 2007). An acid (0.4 to 2.0 weight percent) was added to previously water-degummed oil and was dispersed with a high shear mixer, producing 100 million droplets of aqueous acid per gram of oil. It was claimed that the acid is added to decompose the non-hydratable phospholipids present in the oil. The contact time was no greater than 5 minutes. A base was added to bring the pH greater than 2.5 with a total amount of water to no more than 5%. Once the base and water had been added, the gums were separated from the oil via centrifugation. Oybek Zufarov et al. (2008) carried out a number of experiments on rapeseed and sunflower oil that was either solvent extracted or pressed using a single centrifuge as described in Dijkstra and Van Opstal's patent. Results from the rapeseed experiments may be found in Table 6-J, and sunflower in Table 6-K.

The results obtained by Zufarov et al. do not meet the necessary requirements for a rugged degumming process preparing the oil for physical refining. Dijkstra and Van Opstal (1989) further refined the TOP process by adding an acid retention tank, a second centrifuge, and finally a washing or caustic refining step in order to process oils with high levels of NHPs. In total, four separate centrifuges are required for the TOP degumming process: water degumming centrifuge, initial acid degumming centrifuge, second acid degumming centrifuge, and a caustic refining or water washing centrifuge.

Table 6-J. Extracted and Pressed Rapeseed (Zufarov et al., 2008).

	Rapeseed					
	Extracted			Pressed		
	P (ppm)	Ca (ppm)	Mg (ppm)	P (ppm)	Ca (ppm)	Mg (ppm)
Crude Oil	863.6	181.2	92.2	156.4	78.3	25.6
Water Degummed	70.4	69.3	41.4	60.9	58.4	15.3
Acid Degummed	21.4	19.9	12.3	16.9	9.7	9.7
TOP Degummed	15.1	15.6	3.2	9.6	5.1	1.7

Table 6-K. Extracted and Pressed Sunflower (Zufarov et al., 2008).

	Sunflower					
	Extracted			Pressed		
	P (ppm)	Ca (ppm)	Mg (ppm)	P (ppm)	Ca (ppm)	Mg (ppm)
Crude Oil	293.5	74.9	18.4	95.7	26.0	18.4
Water Degummed	56.9	34.4	6.5	50.1	12.9	6.5
Acid Degummed	11.8	9.3	2.3	7.1	4.1	2.3
TOP Degummed	10.6	5.1	0.5	4.5	0.9	0.5

UF Degumming

Ultrafine (UF) degumming was developed in cooperation of Krupp Engineering and Cereol's research group in Budapest, Hungary (Rohdenburg et al., 1993). Water degummed oil was treated with 100 to 800 ppm of concentrated acid for 15 minutes at 40 to 60°C. The oil was cooled to 30°C, where a stoichiometric amount to an additional 50% excess caustic was added to neutralize the acid in a 5 to 15% solution. The oil mixture was slowly agitated for two hours and then rapidly heated to 80°C. The oil was then washed with 10% water. Analytical results presented in the patent may be found in Table 6-L. The inventors claimed an additional benefit of reducing the wax content via the degumming process. A total of three centrifuges is required for the UF degumming process.

Organic Refining Process

The organic refining process or ORP was developed by Ag Processing. Crude- or water-degummed oil was heated to 92 to 98°C. A citric acid solution containing 1 to 5% acid was prepared and added to the oil from 3 to 20% by weight to the oil with a knife mixer. After a short retention (less than 16 minutes), the gums were separated from the oil. The degummed oil was reported to be less than 10 ppm phosphorus and

Table 6-L. UF Degumming (Rohdenburg et al., 1993).

	Initial Phosphorus (ppm)	Acid Typed	Stoichiometric Amount of Caustic (%)	Washing Water (%)	Final Phosphorus (ppm)
Water Degummed Sunflower	75	Citric	100	10	2.5
Water Degummed Sunflower	80	Citric	110	19	4.5
Pressed Sunflower	150	Phosphoric	100	2 × 10	7.0
Water Degummed Rapeseed	90	Phosphoric	100	10	9.0
Water Degummed Soybean	100	Citric	100	10	6.0
Water Degummed Sunflower	52	Citric	100	Yes, amount not specified	3.0

the citric acid could be recovered and reused in the process (Copeland et al., 2001). The recovered gums contained a large amount of *lyso*-phospholipids generated in the process in comparison to conventional water or acid degumming process (see Table 6-M). The *lyso*-gums were generated due to the acid hydrolysis of the fatty acid in either the *sn*1 or *sn*2 positions.

Dry Degumming

Dry degumming was developed for palm, palm kernel, and coconut type oils containing small amounts of phospholipids. The degumming process combines the acid degumming step with the bleaching process, thus eliminating the water addition and centrifugation of the gums. A 75 to 85% phosphoric acid is added at a rate of 500 to 10,000 ppm weight percent to the oil at 80 to 100°C. Following the acid reaction, 1 to 3% bleaching earth or a combination of bleaching earth and silica is added to oil in the bleaching process. The oil is then filtered using a conventional filtration system. No centrifuges are used in the dry degumming process. Brooks and Hollis (2011),

Table 6-M. ORP Gums (Copeland et al., 2001).

	ORP Gums	Conventional Lecithin
Phosphatidylcholine	14.2	16
Phosphatidylethanolamine	19.9	14
Phosphatidylinositol	8.7	9
Phosphatidic Acid	14.2	5
lyso- forms	6.2	0.5

disclosed that leaving 0.2% moisture in the oil after the vacuum bleaching and filtration reduced the residual phosphorus in the oil by approximately 50%, 4 to 2 ppm.

Data presented at the 102nd AOCS (American Oil Chemists' Society) meeting held in Cincinnati, Ohio, linked high levels of phosphoric acid in the dry degumming process to increased levels of 3-MCPD (3-monochloropropane-1,2-diol) esters in fully processed palm oils (Ramli et al., 2011a). Crude palm oil was physically refined in a 200 kg batch pilot refining plant to determine the role of drying degumming on 3-MCPD esters in the physical refining process. It was reported that the level of 3-MCPD esters was lowered from a high of 3.89 ppm in a dry degumming process using an acid-activated clay to a low 0.25 ppm in a water degumming followed by a neutral bleaching clay process (Ramli, et al., 2011b). It may be assumed that if the link is confirmed, the industrial process of dry degumming with large amounts of phosphoric acid and acid-activated clays may be eliminated in the future in favor of a water or enzymatic degumming process.

Caustic Refining

Caustic refining is the traditional method of removing both phospholipids and free fatty acids from crude, water degummed, or acid degummed oils. Caustic refining takes advantage of the ability to transform free fatty acids (FFAs) present in the oil into a soap, dramatically changing the HLB value of the molecule from 1 to 18, thus making the sodium soap both water soluble and a very strong emulsifier. The process also takes advantage of any remaining hydratable phospholipids to help produce an emulsion for the removal of the NHPs. The downside of the emulsion produced from soaps and hydratable phospholipids is the unintentional removal of neutral oil with the soapstock. Additionally, it must be stated that sodium hydroxide is not a selective chemical. It reacts with all compounds present in oil sources (FFAs, triacylglycerols, diacylglycerols, monoacylglycerols, tocopherols, sterols, etc.) decreasing the total yield. Side reactions or unwanted saponification of neutral oil may be reduced (but not eliminated) by adding a dilute caustic solution and operating the process at lower temperatures.

Oil to be caustically refined is heated to 40 to 90°C, where 350 to 1000 ppm of phosphoric or citric acid is added and mixed. If a process where high temperature and a high-shear mixer are utilized, the retention time may be as little as a few minutes. If a process where low temperature and a static mixer were used, the retention time may be as long as 24 hours, using a "day" tank. The acid reacts with the NHPs to dissociate the calcium, magnesium, and iron salts, allowing the phosphatidic acid to become more hydratable (see Fig. 6.12). Based on the amount of free fatty acids present in the oil, a sodium hydroxide solution would be added to the oil to neutralize the residual acid, the free fatty acids, and form an emulsion for the trapping or encapsulation of the dissociated phosphatidic acid. The soapstock would then be separated from the oil via a "primary" or refining centrifuge. The oil would be washed with 5 to 10% "soft" water or optionally treated with silica to adsorb any remaining soaps. The

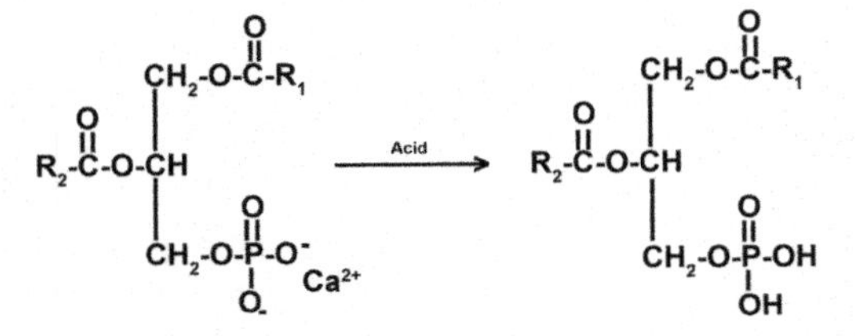

$R_{\#}$ = fatty acid chain X = functional group

Fig. 6.12. Acid dissociation of calcium salt of phosphatidic acid.

composition of the soapstock will depend on the oil being processed and the amount of impurities present. Table 6-N contains 11 different samples of soybean oil soapstock taken from 11 different caustic refining operations. They were analyzed for moisture, gums, and neutral oil. The variation between raw material (crude or water degummed oils), production facilities, and processing conditions yielded dramatically different results compared to the acceptable norm of 30% neutral oil present in soapstock.

The caustic refining process may use as many as three centrifuges and as few as a single centrifuge, depending on the process layout.

Enzymes

Enzymes are nature's catalysts. They are proteins that have a remarkable ability to catalyze very specific chemical reactions of biological importance. They are very specific in both the substances they react with (substrates) and the reaction they catalyze. In an enzyme catalyzed reaction, none of the substrate is diverted into a nonproductive side reaction, so no waste is generated. If enzymes were only 90% specific, then after only 10 cycles, the overall yield would be 35% (Garrett & Grisham, 2010). A yield of only 35% is an unacceptable utilization of resources and accumulation of unwanted by-products which could potentially be toxic for a biological system.

An enzyme's catalytic activity and selectivity are determined by their 3-dimensional (3-D) structure. Their 3-D structure is determined by roughly 20 different naturally-occurring amino acids linked together in a chain-like fashion, forming a macromolecule. These macromolecules fold into 3-D structures (Copeland, 2000). The individual amino acid's side chains enable the chemical reactivities that produce the enzymes' ability to distinguish, orientate, and bind the substrate to the active site. Once the substance has been engaged with the active site of the enzyme, the chemical reaction takes place, followed by the release of the chemical products.

Table 6-N. Soapstock Compositions.

	Moisture	Gums	Neutral Oil
As Is (%)	32–75	10–60	6–50
Dry bases (%)		18–80	18–75

The International Union of Biochemistry and Molecular Biology has developed a nomenclature for naming enzymes, Enzyme Commission number (EC number). Each enzyme is given a four-sequence number to classify it, based on the chemical reaction it catalyzes.

EC 3—Hydrolases are enzymes that break down molecules using water. The chemical equation is as follows:

$$A - B + H_2O \rightarrow A - OH + B - H$$

EC 3.1—Hydrolases that react on ester bonds.
EC 3.1.1—Hydrolases that react on the ester bond located on the carboxylic acid.
EC 3.1.4—Hydrolases that react on the phosphoric diester bonds.

Most enzyme reactions are reversible, but the reaction conditions must be dramatically changed for a reverse reaction to occur. Lipases (EC 3.1.1.3) are enzymes in the presence of water that cleave to the fatty acids present on triacylglycerols. The formations of triacylglycerols are made using the same lipase by removing the excess water generated during the reaction (Yamane, 1987). Typically these lipases are immobilized on a carrier, and the reaction is performed under a vacuum where the only water present is contained in the active site and unavailable for hydrolysis (Cowan, 2011).

Phospholipases

Enzymes that are specific for phospholipids are known as phospholipases. There are three specific types of phospholipase that will be reviewed in this chapter: (1) phospholipase A_1 (PLA_1), phospholipase A_2 (PLA_2), and phospholipase B (PLB); (2) phospholipase C (PLC); and (3) phospholipase D (PLD). Fig. 6.13 shows the location

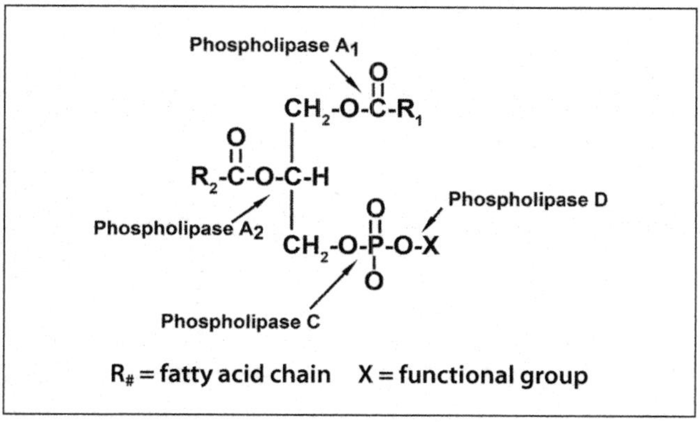

Fig. 6.13. Phospholipase reaction sites.

where PLAs, PLC, and PLD react with a phospholipid. PLA_1 removes the fatty acid attached to the phospholipid in the *sn*1 position as may be seen in Fig. 6.14. A PLA_1 reaction produces a 1 *lyso*-phospholipid where the fatty acid has been cleaved from the *sn*1 position, producing a molecule with fatty acid remaining in the *sn*2 position (International Union of Pure and Applied Chemistry, 2011). PLA_2 removes the fatty acid attached to the phospholipid in the *sn*2 position as may be seen in Fig. 6.15. A PLA_2 reaction produces a 2 *lyso*-phospholipid where the fatty acid has been cleaved from the *sn*2 position, producing a molecule with fatty acid remaining in the *sn*1 position (International Union of Pure and Applied Chemistry, 2011).

PLB, also known as a *lyso*-phospholipase, reacts with *lyso*-phospholipids removing the remaining fatty acid attached to the glycerol backbone producing a glycerophospholipid (Fig. 6.16). PLC and PLD both attack the phosphoric diester bond. PLC cleaves the phosphate group from the phospholipid creating a diacylglycerol (Fig. 6.17) while the PLD cleaves to the functional group creating a phosphatidic acid (Fig. 6.18). Table 6-O lists the various enzymes, substrates, and reaction products involved in enzymatic transformation for most lipids.

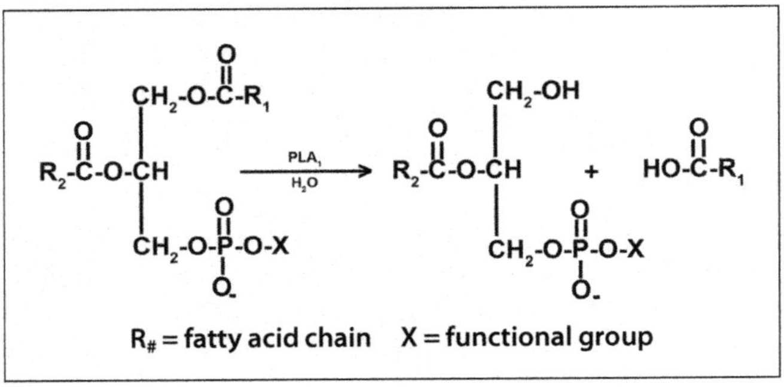

$R_{\#}$ = fatty acid chain X = functional group

Fig. 6.14. Phospholipase A_1 catalyzed reaction.

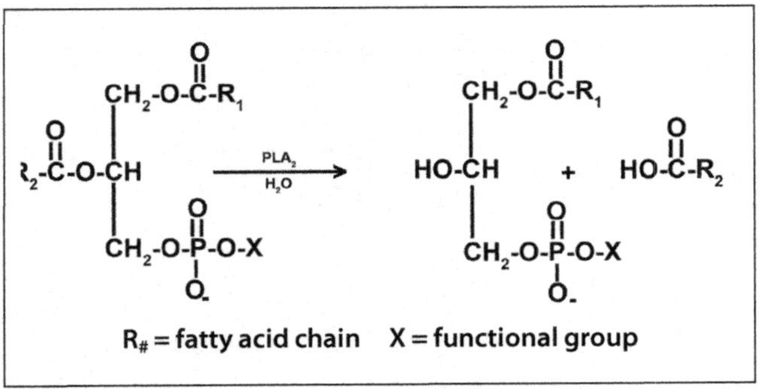

$R_{\#}$ = fatty acid chain X = functional group

Fig. 6.15. Phospholipase A_2 catalyzed reaction.

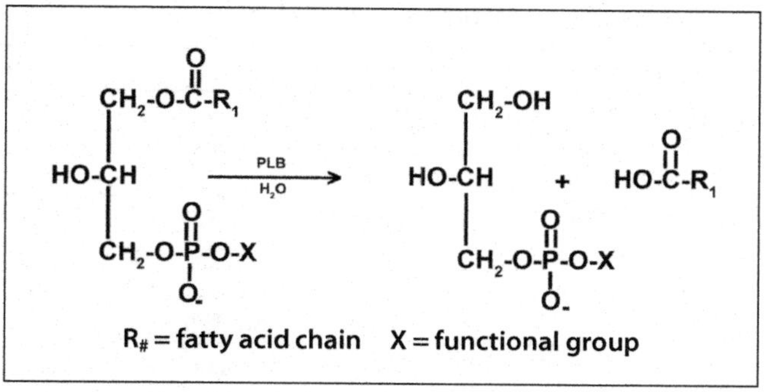

$$\begin{array}{l} CH_2\text{-O-}\overset{O}{\overset{\|}{C}}\text{-}R_1 \\ HO\text{-}CH \\ CH_2\text{-O-}\overset{O}{\overset{\|}{P}}\text{-O-X} \\ \quad\quad\quad\;\; O^- \end{array} \xrightarrow[H_2O]{PLB} \begin{array}{l} CH_2\text{-OH} \\ HO\text{-}CH \\ CH_2\text{-O-}\overset{O}{\overset{\|}{P}}\text{-O-X} \\ \quad\quad\quad\;\; O^- \end{array} + HO\text{-}\overset{O}{\overset{\|}{C}}\text{-}R_1$$

$R_{\#}$ = fatty acid chain X = functional group

Fig. 6.16. Phospholipase B catalyzed reaction.

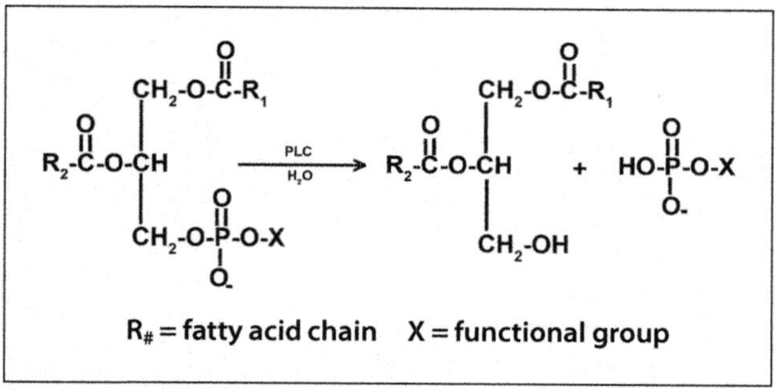

$$\begin{array}{l} \quad\quad\quad\;\; CH_2\text{-O-}\overset{O}{\overset{\|}{C}}\text{-}R_1 \\ R_2\text{-}\overset{O}{\overset{\|}{C}}\text{-O-}CH \\ \quad\quad\quad\;\; CH_2\text{-O-}\overset{O}{\overset{\|}{P}}\text{-O-X} \\ \quad\quad\quad\quad\quad\quad\;\; O^- \end{array} \xrightarrow[H_2O]{PLC} \begin{array}{l} \quad\quad\quad\;\; CH_2\text{-O-}\overset{O}{\overset{\|}{C}}\text{-}R_1 \\ R_2\text{-}\overset{O}{\overset{\|}{C}}\text{-O-}CH \\ \quad\quad\quad\;\; CH_2\text{-OH} \end{array} + HO\text{-}\overset{O}{\overset{\|}{P}}\text{-O-X} \text{ (}O^-\text{)}$$

$R_{\#}$ = fatty acid chain X = functional group

Fig. 6.17. Phospholipase C catalyzed reaction.

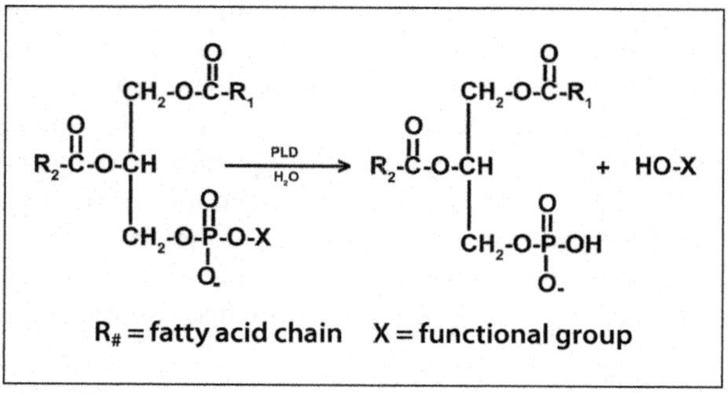

$$\begin{array}{l} \quad\quad\quad\;\; CH_2\text{-O-}\overset{O}{\overset{\|}{C}}\text{-}R_1 \\ R_2\text{-}\overset{O}{\overset{\|}{C}}\text{-O-}CH \\ \quad\quad\quad\;\; CH_2\text{-O-}\overset{O}{\overset{\|}{P}}\text{-O-X} \\ \quad\quad\quad\quad\quad\quad\;\; O^- \end{array} \xrightarrow[H_2O]{PLD} \begin{array}{l} \quad\quad\quad\;\; CH_2\text{-O-}\overset{O}{\overset{\|}{C}}\text{-}R_1 \\ R_2\text{-}\overset{O}{\overset{\|}{C}}\text{-O-}CH \\ \quad\quad\quad\;\; CH_2\text{-O-}\overset{O}{\overset{\|}{P}}\text{-OH} \\ \quad\quad\quad\quad\quad\quad\;\; O^- \end{array} + HO\text{-X}$$

$R_{\#}$ = fatty acid chain X = functional group

Fig. 6.18. Phospholipase D catalyzed reaction.

Table 6-O. Common Names of Lipid Modification Enzymes.

Enzyme	EC number	Substrate	Products	Co-Product
Lipase	3.1.1.3	Triacylglycerol, Diacylglycerol, Monoacylglycerol	Diacylglycerol, Monoacylglycerol, Glycerol	FFA
Phospholipase A_1	3.1.1.32	PC, PE, PI, PA, PS	1 *lyso*-PC, 1 *lyso*-PE, 1 *lyso*-PI, 1 *lyso*-PA, 1 *lyso*-PS	FFA
Phospholipase A_2	3.1.1.4	PC, PE, PI, PA, PS	2 *lyso*-PC, 2 *lyso*-PE, 2 *lyso*-PI, 2 *lyso*-PA, 2 *lyso*-PS	FFA
Phospholipase B	3.1.1.5	*lyso*-PC, *lyso*-PE, *lyso*-PI, *lyso*-PA, and *lyso*-PS	glycerophosphocholine, glycerophosphoethanolamine, glycerophosphoinositol, glycerophosphoric acid, glycerophosphoserine	FFA
Phospholipase C	3.1.4.3	PC, PE	Phosphocholine, Phosphoethanolamine	DAG
PI-Phospholipase C	3.1.4.11	PI	Phosphoinositol	DAG
Phospholipase D	3.1.4.4	PC, PE, PI, PS	Phosphatidic Acid	Choline, ethanolamine, inositol, phosphoric acid, serine

Lipases, including phospholipases, use an amino acid catalytic triad serine-histidine-aspartic acid/glycine (Ser-His-Asp/Glu) for ester bond hydrolysis (Brumlik & Buckley, 1996; Shu et al., 2007). In this triad, histidine acts as a general base, forming a hydrogen bond with a catalytic serine residue which then allows the formation of the first tetrahedral transition state. At low pH conditions, that is, pH 3 and lower, histidine will be protonated. If histidine is already protonated, the hydrogen bond formation with catalytic serine residue will not occur, and reaction will not proceed, thereby reducing or eliminating enzymatic activity (Kerovuo, 2008). Secondly, at a pH less than 3, the enzyme will undergo irreversible denaturation, resulting in loss of activity. Lastly, zinc and calcium ions are required for optimum activity of phospholipase enzymes (Maria et al., 2007). At a low pH, hydrogen ions start competing with zinc and calcium ions, thereby further reducing activity of the phospholipase enzyme. Thus, at low pHs described in the various degumming patents, the activity of phospholipase enzyme will be substantially reduced or eliminated (Kerovuo, 2008).

Enzymatic Degumming Processing

Lurgi (formerly Rohm GmbH) developed the first enzymatic process for degumming edible oils (Aalrust et al., 1993), better known as the EnzyMax® process. The process used an aqueous solution containing a PLA_1, PLA_2, or PLB for the treatment of water-degummed vegetable (preferably soybean) or animal oil. The enzyme was obtained from a ground pancreas pulp containing a PLA_2. The enzyme was dissolved with sodium citrate, sodium dodecyl sulfate (a very strong emulsifier), and water producing an optimum pH of 5. The enzyme/water mixture was then added to 800 grams of oil that was previously heated to 50 to 75°C. An emulsion was formed, and the enzyme was allowed to react with the phospholipids for 3 to 4 hours before centrifugation, and an oil was obtained with a residual phosphorus of 3 ppm. Trials were repeated with a phospholipase B and from *Corticium* species. This enzyme must have contained some PLA_1 or PLA_2; otherwise this *lyso*-phospholipase (PLB) would have been unable to react due to a lack of substrate availability 1 *lyso*- or 2 *lyso*-phosphatide. Aalrust et al., clearly stated in the patent that, "...the object of the process cannot be achieved with phospholipase C or D." Additionally, the inventors limited the scope of their claims to oils with 50 to 250 ppm phosphorus. They reasoned, "Because phospholipase A_1, A_2, and B would attack lecithin, it would make no sense to use the method of the invention on oils having a high content of lecithin, such as raw soybean oil." Very limited amounts of the enzyme were available and made the industrial application of the process difficult (Biotimes®, 2000).

Lurgi provided more detail on the EnzyMax® process as may be seen in Fig. 6.19 (Dahlke, 1998). Dahlke reinforces the idea that the process was designed to only remove non-hydratable phospholipids (NHP) from the oil, and that only a PLA_2 would be able to economically perform a full degumming. The *lyso*-phosphatide resulting from the enzyme reaction is insoluble in oil, and therefore removed in the gums, and the free fatty acid formed remains in the oil. According to the author, the process phases on an enzyme degumming are (Dahlke, 1997):

- adjusting the optimum temperature for the enzyme reaction;
- adding citric acid and retention of the oil with acid for a period of time;
- adding caustic soda to buffer the water phase to the expected optimal pH for the enzyme to be used;
- adding enzyme and water to the oil, via mixers (static + mechanical)—gums were partially re-circulated here when using the porcine enzyme;
- reacting the NHP in the reaction tank (up to six hours);
- heating the oil to the optimal temperature for separation in the centrifuge; and
- separating *lyso*-phosphatides from the oil in the centrifuge

Aalrust et al. and Dahlke disclosed the importance of fine dispersion of the buffered water + enzyme + oil mixture, but no details were provided regarding the nature of such mixers. The original patent and subsequent publication mention only that the enzyme dispersion can be achieved with "conventional equipment." Details of the reaction tank are not provided, except that it consists of a 6-compartment tank, reducing the possibility of short circuiting the reaction time, and the graphics indicate the need for agitation in each compartment. Both soybean (65 ppm P before reaction) and rapeseed (100 to 200 ppm before reaction) oil are claimed by Dahlke to

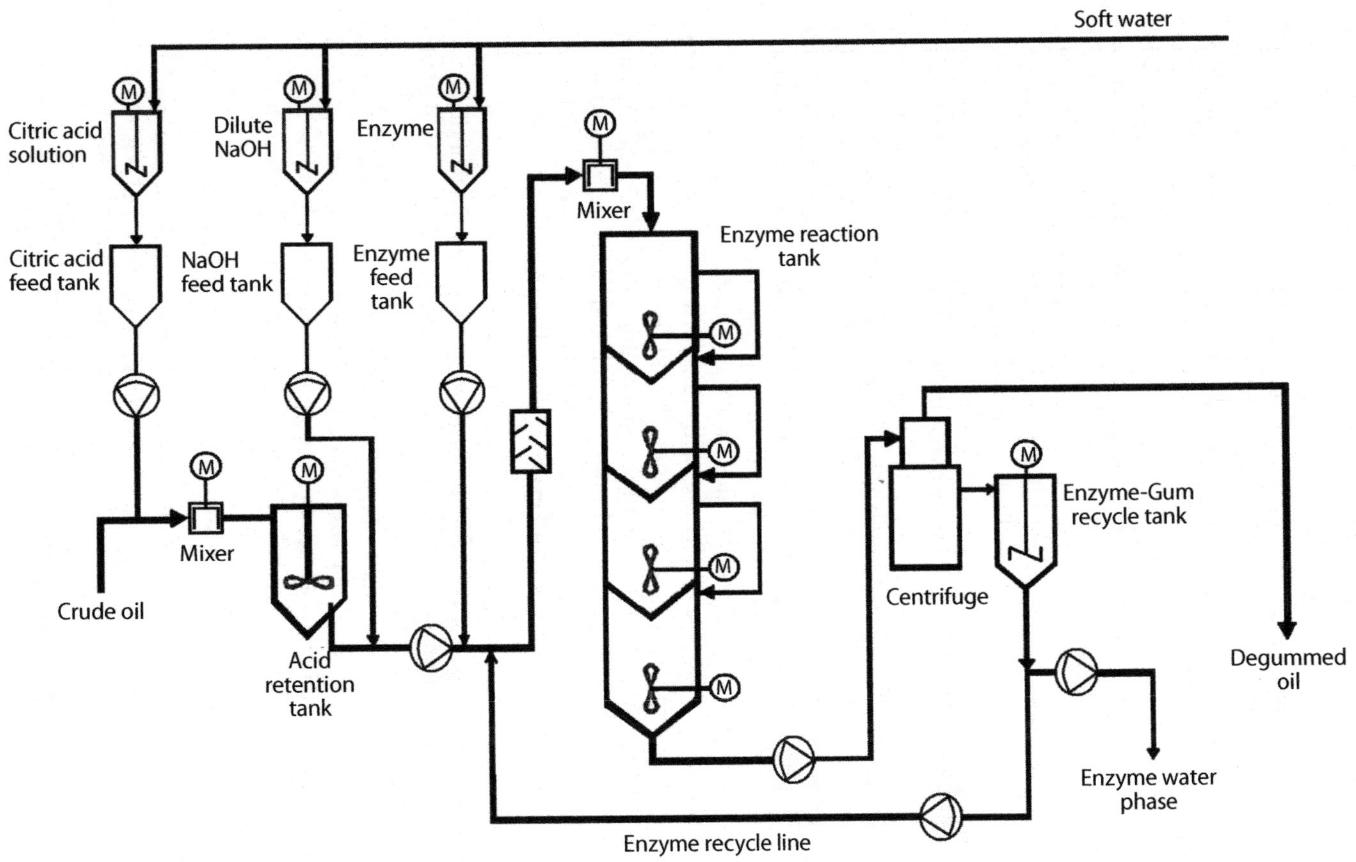

Fig. 6.19. EnzyMax® degumming process flow diagram.

be successfully processed in such manner in a 500 t/d industrial plant, with residual phosphorus levels below 10 ppm in the oil after centrifugation of the reacted oil.

In 1995, Cereol's Mannheim facility worked closely with both Lurgi and Novozymes on the industrial implementation of the EnzyMax® process for rapeseed oil. The facility originally operated on the Lecitase® 10L porcine pancreatic enzyme (Biotimes, 2000). The cost and availability of the enzyme required the recirculation of the enzyme and some of the gums. The plant operated for a period of 5 years until Lecitase® Novo was developed allowing for the elimination of the recirculation. Both enzymes were claimed to have delivered degummed rapeseed oil with less than 10 ppm of phosphorus. Initial phosphorus in the oil was of 100–280 ppm (Münch, 2001). The enzymatic degumming process was discontinued in Mannheim after the implementation of the Exergy process in the rapeseed preparation plant in 2002 (Münter, 2007; Exergy, Unknown)

Showa Sangyo Company developed the second enzymatic process for degumming oil and fat (Yagi et al., 1996). The inventors used a commercial porcine PLA_2 from Novozymes that was calcium dependent (Lecitase® 10L). The process used extremely large amounts of water, 30 to 100% by weight for the enzymatic reaction and additional 30 to 200% water by weight for water washing. The wash water may optionally contain an acid. The very large amounts of water required made the process developed by Yagi et al. industrially impractical.

Roehm GmbH and Metallgesellschaft AG disclosed the use of a microbial PLA_2 from *Aspergillus* strain that works at a pH less than 3 in their patent (Loeffler et al., 1999). It was claimed that both acid degumming and enzymatic degumming could be achieved in a single process. However, as was discussed earlier, phospholipases are unable to work at low pHs. The process was not an enzymatic process, but an acid degumming process that was never industrially utilized.

Novozymes A/S (formerly Novo Nordisk A/S) described the identification and expression of their phospholipase A1 produced from Fusarium oxysporum in two patents which became their commercial product Lecitase® Novo (Clausen et al., 2000). In still another patent, they describe the discovery and activity of phospholipases from Hyphozyma capable of removing both the fatty acyl groups present on a phospholipid molecule when mixed with water degummed oil (50 to 250 ppm phosphorus) with 0.5 to 5% water, pH from 1.5 to 5, temperature from 30 to 45°C, and a reaction time of 1 to 12 hours (Hasida et al., 2000). The optimum activity for vegetable oil degumming was claimed to be a pH 4.5 at 40°C after a 6 hour reaction time. The enzyme had very little activity at pH 3 within an oil matrix (Clausen, 2000), confirming Kerovou's opinions presented earlier in this chapter.

Bunge (formerly Central Soya Company) implemented enzymatic degumming of a mixture of crude non-degummed and water degummed oils using Novozymes Lecitase® Ultra Phospholipase A1/physical refining of soybean oil in its Morristown, Indiana, facility in 2002 (BioTimes, 2004). The facility was able to demonstrate an oil yield increase of 1.2% over the chemical refining process from the same quality soybean oil feedstock. It is believed by the authors that this was the first successful implementation of physical refining of soybean oil.

Edwards et al. (2003) disclosed sequences of complementary DNA's encoding secreted proteins. It was claimed that these proteins could be used with the degumming process disclosed by Loeffler et al.

Verenium Corporation (formerly Diversa) disclosed the identification and expression of phospholipases capable of cleaving the glycerolphosphate ester linkage of a phospholipid (Gramatikova et al., 2007) yielding a phosphate group and a diacylglycerol. It was shown that the DAG would remain in the oil, therefore increasing the yield compared with PLA degumming enzymes that only a fatty acid could be recovered in the distillate. The commercial enzyme was given the name Purifine® PLC. The phospholipase C developed reacted with phosphatidylcholine and phosphatidylethanolamine, but not PI or PA. An additional benefit demonstrated by the producer was the ability to operate at a neutral pH.

Novozymes A/S disclosed variants of enzymes from *Humicola langinose* for enzymatic degumming (Bojsen et al., 2007). The application of their enzyme utilized the degumming processes described by Aalrust et al. Loeffler et al., and Yagi et al.

The Council of Scientific and Industrial Research located in New Delhi, India, disclosed a process using Novozymes Lecitase® Novo for the treatment of rice bran oil (Chakrabarti et al., 2009). The *Aspergillus oryzae* was added to a solution of citric acid and sodium hydroxide. The enzyme mixture was then added to the oil with shear mixing. The retention time ranged from 0 to 110 minutes prior to heating and separation of the gums from the oil. The enzymatically treated oil was then bleached with 2 to 4% bleaching earth and zero to 1% activated carbon. The oil was then de-waxed to achieve a residual phosphorus content of 1 to 3 ppm. Unfortunately, the inventors did not utilize any control experiments and did not analyze the residual phosphorus after the centrifugation prior to the bleaching nor the de-waxing process. It would be expected that, after the enzymatic treatment and separation of the gums via centrifugation, the oil would have had a residual phosphorus of less than 3 ppm without any additional processing for a success enzymatic reaction to have had occurred. The authors stated, "... the oil content of the gums is only lost 30 to 40%" (Chakravarti et al., 2000), suggesting that the enzymatic process did not go to completion, resulting in high oil losses due to emulsification of oil in the removed phospholipids. This would require very large amounts of bleaching earth, activated carbon, a dewaxing process where typically sodium hydroxide is used to form soaps in order to seed the crystallization process.

Bunge discovered operating the enzymatic degumming process, as described by Aalrust et al., produced a detrimental side effect of plating out the salts of calcium and magnesium onto the processing equipment. Bunge disclosed a method for improving the enzymatic degumming process where the calcium and magnesium salts of citric acid were eliminated from fouling the heat exchangers and centrifuge after the enzymatic reaction tank (Dayton et al., 2010). The enzyme was allowed to react at its optimum pH where the formation of insoluble calcium and magnesium citrates is formed. Once the reaction had been completed, an acid was added at the reaction tank's discharge pump to lower the pH, making the calcium and magnesium citrate salts soluble, and thereby reducing or eliminating the fouling. Once the low pH heavy phase had been separated from the oil, sodium hydroxide was added to the heavy phase to bring the pH to neutral.

Utilizing equations developed by Greenwald and Boulet (Greenwell et al., 1940; Boulet et al., 1959), the solubility curve of the citrates based on the total metal content (Ca + Mg + Fe) may be drawn versus the pH of the aqueous system (see Fig. 6.20). The graph demonstrates the precipitation of citrate salts above pH 4.2 where the enzymes are successfully utilized for degumming. Once the enzymatic reaction has been completed, the pH is lowered below 4.2, allowing the salts to remain in solution eliminating the precipitation problem.

Danisco received a patent for a new phospholipase A_2 for degumming edible oils (Søe et al., 2012). The edible soybean oils were supplemented with 1% plant sterols and 2% phosphatidylcholine prior to the degumming reaction. It was claimed that at least 5% of the sterols were converted into sterol esters via an acyltransferase reaction.

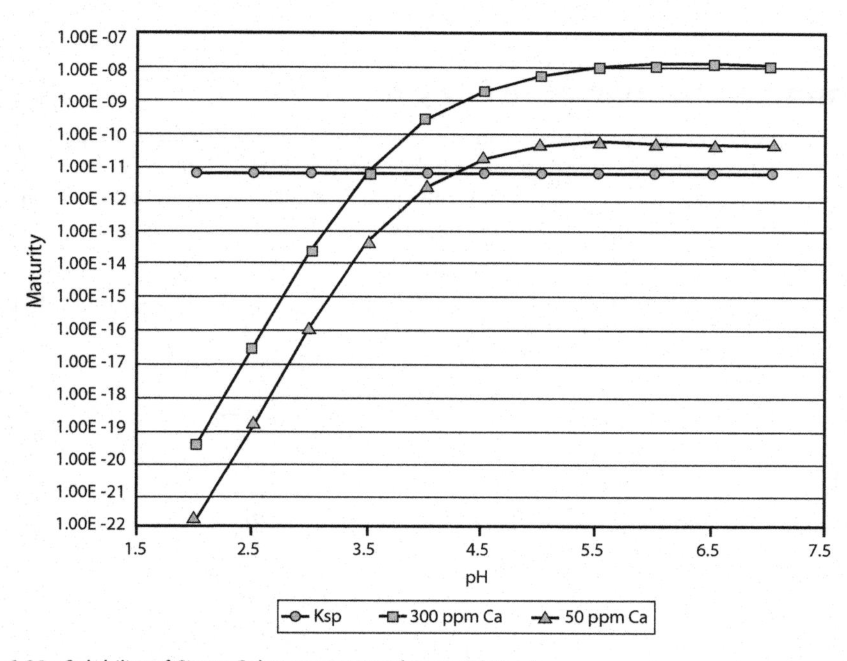

Fig. 6.20. Solubility of Citrate Salts versus pH within an Oil Matrix.

Bunge further disclosed in patent application filings the combination of phospholipase A enzymes and phospholipase C enzymes (Dayton & Galhardo, 2008; Dayton, et al., 2013a). The temperature, pH, water content, reaction time, enzyme dosages, and sequential or simultaneous addition of enzymes were evaluated in 38 experimental trials. It was discovered that PLC and PLAs have a synergistic effect when utilized in combination. In a simultaneous addition of the enzymes, it was disclosed that almost complete conversion and removal of the phospholipids was achieved, yielding an oil with 2.9 ppm phosphorus in as little as 30 minutes. The PLC alone required two hours for complete hydrolysis of PC and PE while the PLA reaction required at least four hours for complete hydrolysis. Additional analysis yielded unexpected results because both the FFA (0.17%) and the DAG (0.42%) were depressed significantly in the enzymatic degummed oil, compared to the theoretical values for their content (0.98% and 1.73%, respectively). It was later disclosed in additional patent filings by Bunge (Dayton, 2012; Dayton 2013b) that the formation of TAGs when the combination of PLAs and PLC were used in a treatment of gums.

DSM (formerly Verenium) and Bunge disclosed the development of a PI-specific PLC (PI-PLC) for hydrolysis of phosphatidyl inositol in related patent filings by both companies (Barton et al., 2011; Dayton et al., 2011). The addition of PI-PLC allows for almost complete hydrolysis of the phospholipids present in oil sources increasing the yield gain by almost an additional one percent over the yield gains provided by PLAs, traditional PLC, and the combinations of PLAs and PLC alone.

Commercially Available Enzymes

Industrial enzymes are produced from the fermentation and expression of various microbial sources that are food grade, Kosher, and/or Halal. Not all enzymes have approval in every country. Below are the commercial producers, product name, expression systems, and EC number of the enzymes.

AB Enzymes

ROHALASE® PL-XTRA—PLA_2 from Trichoderma reesei (AB Enzymes, 2011). EC 3.1.1.4

Danisco

LysoMax®—PLA_2 from Streptomyces violaceoruber (Maria, et al., 2007). EC 3.1.1.4

DSM

Purifine®—PLA_2 from Aspergillus niger (DSM, 2011) EC 3.1.1.4

Purifine®—PLC from Pichia pastoris (Olempska-Beer, 2008) EC 3.1.4.3.

Novozymes

Lecitase® 10L—PLA2 from porcine pancreas (Yang et al., 2006). EC 3.1.14.

Lecitase® Novo—PLA1 from Fusarium oxysporum (Yang et al., 2006) EC 3.1.1.32.

Lecitase® Ultra—PLA1 from Thermomyces lanuginosa/Fusarium oxysporum (Olempska-Beer, 2008). EC 3.1.1.32.

Yields

A crude oil containing 600 ppm and a FFA of 0.6 was analyzed for water degumming (WDG), caustic refining, PLA_1 or PLA_2, and a combination of a PLC plus a PLA_1 (see Table 6-P). Water degumming and PLC degumming do not lower the residual phosphorus to a level where physical refining could be practiced. Caustic refining, enzymatic degumming with PLA, or enzymatic degumming with a combination of PLC and PLA have been validated for their ability to completely remove the phospholipids in either crude non-degummed or water degummed oils.

The losses during the degumming and subsequent processing of these oils demonstrate the yield advantages over the conventional processes of degumming and caustic refining. Crude non-degummed oil may be degummed with PLC, producing an oil with roughly the same level of residual phosphorus as the water degummed oil, but with an increase in the total amount of oil due to the generation of DAG and the reduced emulsification properties of the reacted phospholipids when they were

Table 6-P. Refining Yields for a Crude Oil Containing 600 ppm Phosphorus and 0.6% FFA.

	Water[a]	PLC[b]	Caustic WDG[c]	PLA WDG[d]	Caustic[e]	PLA[f]	PLC + PLA[g]
Initial Phosphorus (ppm)	600	600	150	150	600	600	600
Final Phosphorus (ppm)	150	150	1	1	1	1	1
Dry Heavy Phase Generated (kg)	20.3	11.2	17.3	5.0	37.2	16.3	13.1
Lipids Lost (kg)	20.3	6.2	16.3	3.7	35.9	15.8	11.9
Oil lost in Bleaching (kg)	—	—	0.8	0.8	0.8	0.8	0.8
Distillate (kg)							
Fatty Acids	—	—	1	6.7	1	9	9
Unsaponifiables	—	—	2	2	2	2	2
Oil Yield (kg)	980	991	981	988	961	974	977

[a]Water—2% water used in the water degumming process
[b]PLC—2% water used in the PLC degumming process
[c]Caustic WDG—500 ppm phosphoric acid pretreatment, 0.03% excess sodium hydroxide
[d]PLA WDG—500 ppm citric acid, 90 ppm sodium hydroxide (pH 4.5), 2% hydration water
[e]Caustic—500 ppm phosphoric acid, 0.03% excess sodium hydroxide
[f]PLA—500 ppm citric acid, 90 ppm sodium hydroxide (pH 4.5), 2% hydration water
[g]PLC plus PLA—500 ppm citric acid, 90 ppm sodium hydroxide (pH 4.5), 2% hydration water

converted into their water soluble phosphates. PLA should not be used as a water degumming aid, if the degummed oil is to be sold or exported due to the limitations on the amount of FFA present in the oil. As an example, if the PLC enzyme was replaced with PLA and all things were kept equal, the FFA in the PLA degummed oil would be 1.05% with a residual phosphorus of 150 ppm without the DAG gain. Most processing plants where oil is exported utilize expanders where the initial phospholipid content is much higher and the final FFA after PLA treatment would be out of specification.

If the crude oil was water degummed for lecithin production or the oil was imported, an increase in yield of 7 kg was observed for the PLA degumming process over the caustic refining process. Besides the increase in oil yield, one of the major benefits was the reduction of heavy phase generated. In the example above, the amount of dried heavy phase was 17.3 kg compared to only 5 kg in the enzymatic process.

The analysis of the data generated from the treatment of crude non-degummed oil with the caustic refining process compared to either PLA or PLC plus PLA treatment, and physical refining demonstrates the following:

- The amount of lipids lost in the removal of the phospholipids was almost 50% in the comparison to PLA to caustic refining. Thus, the amount of soapstock or gums that may be returned to the meal was reduced by 21 kg.
- The amount of lipids lost in the removal of the phospholipids was reduced by nearly two-thirds in the combination of PLC plus PLA to caustic refining. The amount of soapstock required or gums returned to the meal was reduced by 24 kg.
- The physical refining process allows for the recovery of the fatty acids present in the crude oil without the need of acidulation or the discharge of large amounts of water. An additional gain occurs in the recoverable fatty acids that are cleaved from the phospholipids with PLA.
- The overall yield was improved over the caustic refining process by 13 kg for PLA alone and 16 kg for the combination of PLA plus PLC.

The combination of PLA and PLC enzymes allows the reaction all of the phospholipids present in crude oil while still maximizing the yield gain from PLC for physical refining of oils, independent of the level of non-hydratable phospholipids present in the feed material. The additional benefit discovered was the synergist effect of the two enzymes in combination. The presence of PLC allowed for a significant increase in the reaction rate of either enzyme alone, regardless of the concentration of the single enzyme. As mentioned previously, complete hydrolysis of all of the phospholipids was accomplished in as little as 30 minutes when a combination of PLA and PLC were utilized together (Dayton et al., 2013a).

The most important conclusion from the data is the benefit generated when greater amounts of phospholipids present in the oil, the greater the potential yield gain from the enzymatic process over the conventional processes of water or acid degumming followed by caustic refining. This is not intuitive, oils processed by conventional means have losses that are always multiples of the amounts of phospholipids present in the oil.

Bleaching

Enzymatically degummed oil with either PLA or a combination of PLA and PLC may be treated with 0.01 to 0.05% neutral silica to assist in the removal of any trace metals present in the oil. The oil/silica is dried and the silica is removed via filtration. The trace metal content should be at or below the detection limit after the silica removal in order for successful physical refining. The chlorophyll is removed using the industry standard amounts of bleaching earth and process conditions.

Deodorization

The deodorizer must be either built for physical refining or must be modified to handle the additional load caused by the fatty acids present in the feed material. The deodorizer should be equipped with an ice condensation system and a dual temperature scrubber in order to maximize the yield gain and value of the by-products. Typically, bleached oil from a caustic refining operation will enter the deodorizer with 0.05 to 0.15% FFA, while bleached oil from an enzymatic process may contain from 0.4 to 1.6% FFA. The fatty acids are recovered in a separate condenser from the unsaponifiables in order to maintain the value of both co-products.

Environmental Impact

A life cycle assessment addresses the "cradle to grave" production of raw materials through the handling of waste materials generated in the process. An LCA was made comparing a caustic refining process with soapstock acidulation versus a PLA_1 degumming/physical refining plant with the *lyso*-phospholipids placed on the meal for 750 metric tonnes per day of water-degummed soybean oil in the United States. The environmental load per 1000 kg of refined oil may be found in Table 6-Q.

Industrial Implementations (public announcements)

December 1995—EnzyMax® implemented in Mannheim, Germany (Inform, 1995)

December 2000—EnzyMax® implemented in Mannheim, Germany with Lecitase® Novo (Biotimes®, 2000).

March 2004—Enzymatic degumming and physical refining of soybean oil using Lecitase® Ultra in Morristown, Indiana (Biotimes®, 2004a)

Table 6-Q. Life Cycle Assessment (Oxenboll, 2005).

Impact Potential*	Savings by replacing caustic refining with PLA_1/physical refining per 1000 kg refined oil
Fossil Energy (MJ)	400
Global Warming (kg) CO_2 equivalents	44
Acidification (g) SO_2 equivalents	527
Nutrient Enrichment (g) PO_4 equivalents	375
Smog Formation (g) Ethylene equivalents	18

*LCA was conducted in accordance with ISO 14040.

April 2004—Enzymatic degumming in Cairo, Egypt (Biotimes®, 2004b)

February 2005—Enzymatic degumming in Bangalore, India (Biotimes®, 2005)

March 2010—Enzymatic degumming at Molinos Rio de la Plata's San Lorenzo plant, Argentina (Verenium, 2010)

June 2010—Enzymatic degumming at A.P. Solvex Ltd., India (Biotimes®, 2010)

December 2010—Enzymatic degumming at Camera in Lijuí, Brazil (Wheatley, 2011)

February 2011—Enzymatic degumming using Purifine® in Terminal 6 Puerto San Martin, Argentina (Verenium, 2011)

Conclusions

The authors of this chapter believe that enzymatic degumming is an important technology for today and the future. The implementation of phospholipases, in order to reduce neutral oil losses, transforms the native phospholipids into recoverable DAGs and fatty acids, while reducing the amount of material lost in the soapstock or sent back to the meal is an important part of profitability, while reducing harsh chemicals and increasing the sustainability of a processor today.

References

Aalrust, E.; Beyer, W.; Ottofrickenstein, H.; Penk, G.; Plainer, H.; Reiner, R. Enzymatic treatment of edible oils. United States Patent 5,264,367 issued in November 1993.

AB Enzymes Technical Data Sheet, Rohalase® PL-XTRA, 2011-04-18 Rev. Nr. 00.

ALC, American Lecithin Company product brochure "A Simple Guide to Use and Selection." Date and author unknown.

Anderson, D. Chapter 1: A primer on oils processing technology. Bailey's Industrial Oil and Fat Products, Sixth ed., Vol. 6, John Wiley & Sons, Inc. (2005).

Athenstaedt, K.; Daum, G. Phosphatidic acid, a key intermediate in lipid metabolism, Eur. J. Biochem. 1999, 266, 1–16.

Barton, N.; Hitchman, T.; Lyon, J.; O'Donoghue, E.; Wall, M. Phospholipases, Nucleic Acids Encoding Them and Methods for Making and Using Them. WO 2011/046812 published April 2011.

Benson, A.A.; Maruo, B. Plant Phospholipids. Identification of the phosphatidyl glycerols. Biochimica et Biophyisca Acta, 1958, 27, 189–195.

BioTimes®, Enzymes degumming is coming! 2000.

BioTimes®, Enzymatic degumming at Bunge and the bottom line. March 2004a.

BioTimes®, Higher Vegetable oil yields for United. April 2004b.

BioTimes®, New treatment improves "heart oil" from India. February, 2005.

BioTimes®, Well-oiled partnership increases yields. June 2010.

Bojsen, K.; Svendsen, A.; Fuglsang, C.C.; Patkar, S.A.; Borch, K.; Vind, J.; Petri, A.; Glad, S.O.S.; Glad, S.; Budolfsen, G. Lipolytic enzyme variants. Unites States Patent 7,312,062 issued December 2007.

Boulet, M.; Marier, J.R. Solubility of tricalcium citrate in solutions of variable ionic strength and in milk ultrafiltrations. Issued as N.R.C. No. 5543 (1953).

Branton, D. Membrane Structure. Annu. Rev. Physiol. 1969, 20, 209–238.

Brooks, D.; Hollis, R. The effect of bleached oil moisture in bleaching dry and semi-dry degummed crude palm oil, presented at the 14th AOCS Latin American Congress and Exhibition on Fats and Oils (2011), Cartagena, Colombia.

Brumlik, M.J.; Buckley, J.T. Identification of the catalytic triad of the lipase/acyltransferase from Aeromonas hydrophila, J. Bacteriol. 1996, 2060–2064.

Chakrabarti et al. Process for the pre-treatment of vegetable oils for physical refining. United States Patent 7,494,676 issued February 2009.

Cherry, J.P.; Kramer, W.H. Plant Sources of Lecithins. Lecithins: Sources, Manufacturing & Uses; Szuhaj, B.F., Ed., AOCS Press: Champaign, Illinois, 1989; pp. 16–31.

Clausen, I.G.; Patkar, S.A.; Borch, K.; Barfoed, M.; Clausen, K.; Fuglsang, C.C.; Dybdal, L.; Halkier, T. Method for reducing phosphorus content of edible oils. United States Patent 6,103,505 issued August 2000.

Clausen, I.G.; Patkar, S.A.; Borch, K.; Barfoed, M.; Clausen, K.; Fuglsang, C.C.; Dybdal, L.; Halkier, T. Method for reducing phosphorus content of edible oils. United States Patent 6,143,545 issued November 2000.

Copeland, D.; Belcher, W.R. Methods for refining vegetable oils and byproducts thereof. United States Patent 6,172,248 issued January 2001.

Copelane, R. Chapter 3. Structural Components of Enzymes, Enzymes, 2nd ed., Wiley VCH, New York (2000).

Cowan, D. New applications for enzymes in oil processing. Enzymatic processing and modification—current and future trends. Ghent Belgium, June 2011.

Dahlke, K. The enzymatic degumming—EnzyMax®, Oleagineux, Corps Gras Lipids, 1997, 4(1) 55–57.

Dahlke, K. An enzymatic process for the physical refining of seed oils, Chemical Engineering Technology, 1998, 21, 278–281.

Davis, H.T. Factors determining emulsion type: Hydrophile—lipophile balance and beyond. Colloids and Surfaces A: Physiochemical and Engineering Aspects. Vol. 91, 3 November 1994. P 9–24. A selection of papers presented at the First World Congress on Emulsions.

Dayton, C.L.G.; Berkshire, T.L.; Staller, K.L. Process for improving the enzymatic degumming of vegetable oils and reducing fouling of downstream processing equipment. United States Patent 7,713,727 issued May 2010.

Dayton, C.L.G.; Galhardo, F. Enzymatic degumming utilizing a mixture of PLA and PLC phospholipase. United States patent application US 2008/0182322 published July 2008.

Dayton, C.L.G.; Rosswurm, E.M.; Galhardo, F.S. Enzymatic degumming utilizing a mixture of PLA and PLC phospholipase with reduced reaction time. United States Patent 8,460,905 issued June 2013a.

Dayton, C.L.G. Generation of Triacylglycerols from gums. United States Patent 8,241,876 issued August 2012.

Dayton, C.L.G., Generation of Triacylglycerols. United States Patent 8,541,211 issued September 2013b.

Dayton, C.L.G.; Galhardo, F.; Barton, N.; Hitchman, T.; Lyon, J.; O'Donoghue, E.; Wall, M. Oil Degumming Methods, WO 2011/046815 published April 2011.

Dijkstra, A.J.; Van Opstal, M. Process for producing degummed vegetable oils and gums of high phosphatidic acid content. United States Patent 4,698,185, October 1987.

Dijkstra, A.J.; Van Opstal, M. The total degumming process, JAOCS, 1989, 66(7), 1002–1009.

Domaille, P., et al., Analytical profiling of small scale reactions of phospholipase-C mediated vegetable oil degumming, presented at the 2007 American Oil Chemists' Society Meeting.

DSM Application Data Sheet for GumZyme™, 2011.

Edwards, J.B.D.M.; Bougueleret, L.; Jobert, S. Complementary DNA's encoding proteins with signal peptides. United States Patent 6,548,633 issued April 2003.

Exergy Consulting, Plant design, Project Management, and Engineering Services for Process Industry and Bio-Energy Sector. 2004.

Folch, J. Letter to the editor, J. of Biol. Chem. 1941, 139, 973–974.

Folch, J. The chemical structure of phosphatidyl serine, J. of Biol. Chem. 1948, 174, 439–450.

Forster, A.; Harper, A.J. Physical Refining. JAOCS, 1983, 60, 265–271.

Garrett, R.A.; Grisham, C.M. Chapter 13. Enzymes—Kinetics and Specificity, Biochemistry 4th ed. Brooks/Cole Boston Mass (2010).

GEA Evaporation Technologies AB brochure, Exergy® Steam Processor—Heat treatment of oil seeds/beans and protein, Date unknown.

Gramatikova, S.; Hazelweed, G.; Lam, D.; Barton, N. Phospholipases, nucleic acids encoding them and methods for making and using them. United States Patent 7,226,771 issued June 2007.

Greenwald, I.; Redick, J.; Kibrick, A.C. The dissociation of calcium and magnesium phosphates. Journal Biological Chemistry, 1940, 135(65).

Gunstone, F.D.; Harwood, J.L.; Dijkstra, A.J. The Lipid Handbook; third ed., CRC Press: Boca Raton, Florida, 2007; Section 3.4.

Harwood, J.L. Fatty acid biosynthesis. Plant Lipids: Biology, utilization and manipulation; Murphy, J.P., Ed. Blackwell Publishing, Oxford, 2005; pp. 27–66.

Hasida, M.; Tsutsumi, N.; Halikier, T.; Stringer, M.A. Acidic phospholipase, production and methods using thereof. United States Patent 6,127,137 issued October 2000.

Higgins, N.W. Low trans-stereoismoer shortening system. United States Patent 7,718,211 issued May 2010.

Hsu, F.F.; Turk, J. Studies on phosphatidylglycerol with triple quadrupole tandem mass spectrometry with electrospray ionization: Fragmentation processes and structural, J. Am. Soc. Mass Spectrom. 2001, 12(9) 1036–1043.

Hvolby, A. Removal of nonhydratable phospholipids from soybean oil, JAOCS, 1971, 48, 503–509.

International Union of Pure and Applied Chemistry, Nomenclature of Lipids—Recommendations Lip-1 and Lip-2, 2011.

Jamil, S.; Dufour, J.P.G.; Deffense, E.M.J. Process for degumming a fatty and fatty substance thus obtained. United States Patent 6,015,915 issued January 2000.

Jiang, F.; Wang, J.; Kaleem, I.; Dai, D.; Zhou, X.; and Li, C. Degumming of vegetable oil by novel phospholipase B from Pseudomonas fluorescens BIT-18. Bioresource Technol. 2011, doi10.1016/j.bioretech.2011.05.050.

Kanamoto, R.; Wada, Y.; Miyajima, G.; Kito, M. Phospholipi-phospholid interaction in soybean oil. JACOS, 1981, 12, 1050–1053.

Kantor, M. Refining Drying Oils. JAOCS, 1950, 27, 455–4462.

Kerovuo, J. Declaration of Janne Kerovuo under 37 CFR § 1.132 for 10/556,816 December 2008.

Kövári, K.; Münter, C.; Münch, Ernst-Wilhelm; Denise, J. Exergy Process, a possibility of short heat treatment of oils seeds. Part II Presented at the 91st AOCS Annual Meeting and Expo, 2000. San Diego, California.

List, G.R.; Avellaneda, J.M.; Mounts, T.L. Effect of Degumming Conditions on Removal and Quality of Soybean Lecithin. JAOCS, 1981, 58, 892–898.

List, G.R.; Mounts, T.L.; Lanser, A.C. Factors promoting the formation on nonhydratable soybean phosphatides. JAOCS, 1992 (69), 443–446.

Loeffler, F.; Plainer, H.; Sproessler, B.; Ottofrickensten, H. Vegetable oil enzymatic degumming process by means of aspergillus phospholipase. United States Patent 6,001,640 issued December 1999.

Logan, A. Degumming and Centrifuge Selection, Optimization and Maintenance. IUPAC-AOCS Workshop on Fats, Oils and Oilseeds Analysis and Production. Tunis, Tunisia 2004.

Maria, L. de; Vind, J.; Oxenboll, K.M.; Svendsen, A.; Patkar, S. Phospholipases and their industrial applications. Appl. Microbiol Biotechnol. 2007, 74, 290–300.

McClements, D.J. Food Emulsions: Principles, Practices, and Techniques; second edition, CRC Press: Boca Raton, Florida, 2005; Chapter 4.

Münch, E. Practical experience on enzymatic degumming, Proceedings of the World Conference on Oilseed Processing and Utilization, edited by Richard Wilson, 2001, 17–20.

Münter, C. Conditioning of flaked rapeseed using Exergy steam processor—a report from full scale test runs in a Swedish crushing plant. Lipid Technology, 2007, 19(5) 104–107.

Murzyn, IK.; Rog, T.; Pasenkiewicz-Gierula, M. Phosphatidylethanolamine-Phosphatidyl bilayer as a model of the inner bacterial membrane, Biophys J. 2005, 88, 1091–1103.

Nielsen, K.J. The composition of difficultly extractable soybean phosphatides, JAOCS, 1960, 37(5), 217–219.

Olempska-Beer, Z. Phospholipase C expressed in Pichia pastoris: Chemical and Technical Assessment, Joint Expert Committee on Food Additives, Rome Italy, 2008.

Oxenboll, K.M. Environmental Assessment of the Enzymatic Degumming process at a Vegetable Oil Refinery, Nordic Symposium, Copenhagen, Denmark 2005.

Pan, L.G.; Campana, A., Tom, M.C.A. A kinetic study of phospholipid extraction by degumming process in sunflower oil. JAOCS, 2002, 12, 1273–1277.

Pangborn, M. Isolation and purification of serologically active phospholipids from beef heart. J. Biol. Chem. 1942, 143, 247–256.

Pardun, H. Die Pflanzenlecithine, Verlag für chem., Industrie H. Ziolkowsky KG, 80900 Augsburg 1, Germany, 1988, 181–194.

Racicot, L.D.; Handel, A.P. Degumming of soybean oil: Quantitiative analysis of phospholipids in crude and degummed oil, JAOCS, 1983, 60(6) 1098–1101.

Ramli, M.R.; Ibrahim, N.A.; Hussein, R.; Kunton, A.; Razak, R.A.A.; Nesaretnam, K. Effects of degumming and bleaching on 3-MCPD esters formation during physical refining, JAOCS, 2011, 88, 1839–1844.

Ramli, M.R.; Siew, W.L.; Ibrahim, N.A.; Razak, R.A.A.; Kuntom, A.; Nesaretnam, K. Monitoring of 3-MCPD Esters Formation in Palm Oil on a Pilot Scale Refining. Presented at the 102nd AOCS Annual Meeting in Cincinnati, OH. (2011b).

Ringers, H.J.; Segers, J.C. Degumming process for triglyceride oils, United States Patent 4,049,868 issued September 1977.

Rohdenburg, H.L.; Csernitzky, K.; Chikany, B.; Peredi, J.; Borodi, A.; and Ruzics, A.F. Degumming Process for plant oils, United States Patent 5,239,096 issued August, 1993.

Rooney, S.A.; Page-Roberts, B.A.; Motoyama, E.K. Role of lamellar inclusions in surfactant production: studies on the phospholipid composition and biosynthesis in rat and rabbit lung subcellular fractions. J. Lipid Res. 1975, 16, 418–425.

Sambanthamurthi, R.; Sundram, K.; Tan, Y-A. Chemistry and biochemistry of palm oil, Prog. Lipid Res. 2000, 39, 507–558.

Schlame, M.; Brody, S.; Hosteltler, K.Y. Eur. J. Biochem. 1993, 212, 727–735.

Schneider M. Phospholipids. In Gunstone F.D., Padley F.B., Eds. Lipid technologies and applications. New York: Marcel Dekker; 1997, 62.

Segers, J.C. Superdegumming, a new degumming process and its effect on the effluent problems of edible oil refining, Fette Seifen Anstrichem, 1982, 84, pp. 543–546.

Segers, J.C.; Van de Sande, R.L.K.M. Degumming—Theory and Practice. Edible fats and oils processing: basic principles and modern practices: World Conference Proceedings; Erickson, D., Eds; AOCS Press: Champaign, Illinois, 1990; pp. 88–93.

Segers, J.C. Private communication with C.L.G. Dayton 12/13/2010.

Sen Gupta, A.K. Micellar Structures and their Implication in the Chemistry and Technology of Fats and Other Lipids. Fette Seifen Anstrichmittel. 1986, 88, pp. 79–86.

Shu, Z.Y.; Yan, Y.J.; Yang, J.K. *Aspergillus niger* lipase: gene cloning, over-expression in *Escherichia coli* and in vitro refolding. Biotechnol Lett, 2007, 29 pp. 1875–1879.

Sinram, R.D. The added value of speciality lecithins. Oil Mill Gazetteer; September 1991, 22–26

Søe, J.B; Turner, M. Enzymatic Oil-Degumming Method, United States Patent 8,192,782 issued June 2012.

Szuhaj, B.F; van Nieuwenhuyzen, W. Nutrition and biochemistry of phospholipids. Illinois: AOCS Press, 2003.

Tosi, E.A.; Cazzoli, A.F.; Ré, E.D.; Tapiz, L.M. Phosphatides content in soybean oil as function of bean moisture content-at-harvest and storage-time. Grasas y Aceites. 2002, 53(4), 400–402.
Verenium Press Release, World's largest soybean processing plant converts to usage of Verenium's Purifine® PLC Enzymatic degumming process, March 3, 2010.
Verenium Press Release, World's Largest Soybean Processing Industrial Complex Converts Half of Capacity to Usage of Verenium's Purifine® PLC Enzymatic Degumming Process Engineered and Installed by Alfa Laval, February 28, 2011.
Wheatley, J. Biodiesel upstart. In Alfalaval customer stories. 2011.
Yagi, T.; Higurashi, M.; Tsuruoka, H.; Nomura, I. Process for refining oil and fat. United States Patent 5,532,163 issued July 1996.
Yamane, T. Enzyme technology for the lipids industry: An engineering overview, JAOCS, 1987, 64, 1657–1662.
Yang, B.; Wang. Y.H.; Yang, J.G. Optimization of enzymatic degumming process for rapeseed oil, JAOCS, 2006, 83(7).
Yang, J.G.; Wang, Y.H.; Yang, B.; Mainda, G.; Guo, Y. Degumming of vegetable oil by a new microbial lipase. Food Technol. Biotechnol, 2006, 44(1) 101–104.
Young, F. Physical refining published by AOCS in Edible fats and oils processing: basic principles and modern practices: World conference proceedings, edited by David Erickson 1990, pp. 124–135.
Zufarov, O. Schmidit, S. and Sekretar, S. Degumming of rapeseed and sunflower oils. Acta Chimica Slovaca, 2008, 1(1) 321–328.

7

Nano Neutralization™

Eric Svenson and Jim Willits
Desmet Ballestra North America, Marietta, Georgia, USA

Introduction

Neutralization is the first process to chemically refine soft seed oils, and depending on the market being served, may include a separate degumming operation. The purpose of neutralization is to reduce the hydratable and non-hydratable calcium and magnesium phospholipids, and the free fatty acids to acceptable levels. It is an important first step that must be accomplished before the oil can proceed to bleaching and deodorization.

The use of computer-controlled instruments has helped vegetable oil refiners to better control their operation. Using these instruments has decreased operating costs and improved oil yields. However, the basic process has remained unchanged for several decades.

Nano Neutralization™ is a novel new technology. It has proven to increase oil yields and drastically reduce the chemicals required to produce once-refined vegetable oil while maintaining oil quality.

Traditional Oil Neutralization Methodology

Traditional neutralization generally consists of an acid treatment prior to treating the crude vegetable oil with a dilute caustic solution. The purpose of the acid treatment is to convert the nonhydratable phospholipids (NHP), which are typically Ca/Mg phosphatidates to hydratable phospholipids (HP). A Ca and Mg salt is generated in the chemical reaction with the acid. To ensure that the Ca and Mg are below the required levels, an *Excess* amount of acid is added. Both the HP and the Ca/Mg salt can easily be removed through centrifugal separation after the caustic solution is mixed with the vegetable oil.

Two acids that are typically used for acid treatment are phosphoric or citric acid. Phosphoric acid is generally preferred since it is a stronger acid requiring a lesser amount on a dry basis than citric acid. Another reason for the preference is that it usually costs less than citric acid. However, processors may elect to treat with citric acid if the crude vegetable oil has unusually high levels of phosphorus. In those cases, sometimes it is difficult to distinguish between the phosphorus if the neutralized oil is from the NHP or the residual acidity from the acid treatment itself. Another benefit of citric acid is that it disassociates at deodorization temperatures unlike phosphoric acid. Also if the crude oil is overtreated with phosphoric acid and not completely neutralized, the excess phosphoric acid may cause color and flavor reversion in the finished deodorized oil.

Once the crude vegetable oil has been treated with acid to convert the NHP to HP, it is treated with a dilute alkali solution to neutralize the free fatty acid and the excess acidity. The processor typically adds more caustic than is theoretically necessary for neutralizing the acid. This additional amount of caustic is commonly referred to as the Excess. The processor runs the risk of saponification of the neutral oil into soap if the Excess is too great.

Formulas for dosing the oil with acid and caustic are available from numerous handbooks on edible oil processing. These formulas enable the oil refiners to calculate the dosage required for refining the oil. More important, however, is the experience of individual refiners and the quality of the crude oil that dictates the optimum dosage for any particular oil.

Neutralization Chemistry

The law of conservation of mass states that the mass of a closed system will remain constant over time. Mass can neither be created nor destroyed. However, mass can be rearranged in a chemical reaction, which is what takes place when converting the NHP to HP and free fatty acid to soap.

It requires 2 moles of phosphoric acid (H_3PO_4) to react with 3 moles of calcium type NHP. The product is 3 moles of HP and 1 mole of calcium phosphate salt [$Ca_3(PO_4)_2$]. It also requires 2 moles of phosphoric acid to convert 3 moles of magnesium type NHP to 3 moles of HP and 1 mole of magnesium phosphate salt [$Mg_3(PO_4)_2$]. The two chemical equations are as follows:

$$3\text{NHP Ca} + 2H_3PO_4 \Rightarrow 3\text{HP} + Ca_3(PO_4)_2$$

$$3\text{NHP Mg} + 2H_3PO_4 \Rightarrow 3\text{HP} + Mg_3(PO_4)_2$$

The molar mass of the various compounds is found in Table 7-A.

Calcium NHPs

$$3\text{NHP Ca} + 2H_3PO_4 \Rightarrow 3\text{HP} + Ca_3(PO_4)_2$$

$$(3 \text{ mol} \times 840 \text{ g/mol}) + (2 \text{ mol} \times 98 \text{ g/mol}) \Rightarrow (3 \times 802 \text{ g/mol}) + 310 \text{ g/mol}$$

The theoretical amount of phosphoric acid (on a dry basis) necessary to convert NHP Ca to HP can be calculated as follows:

$$\text{ppm } H_3PO_4 = \text{ppm Ca} \times (2 \text{ mol} \times 98 \text{ g/mol}) \;/\; (3 \text{ mol} \times 40 \text{ g/mol})$$

$$\text{ppm } H_3PO_4 = \text{ppm Ca} \times 1.63$$

Table 7-A. Molar Mass of Various Elements and Chemical Compounds.

Elements & Chemical Compound	Molar Mass (g/mol)
NHP Ca	840
NHP Mg	824
HP	802
Ca	40
Mg	24
H_3PO_4	98
$Ca_3(PO_4)_2$	310
$Mg_3(PO_4)_2$	263
NaOH	40
Na_3PO_4	164
H_2O	18

Magnesium NHPs

$$3\text{NHP Mg} + 2H_3PO_4 \Rightarrow 3\text{HP} + Mg_3(PO_4)_2$$

$$(3\text{ mol} \times 824\text{ g/mol}) + (2\text{ mol} \times 98\text{ g/mol}) \Rightarrow (3\text{ mol} \times 802\text{ g/mol}) + 263\text{ g/mol}$$

The theoretical amount of phosphoric acid (on a dry basis) necessary to convert NHP Mg to HP can be calculated as follows:

$$\text{ppm } H_3PO_4 = \text{ppm Mg} \times (2\text{ mol} \times 98\text{ g/mol}) / (3\text{ mol} \times 24\text{ g/mol})$$

$$\text{ppm } H_3PO_4 = \text{ppm Mg} \times 2.72$$

As an example, let's say crude degummed soybean oil will be processed at 45,000 lbs/hr of with 100 ppm Ca and 80 ppm Mg, and it will be treated with 85% phosphoric acid plus 50% *Excess.* The theoretical minimum amount of 85% phosphoric acid needed to treat can be calculated by the following formula:

$$\text{Min } 85\%\ H_3PO_4 \text{ lbs/hr} = \text{oil flow rate lbs/hr} \times (\text{Ca ppm} \times 1.63 + \text{Mg ppm} \times 2.72) / (1{,}000{,}000 \times 0.85)$$

$$\text{Min } 85\%\ H_3PO_4 \text{ lbs/hr} = 45{,}000 \text{ lbs/hr} \times (100 \text{ ppm} \times 1.63 + 80 \text{ ppm} \times 2.72) / (1{,}000{,}000 \times 0.85)$$

$$\text{Min } H_3PO_4 \text{ lbs/hr} = 20.149 \text{ lbs/hr}$$

The next step is to determine how much *Excess* H_3PO_4 to use. The purpose of the *Excess* is to ensure that the NHP Ca and NHP Mg are converted to HP with the understanding that a complete reaction may not be possible. There are several factors that will play a role in determining the amount of *Excess* that is required. These include the temperature of the oil being treated, the duration of acid mixing, the intensity of the mixing, and the final phosphorus level desired. It is important to keep in mind that the higher the *Excess* H_3PO_4 will require a higher dosage of the alkaline solution to neutralize it. The amount to include for 50% *Excess* would be the following:

$$\textit{Excess}\ H_3PO_4\ \text{lbs/hr} = \text{Minimum lbs/hr} \times 50\%$$

$$\textit{Excess}\ H_3PO_4\ \text{lbs/hr} = 20.149\ \text{lbs/hr} \times 0.50$$

$$\textit{Excess}\ H_3PO_4\ \text{lbs/hr} = 10.074\ \text{lbs/hr}$$

The total amount of H_3PO_4 for treat is the sum of the minimum plus the *Excess.*

$$\text{Total } 85\%\ H_3PO_4\ \text{lbs/hr} = \text{Min } 20.149\ \text{lbs/hr} + \textit{Excess}\ 10.074\ \text{lbs/hr}$$

$$\text{Total } 85\%\ H_3PO_4\ \text{lbs/hr} = 30.223\ \text{lbs/hr}$$

Once the crude soybean oil has been acid treated to convert the NHP to HP, it needs to be treated with an alkaline solution to neutralize the free fatty acid (FFA). A higher dosage above the stoichiometric or theoretical amount is necessary to neutralize the excess phosphoric acid and to ensure most of the FFA has been neutralized. A complete reaction and removal of FFA is not possible. The predominant alkali compound used in neutralizing crude soybean oil is sodium hydroxide (NaOH), which is commonly known as caustic soda. Other alkali compounds such as potassium hydroxide (KOH) and sodium bicarbonate ($NaHCO_3$) have been tested, but they are rarely used. Sodium hydroxide is typically expressed in either degrees Baumé (°B) or normality in vegetable oil refining. Table 7-B shows the relationship between °B, normality, and % concentration.

Three moles of NaOH are required to neutralize one mole of H_3PO_4 as shown in the following formula:

$$3NaOH + H_3PO_4 \rightarrow Na_3PO_4 + 3H_2O$$

$$(3 \times 40\ \text{g/mol}) + (98\ \text{g/mol}) = (164\ \text{g/mol}) + (3 \times 18\ \text{g/mol})$$

Table 7-B. Sodium Hydroxide Concentration.

°B at 15°C	Normality	% NaOH Content	°B at 15°C	Normality	% NaOH Content
10	1.64	6.57	22	4.02	16.09
12	2.00	8.00	24	4.47	17.87
14	2.38	9.50	26	4.92	19.70
16	2.76	11.06	28	4.40	21.58
18	3.17	12.68	30	5.88	23.50
20	3.59	14.36			

The amount of NaOH on a dry basis required to neutralize the excess H_3PO_4 (85% solution) is determined by the following formula:

$$\text{NaOH lbs/hr} = \textit{Excess}\ H_3PO_4\ \text{lbs/hr x } 0.85 \times (3 \times 40 / 98)$$

$$\text{NaOH lbs/hr} = \textit{Excess}\ H_3PO_4\ \text{lbs/hr} \times 0.85 \times 1.22$$

Using the previous example, there was 10.074 lbs/hr excess H_3PO_4 added to the treated crude oil. The required amount of NaOH needed to neutralize the excess H_3PO_4 would then be 10.447 lbs/hr NaOH dry basis.

Traditionally the NaOH solution added to the crude oil is referred to as the *Treat* and is indicated as a percentage of the oil flow rate. A critical decision is required in determining the NaOH concentration and amount needed to treat the crude oil. The free fatty acid (FFA) concentration of the crude oil is used to calculate the required amount needed to neutralize the FFA plus an *Excess* amount, which is derived from experience. As previously stated, the processor runs the risk of saponifying the neutral oil into soap if too much NaOH is used. The traditional formula used to calculate the % Treat is as follows:

$$\%\ \text{Treat} = [(\%\ \text{FFA} \times 0.142 + \%\ \text{Excess}) / \%\text{NaOH}] \times 100$$

In the above formula, the molecular weight of NaOH (40 g/mol) is divided by the molecular weight of oleic acid (282 g/mol) resulting in a factor of 0.142. As an example, a processor wants to treat crude nondegummed soybean oil which has 0.55% FFA with 16°B NaOH with an Excess of 0.15%, and this will produce the following:

$$\%\ \text{Treat} = [(0.55 \times 0.142 + 0.15) / 11.06] \times 100$$

$$\%\ \text{Treat} = [(0.55 \times 0.142 + 0.15) / 11.06] \times 100$$

$$\%\ \text{Treat} = 2.06\%$$

Using the previous flow rate of 45,000 lbs/hr, the amount of 16°B NaOH solution needed for treat would be 927 lbs/hr.

$$\text{NaOH lbs/hr} = 45{,}000\ \text{lbs/ hr} \times 2.06 / 100 = 927\ \text{lbs/hr}$$

Neutralization Process

A combination of modern lab analytical capabilities with process automation and ultra-high shear mixing technology has resulted in the possibility of treating most oils with a very low dosage (near stoichiometric quantity) of acid for treatment of non-hydratable phospholipids. Since excess acid adds to consumables' expense and requires added caustic for the neutralization step, a highly automated system will minimize ongoing operating costs. A typical modern process flow is shown in Fig. 7.1.

The crude vegetable oil is stored in well agitated day tanks to ensure a consistent feedstock supply to the oil refinery. Once the crude oil has been analyzed for P, Ca,

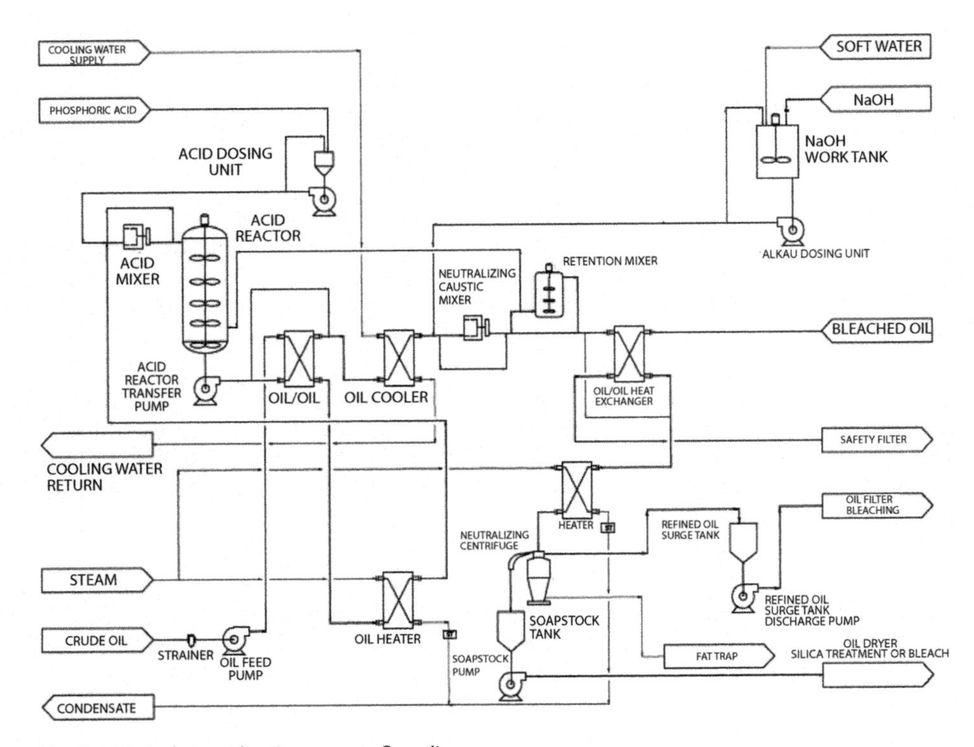

Fig. 7.1. Typical neutralization process flow diagram.

Mg, and FFA and the proper amounts of H_3PO_4 and NaOH have been calculated and entered into the control system, the crude oil is heated to 194°F (90°C). The acid is added to the hot crude oil ahead of intensive mixing in an ultra-high shear mixer. Heating the crude oil and the intimate mixing enable the acid molecules to be well dispersed in the oil very quickly and reduce the downstream reaction time.

The mixture of hot crude oil and acid then enters a reactor vessel where it is slowly stirred for approximately 30 minutes. This allows the acid to react with all of the oil-soluble calcium and magnesium salts of phosphatidic acids and converts them from being oil soluble to being more water soluble. This enables them to be hydrated, and to be centrifugally separated downstream.

The acid treated hot crude oil is then cooled to 122°F (50°C). The NaOH solution is then added to neutralize the free or excess H_3PO_4 and FFA. It is important that the water used in making the NaOH solution be softened to prevent adding Ca and Mg ions into the acid-treated oil. There must be at least enough water to hydrate the phospholipids, and enough caustic to neutralize the free fatty acids, and the acidity from H_3PO_4. Once the NaOH solution has been added, the mixture passes through an ultra–high shear mixer before entering a retention mixer. The acid treated crude oil and NaOH solution are gently mixed for 5–7 minutes in the retention mixer to ensure the reaction reaches completion while keeping the gums and soaps in suspension.

The next step is the separation of the gums and soaps from the oil. The acid and caustic treated oil with the gums and soaps is heated to approximately 167°F (75°C) before it enters the neutralizing centrifuge, which is the most critical component in the refining process. The centrifugal force quickly separates the heavier gums and soaps from the lighter vegetable oil. The centrifugal separator is adjusted to obtain an oil stream (light phase), and a gums/soaps stream (heavy phase) with minimal entrained neutral oil. The vegetable oil off the neutralizing centrifuge should have phosphorus less than 10 ppm and soaps below 200 ppm.

The neutralizing centrifuge requires preventive maintenance and periodic cleaning to ensure consistent oil quality. These actions also ensure that the oil losses are minimized. Oil losses are also greatly impacted by the amount of excess acid and caustic treatment. Enough acid must be added to convert the NHP to HP and enough caustic to neutralize the FFA. An overtreatment of acid in itself does not impact the oil quality, but will require additional caustic to neutralize the excess acid, which in turn will increase operating costs.

The processor also runs the risk of overdosing with caustic. Should this occur, saponification of oil into soap will be the result in higher oil losses. The efficiency of the neutralizing centrifuge is calculated from the following formula:

$$\% \text{ Efficiency} = \text{product output (flow rate)} / \text{crude oil input (flow rate)} \times 100$$

The oil losses are then determined by subtracting the % Efficiency from 100.

$$\% \text{ Oil loss} = 100 - \% \text{ Efficiency}$$

Understanding Nano Reactor Technology

The use of a Nano Reactor® in refining vegetable oil is a new technology that has only been commercially available since early 2011. A Nano Reactor® is a static device with no moving parts. Its name is derived from the nanometer-sized bubbles that are created when the fluid is pumped through the Nano Reactor® under high pressure.

What actually takes place in the Nano Reactor® is not as yet fully understood. What is known is that a Nano Reactor® creates intense hydrodynamic cavitation. It is believed that the shockwaves created by the hydrodynamic cavitation physically break the weak bonds of the Ca and Mg non-hydratable phospholipids in vegetable oils in milliseconds.

Crude vegetable oil is pumped through a series of scientifically-designed geometries at high pressure. As the oil goes through each stage, there is a dramatic pressure drop. The water molecules in the oil vaporize and recompress to a liquid at each stage, creating shockwaves that break the weak metal bonds of the non-hydratable phospholipids to make them more hydratable. In addition to physically breaking the Ca and Mg bonds, the Nano Reactor® thoroughly mixes the acid and caustic solutions that have been added to the crude vegetable oil.

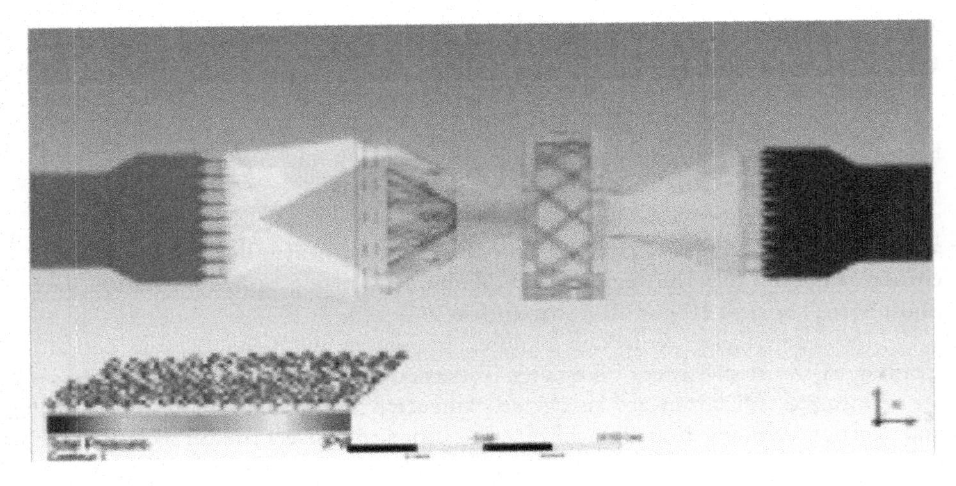

Fig. 7.2. Thermal image of a Nano Reactor® from Cavitation Technology, Inc.

The Nano Neutralization™ Process

A Nano Reactor® can be installed in an existing neutralization process with relative ease and minimal cost. The crude vegetable oil is diverted to the Nano Neutralization™ system after the standard acid treatment. After passing through the Nano Reactor®, the oil is piped back to the standard neutralization line ahead of the retention mixers. The processor has the ability to operate the standard neutralization process if they so choose by opening and closing a few valves. A typical Nano Neutralization™ process flow diagram is shown in Fig. 7.3.

The crude vegetable oil is pumped from the day tank to a small surge tank after treatment with acid. However, the temperature of the crude vegetable oil only needs to be at 120–140°F (50–60°C) instead of the 194°F (90°C) typical for acid treatment. The purpose of the surge tank, which has approximately 5 minutes of retention, is to ensure the high-pressure pump has a constant head of material on its suction side. As the pump pulls the acid treated oil from the surge tank, a caustic solution of appropriate concentration and dosage is added to the oil flow before it enters the Nano Reactor™. The fluid pressure before the Nano Reactor® can reach up to 1200 psig (83 barg), but typically it operates between 900–1000 psig (62–69 barg).

Operation of the Nano Neutralization™ system is simple and straightforward. There are only two control loops necessary for an efficient operation. The first is the variable frequency drive on the high-pressure pump. The speed of the pump is adjusted to maintain the desired pressure differential, which is critical to the process. The other control is the level transmitter installed in the surge tank feeding the high-pressure pump. A flowcontrol valve installed on a recycle line opens and closes to maintain the desired level in the surge tank. This ensures that the high-pressure pump will never operate without fluid.

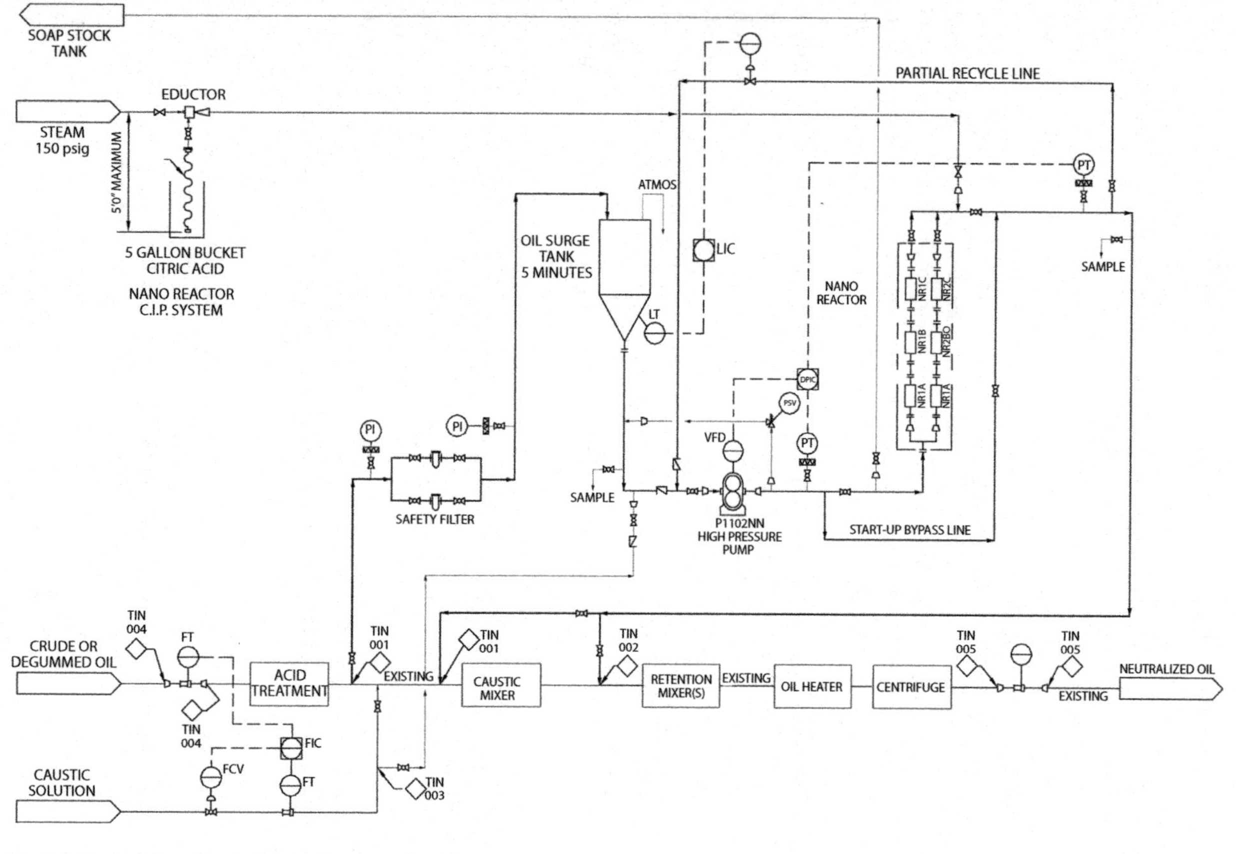

Fig. 7.3. Typical Nano Neutralization™ process flow diagram.

Nano Neutralization™ Results and Benefits

Nano Neutralization™ has been found to drastically reduce chemical consumption. Operation of a plant scale Nano neutralization™ system has indicated the following: drastic reduction in acid (by over 90%) that was previously used to chemically cleave the metals from the non-hydratable phospholipids. Without the need to neutralize excess acid, and with the intensive mixing in the Nano Reactor®, the quantity of caustic required to neutralize the FFA has been reduced significantly.

Also because of reduction in caustic dosage, there is less saponification of oil and a much cleaner separation in the centrifugal separator. This has resulted in an increase in oil yield with less neutral oil loss with the soapstock. The cleaner separation also results in less soap remaining in the oil. Less soap in the oil has resulted in reduction of the amount of silica or wash water needed to remove the soaps prior to bleaching.

Trials at two separate commercial installations have indicated that it is possible to significantly reduce the amount of acid (by almost 90%) for acid pretreatment when processing either crude nondegummed or degummed soybean oil. In both cases, it was possible to achieve the phosphorus levels of <10 ppm in the refined oil. Also, the caustic dosage was reduced by 30–50%. The soaps off the neutralizing centrifuge were below 150 ppm.

In addition to the acid and caustic savings, the two trials indicated improvements in oil yield. The oil yield improvement has been documented to be greater than 1.0%.

A commercial installation that is processing approximately 40,000 pounds per hour of crude water degummed soybean oil has been operating a Nano Reactor® system since February 2011. The plant operating logs indicate savings in phosphoric acid, caustic soda, as well as silica. The plant is currently operating their refining line using approximately 90% less phosphoric acid, and 50% less caustic, compared to usage prior to installation of the Nano Reactor® system. The silica addition has also been reduced, from 0.1% in August 2010 to 0.01% in August 2011. Analytical data of once refined oil before and after installation of the Nano Reactor® is shown in Table 7-C.

Because of low level of soap (<100 ppm) in the refined, this plant has conducted trials to eliminate silica dosage. Trials using only bleaching clay have been successful. The processor is now contemplating eliminating silica addition altogether.

Table 7-C. Once Refined Soybean Oil at a Commercial Installation.

Component	Before Nano Reactor™	After Nano Reactor
P	7–15 ppm	<3 ppm, 1 ppm Typical
Ca	1–3 ppm	<1 ppm
Mg	1–3 ppm	<1 ppm
Soap	300 ppm	<150 ppm, 50–100 ppm Typical

Table 7-D. Nano Neutralization Monthly Savings (24 Hr/Day, 27 Day/Mth).

Component	Change	Savings (lbs/hr)	Price / Lbs	Monthly Savings
Oil Yield	0.2% Increase	90.0	$0.50	$29,160
H_3PO_4	90% Reduction	23.1	$0.75	$11,227
NaOH	30% Reduction	33.9	$0.18	$3,953
Silica	90% Reduction	40.5	$0.68	$17,846
Total				$62,186

Using the previous example of refining 45,000 lbs/hr of crude degummed soybean oil with 100 ppm Ca, 80 ppm Mg, and 0.55% FFA being treated with 85% phosphoric acid and 16°B NaOH, the following chemical consumptions were calculated:

- H_3PO_4: 25.690 lbs/hr (dry basis)
- NaOH:112.973 lbs/hr (dry basis)

The monthly savings of a refinery operating 24 hours per day, 27 days per month with a Nano Reactor® are shown in Table 7-D.

Using a Nano Reactor® in a vegetable oil refinery provides several benefits. There is a substantial chemical savings that is better for the environment at a significant economic benefit. The chemical savings allows for less storage of the raw material, which in turn reduces the environmental impact potential. Crude vegetable oil processed through a Nano Reactor® also allows the neutralization centrifuge to operate smoother, since the separation line between the light and heavy phase are more defined.

8

Physical Refining of Vegetable Oils

Walter E. Farr
The Farr Group of Companies, Memphis, Tennessee, USA

Introduction

In the United States, "refining" refers to the removal of free fatty acids by either a chemical or physical processing of fats and oils. The chemical refining process may include a complete degumming process followed by bleaching, hydrogenation, and deodorization. The chemical refining process has been the dominant process for oils containing high levels of phosphatides for over 70 years.

Definition of Physical Refining

Free fatty acids (FFAs) are neutralized with caustic soda, but another important part of this is removal of all other impurities in the crude oil, carried out with the soapstock. After this oil is properly bleached (reducing color and chlorophyll, and removing any remaining trace metals), the oil passes through the final processing step—deodorization.

A physical refining deodorizer is the same as a conventional deodorizer, but modifications are made to handle the higher FFA load, owing to the fact that the FFAs were not neutralized. A brief definition of physical refining is: FFAs are removed in a physical refining deodorizer, with no neutralization of the FFAs in a previous processing step.

A Review of Current Practices in the Refining of Vegetable Oils

Current refining processes are described as follows:

1. The high-temperature short mix caustic refining process
2. The low-temperature long mix caustic refining process.

While the low-temperature long-mix process is preferred in North America, the high-temperature short-mix process is preferred in Europe and the Middle East.

The critical components of the refining system (long-mix caustic refining process) are as follows:

1. the acid pre-treatment system
2. the caustic addition system
3. the caustic/oil mixer
4. the retention mixers
5. the primary centrifuge
6. the wash water treatment system
7. the water wash centrifuge
8. the vacuum dryer for refined oil.

New Developments Offer Significant Changes in Vegetable Oil Refining

In the last 20 years, there have been new developments in the way that vegetable oils are refined. In soybean oil, for instance, advances that have been implemented include:

- Close coupling of refining and bleaching
- Modified caustic refining or silica refining (elimination of water washing)
- Bleaching improvements (operating filters in packed bed, and lead-trim mode)
- Membrane Degumming and Refining
- Organic refining process (ORP)
- Enzymatic degumming and refining
- Physical refining of soybean oil and other soft oils

Close Coupling of Refining and Bleaching

It has been common practice to send vacuum-dried refined oil to a large intermediate tank, and then bring the oil back in for bleaching. Since this caustic refined oil cools (particularly in cold climates), it must be heated back up to bleaching temperature, wasting a lot of energy. This is counter to the idea of "green processing." However, if the refined oil goes directly to bleaching (close coupling), this energy is saved. A prerequisite for this technique to work well is that when the refinery starts up, it must be in perfect control; otherwise, impure refined oil would immediately blind out the bleach filters.

Modified Caustic Refining or Silica Refining (Elimination of Water Washing)

In this technique, silica instead of water washing is used to remove soap and other trace metals from the refined oil. This was a major step in "greening" the refining system, in that there is a big reduction in fresh water usage. Elimination of refining wash water reduces the brochemical oxygen demand (BOD) load from a typical refinery by 75%. Also, shutting down the water-wash centrifuge saves electrical energy and oil loss. Eliminating water washing can improve yield by 0.20% in a typical refinery.

Bleaching Improvements: Operating Filters in Packed Bed and Lead-Trim Mode

Typical bleaching systems have two filters. The first one is on-line until spent, then the second is put on-line. The former filter is cleaned to be ready for the next cycle. Bleaching clay is added continuously throughout the bleaching cycle.

A new technique introduced about 15 years ago is called *packed bed bleaching.* The filter is precoated with filter-aid; then all bleaching clay needed for the complete filter cycle is added to the filter. (It is added the same way the filter-aid was "pre-coated.") With proper pump sizing, this is done in about 15 minutes. For example, a 600 sq. ft. filter may require 2000 lbs/hr of clay for the proper packed bed. The filter stays on line until the filtrate goes out of specification on chlorophyll, the packed bed is built on the second filter, and the flow is switched into the second filter.

The advantage of this approach lies with the improved bleaching efficiency and longer life of the filter, typically 12–24 hours, depending on filter design and the type of bleaching clay.

A later development/improvement is called the *lead-trim process.* The first filter is started up in packed bed mode and runs its complete cycle. When this filter goes out of specification on chlorophyll, the second filter receives its packed bed. A critical new step is that the flow continues through the first filter, and the filtrate goes directly to the newly packed bed filter. The advantage is that the first filter still has a lot of chlorophyll retention or adsorption capability. This allows a reduction in bleaching clay usage and reduces oil lost in spent bleaching clay. Almost all new plants built in the last 10 years have been built this way, and have a bleaching clay dosage as low as 0.10%. This is truly "greening of the bleaching process." In 1995, the Owensboro Grain Edible Oils Refinery, Owensboro, Kentucky, was the first new plant to be designed to operate this way.

Membrane Degumming and Refining

Membrane separation of soybean oil by ultrafiltration (UF) in miscella (the mixture of extracted oil and hexane that is produced during solvent extraction) has been developed and is in the pilot plant stage. UF membranes work by separating two components of different molecular weights, in this case the refined oil from the phospholipids. The separation is made in the oil mill extraction plant in miscella, due to the drastic reduction in viscosity required for sufficient flux rate through the membranes.

Development work on degumming soybean oil in miscella began about 1985 at Texas A&M University and Anderson, Clayton & Co. Both super-degumming—acid degumming enhanced by temperature reduction and extended retention time (invented by J. Segers and patented by Unilever)—and refining in the miscella were investigated. Several pilot plant trials yielded low P levels. To make RBD oil, only a light bleaching (0.10% clay) with no silica, followed by physical refining, was needed.

Organic Refining Process (ORP)

The "organic refining process" was named for the organic citric acid used in the process. The process utilized a high dosage of dilute citric acid. It was stated that the citric acid usage was reasonable because it was recirculated. The contaminated citric acid went into a large settling tank, the impurities were supposed to separate, and the cleaner citric acid solution would be recycled. When the yield of the process was examined, however, the yield was no better than conventional caustic refining. Only one plant adapted to this process, and it is unlikely another with this technology will ever be built.

Enzymatic Degumming and Refining

Enzymatic degumming and refining technology is covered in more detail in Chapter 6, "Enzymatic Degumming." As mentioned earlier, a good physical refining system must be able to handle crude oils of all quality, that is, oil from field-damaged beans, mishandled, or multiple-handled beans. The three plants built for enzymatic degumming by a large international company are (1) in Indiana (built several years ago); (2) Alabama (under construction; and (3) Argentina. (This very large plant was in the final stages of start-up and optimization in 2011.)

Physical Refining of Soybean Oil and Other Soft Oils

The actual process of physical refining of soybean oil and other soft oils is covered in more detail elsewhere in this chapter.

Development of Physical Refining of Soybean Oil

Physical refining of palm oil and other lauric oils has been practiced for many years. Palm oil is low in phosphotides (P), and very low in NHPs (nonhydratable phosphotides), allowing for a simple acid water wash and bleaching to reduce color and remove all trace metals, and physical refining. Not only does this work well, but it saved the palm oil industry in that caustic refining (neutralization) of high FFA palm oil would have resulted in very high refining losses, making refining palm oil totally unprofitable.

Physical refining of soybean oil and canola oil is much more challenging, as these oils have the highest level of P, and can have the highest level of NHPs of all vegetable oils.

EMI and Desmet installed physical refineries for soybean oil in Mexico and South America nearly 35 years ago. None of these plants could make a quality product, regardless of any possible yield advantage, and each plant had to revert back to conventional refining.

Physical Refining of Soybean Oil

A prerequisite for physical refining of soybean oil was the development of degumming techniques that would lower P and other trace elements to a level that would allow reasonable dosages of silica and bleaching clay, bringing these components to a level equal to totally neutralized oil. All of these special degumming techniques utilized acid (preferably organic citric acid) with high shear mixers.

For physical refining of soybean oil to be commercially viable, the system must be capable of handling crude oils from seed of various qualities. It must be able to handle oil extracted from field-damaged seed, seed that was harvested before full maturity, seed stored improperly, and seed that receives excessive handling. All of these conditions can create some cracked beans, which can start the enzymatic process that creates NHP. The higher the level of NHP, the more challenging it is to obtain adequate degumming.

Having a system for physical refining only for good quality, and using neutralization for poor quality crude oil could not be justified. This resulted in the development of hybrid systems as follows:

- Physical Refining
- Semi-Physical Refining™
- Modified Physical Refining.™

In order to predict the best process for a particular lot of oil, a laboratory test was developed, termed "Degumming Efficiency." Degumming efficiency is determined via the formula:

Degumming efficiency (%) =
100 × (P ppm crude oil – P ppm laboratory degummed oil) / P ppm crude oil

The degummability test for soybean oil is explained in Fig. 8.1.

An Example for Computing Degumming Efficiency

Say the P in the crude oil is 600 ppm, and the P in the degummed oil (via above procedure) is 35 ppm, then:

$$\text{Degumming efficiency} = 100 \times (600 - 35) / 600 = 94.2\%$$

After extensive trials made on many different oils, it was proven that if an oil had a degumming efficiency of 93% or higher, it could be properly pretreated and physical refined with reasonable dosages of silica and bleaching clay. The economics of this process are astonishing.

The Multi-purpose Refinery

The three processes will be thoroughly explained while demonstrating how the degumming efficiency test is used to select the best process for each particular lot of oil.

DEGUMMABILITY TEST FOR SOYBEAN OIL

To determine how well crude oil will degum or refine / To determine which refining process is best suited for a particular oil

TEST PROCEDURE
Analyze the crude oil for P, Ca, Mg.
Add 400g crude oil to 800 ml beaker.
Place on stirrer/hotplate with stirrer bar set at medium speed.
Heat to 130°F (do not overheat).
Slowly add water (deionized water preferred).
Water treat = $(P\ ppm \times 31.7 \times 10^{-4}) \times 0.7$
Example only, if $P = 800$ ppm → $(800 \times 31.7 \times 10^{-4}) \times 0.7 = 1.78\%$
400 g oil x 1.78% = 7.1 g water.
Agitate, low speed, for 20 minutes, no additional heat needed.
Turn off agitator and let sit idle for 5 minutes.
Filter through filter paper, by gravity.
Analyze filtrate for P, Ca, and Mg.

Fig. 8.1. Degummability test for soybean oil.

Physical Refining

Physical refining will be used when the degumming efficiency is 93% or higher. This process is depicted in the simplified block diagram, Fig. 8.2.

In this process, citric acid is injected at the inlet of the ultra-shear mixer; water (for hydration) is injected at the inlet of the Farr Retention Mixer (replacing the conventional hydration tank used in degumming), then through the centrifuge. Oil exits the centrifuge, goes through silica treatment and packed bed bleaching, then through the physical refining deodorizer. Since no caustic is used, and the FFAs are removed in the physical refining deodorizer, this is physical refining in its purest form.

Semi-physical Refining

Semi-physical refining, depicted in Fig. 8.3, is used when the degumming efficiency is 85 to 92%.

In this process, as in physical refining, citric acid is injected at the inlet of the ultra-shear mixer, followed by a light caustic dosage injected in front of the Farr Retention Mixer, then through the centrifuge. Typically, this caustic dosage can be 50% of neutralizing dose. The oil is then bleached before going to the deodorizer designed for physical refining. Since the FFAs were only partially neutralized, this is called Semi-physical refining™.

Since only a small amount of caustic was used, much less soapstock is produced, residual soap in the refined oil exiting the centrifuge is very low, still offering improved yield and quality over conventional refining (total neutralization).

Note: In this process, the ultra-shear mixer is required for the citric acid, but a simpler, lower HP, and less expensive mixer can be used for the caustic. Examples are IKA single stage mixer, Silverson mixer, or an Alfa-Laval MX mixer.

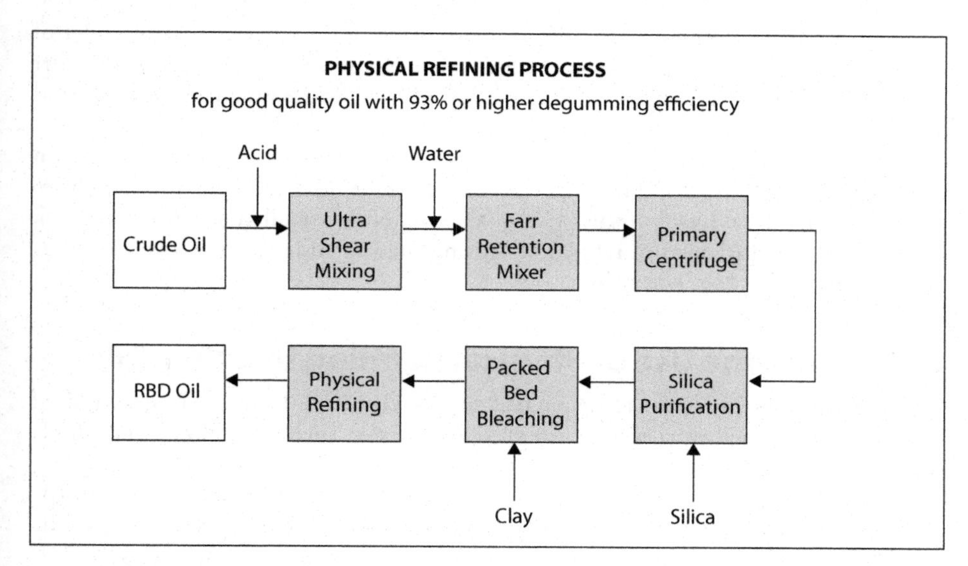

Fig. 8.2. Physical refining process.

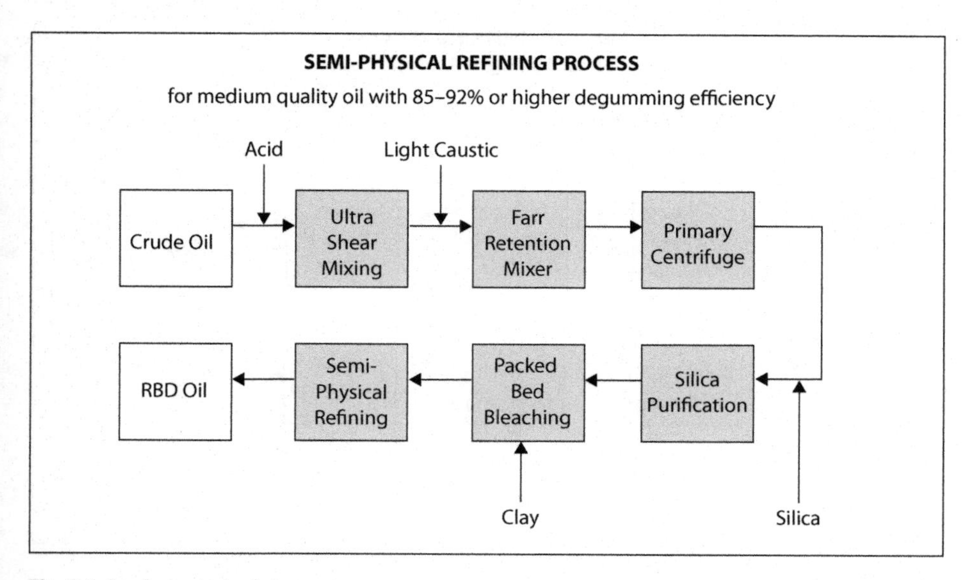

Fig. 8.3. Semi-physical refining process.

Modified Physical Refining™

Modified Physical Refining, depicted in Fig. 8.4, will be used when the degumming efficiency is 75 to 84%.

In this process, citric acid is injected at the inlet of the ultra-shear mixer, then light caustic (typically less than 50% of neutralizing dose) is injected at the inlet of the second ultra-shear mixer, water of hydration is injected at the inlet of the Farr Retention Mixer, then to the centrifuge followed by silica treatment, bleaching, and semi-physical refining.

The critical difference in this process lies with the use of an ultra-shear mixer on caustic. By use of ultra-shear mixing, less caustic can be used, typically less than 50% of neutralizing dose. Even on this poor-quality oil, only a small amount of soapstock is produced, residual soap in refined oil exiting the centrifuge is very low, and silica and bleaching clay dose is very low.

Expanding the Use of Physical Refining to Other Oils

For brevity, all the processes above are related to soybean oil. It should be emphasized that the processes are equally good for all vegetable oils, such as sunflower oil, canola oil, cottonseed oil, peanut oil, safflower oil, etc.

The processes are applicable for all the high FFA vegetable oils. Just as in the case of palm oil, these oils need to be physically refined to dramatically improve the

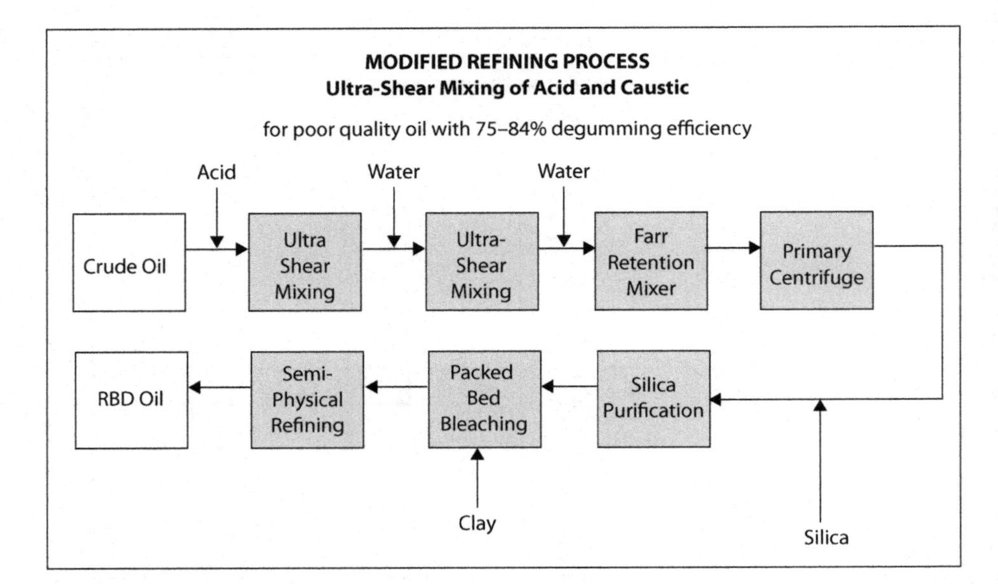

Fig. 8.4. Modified physical refining.

refining yield. High FFA oils conventionally refined (neutralized) create high losses and very large volumes of troublesome soapstock.

High FFA oils that have proven to work well by this physical refining process are wet-milled corn oil, corn oil from syrup in the ethanol process, rice bran oil, wheat germ oil, oil from spent barley, olive oil, avocado oil, and others.

Quality of Physical Refined Vegetable Oils

A paradigm that physical refined oils would never have the quality of conventionally-refined oils (totally neutralized) existed for many years, and hampered many from the pursuit of developing physical refining.

Of course the pre-requisite for high-quality physical refined oil was the development of ultra-shear degumming, improved bleaching techniques, and use of deodorizers built for physical refining.

With the advent of the need for zero trans fat in salad and cooking oils, it was proven that physical refining actually contributed to delivering zero trans fat, if performed in the proper way. High temperature/short-time deodorizers need to operate at 500°F to remove FFA in the short time, which could have been a challenge for physical refining of high FFA oils.

The next myth was that one must deodorize soybean oil at 500°F to get the maximum heat bleach for low color (Wesson Oil color). This myth was exploded at a new plant in Mexico (Ragasa, Monterrey, Mexico). Their salad oil, Nutrioli™, has true zero *trans* fat and color of 0.2–0.3 red (Wesson color). This oil is semi-physical refined, and the best oil in Mexico, Central and South America, equal to the best oils in the United States.

Plants Built with Physical Refining or Semi-Physical Refining of Vegetable Oils

Owensboro Grain Edible Oils Co., Owensboro, Kentucky

Owensboro Grain is a fully integrated (crusher/refiner) facility for soybeans. This is the only plant world-wide than uses all expanders in preparation for solvent extraction. All of the crude oil is water degummed to make lecithin. This degummed oil, which is actually super-degummed oil, is transferred to the refinery. It is ultra-degummed, bleached, and physical refined. This refinery was started in 1995, and even today is still the most efficient and highest quality of any new refinery built in the last twenty years. The capacity is 500 million lb/year.

Ragasa (Raul Garcia), Monterrey, Mexico

Ragasa has operated extraction plants for cottonseed and soybeans for over 50 years. In 1999, the decision was made to build a new refinery for soybean oil in Guadalupe, Mexico. Due to excessive handling of the soybeans exported from the United States, cracked beans are prevalent. This causes a relatively high level of NHP. The degumming efficiency of the crude oil is about 80%. Thus, this process operates with the Modified Physical Refining™ process. That means an IKA ultra-shear mixer is used on both citric acid and on caustic.

The resultant RBD (UDBPR) oil is of excellent quality, true zero trans fat (not rounded off to zero), low color (equal to Wesson Oil), 17 hr AOM, and very good shelf life for packaged oil. Their premium line of salad oil was named Nutrioli™, and has received wide acceptance. In 2011, Nutrioli was exported to the United States, and it will soon be in all Walmart stores in the United States. The capacity of the Monterrey Plant is 300 million lb/year, with a possible capacity increase next year.

Whole Harvest Foods (formerly Carolina Soy Products Inc.), Warsaw, North Carolina

Whole Harvest Foods had operated an extruder/expeller plant for soybeans for several years, marketing the crude oil to others. In 2000, they saw their market for the crude oil disappearing, and on a crisis basis decided to install their own RBD refinery.

Working with Desmet, a physical refining plant was installed. This was the first plant ever installed to operate with physical refining and no back-up with partial neutralization. No caustic is used at all, making this one of the first physical refining plant for soybean oil. The process is quite simple: Extruder/expeller crude oil (no solvent extraction) > IKA ultra-shear mixer (with citric acid) > centrifuge > bleaching > physical refining. As can be seen, the plant has a small carbon footprint, low energy use, no fresh water usage, very little water discharge, and no soapstock produced, making it a shining example of "green oil processing" and a contributor to sustainability.

A significant synergy evolved from combining expeller (no solvent extraction) technology with physical refining. For reasons not yet totally understood, the RBD (UDBPR) oil has an improved fry life, at least twice that of conventionally-processed oils. Carolina Soy produces cooking oil for institutions, food service, and restaurants, but not for retail purposes. The oil is zero *trans* fat, non-GMO, and contains all of the naturally-occurring omega-3 fatty acids and natural antioxidants. Since very low color is not needed for nonretail oil, a minimum amount of silica and bleaching clay can be used.

This oil is marketed under the brand name Whole Harvest™. Canola oil was added to the line, processed in the same manner as the soybean oil. This oil can be labeled "all natural" in good conscience, due to the absence of solvents and/or harsh chemicals.

Conclusions

Physical refining of vegetable oils offers great opportunity for green vegetable oil processing, and is a contribution toward sustainability of oil mills and refineries, as supported by the following.

- The process has a much smaller carbon footprint, less equipment required, considerably less capital cost, and lower cost of automation.
- There is reduced energy and less usage of fossil fuels.
- Eliminating water-washing reduces fresh water use and wastewater discharge (with low BOD).
- Only organic acid, citric acid, is used. No caustic is used. This allows for "all natural" labeling.
- Due to mild processing, all naturally occurring omega-3 fatty acids are retained, and a higher level of natural antioxidants is retained.
- With no caustic neutralization, no soapstock is produced. This is the single most important feature of physical refining. Soapstock is a by-product with little value, and is quite costly to dispose of. Very few processors acidulate soapstock any more due to the environmental nightmare the process causes. Thus, soapstock may be transported long distances, requiring fuel and rolling stock, and the environmental problem is just transferred to another area.
- With physical refining, all the FFAs are removed in the physical refining deodorizer, and recovered in the fatty acid scrubber. With soybean and canola oils, this deodorizer distillate is sold for its tocopherol (vitamin E) content. With an FFA content of <0.30 (as with ultra-degummed soy and canola), the tocopherol content is not diluted. For high-FFA oils, say up to 9%, the tocopherol level is drastically reduced. This can be sold as vegetable fatty acids (which is still a value-added by-product). The quality of physical refined oil is equal to, or better than conventionally refined (neutralized) oils.

References

Bailey's Industrial Oil & Fat products, Sixth edition, Fereidoon Shahidi, Ed. 2005, John Wiley & Sons, Inc., Volume 5, Edible Oil and Fats Products: Processing.

Inform 4: 1273, 1993

Introduction to Fats and Oils Technology, Second Ed., 2000, AOCS Press, R. O'Brian, W. Farr, P. Wan, Eds.

Practical Short Course on Vegetable Oil Processing and Products of Vegetable Oil/Biodiesel, Texas A&M University, October 9–13, 2011, "Latest Technologies in Edible Oil Refining and Biodiesel," Walter Farr.

Revisit ANIAME, Anon XX, Vol. 11, Num. 55 Enero/Marzo 2007, pp. 47–54, "Los ultimas tecnologias en el procesamiento de aceites vegetales como materias primas para aceites comestibles y Biocombustibles," Walter Farr.

9

Conservation of Energy and Resources in Hydrogen Generation and in Hydrogenation

Nancy C. Easterbrook and Walter E. Farr
Air Products & Chemicals Inc., Allentown, Pennsylvania, USA and
The Farr Group of Companies, Memphis, Tennessee, USA

Introduction

Hydrogenation continues to be a standard unit operation in the production of fats and oils for food and industrial uses. Edible oils are long-chain fatty acids attached to a glycerol backbone. The carbon-to-carbon linkages in the fatty acid chain can be either single or double bonds. Stearic or saturated oils or fats have no double bonds and are naturally stable and solid at room temperature. In the 60s and 70s, saturated fats/oils such as tropical oils and butter were publicly linked to health risks. This caused a widespread shift to unsaturated edible oils, of which there are three types:

- monounsaturated oils with one double bond, called oleic oils (18:1)
- polyunsaturated oils with two double bonds, called linoleic oils (18:2)
- polyunsaturated oils with three double bonds, called linolenic oils (18:3)

Natural polyunsaturated oils are subject to auto-oxidation upon contact with air, which reduces quality and shelf life, and ultimately leads to rancidity. This problem was resolved through partial hydrogenation by selective conversion of C18:3 to C18:2, slight conversion of C18:2 to C18:1, and no conversion of C18:1 to C18:0 (although a small amount occurs). After winterization, a good cold test is achieved. The rate of oxidation of 18:3 is 15-fold greater than that of oleic acid (18:1). The relative oxidation rate of 18:2 to 18:1 is 10-fold. The oxidation rate of 18:1 is 10-fold greater than 18:0 (no oxidation for stearic acid). As the degree of saturation increases, the melting point also increases.

Partially hydrogenated oils were standard for several decades until the early 2000s when dietary health risks with the *trans* isomer similar to those of saturated fats became widely known. In 2005, labeling of *trans*-fat content became mandatory. Hence, partial hydrogenation has largely gone out of favor, causing an industry-wide reduction in the number of manufacturing sites that hydrogenate as well as the total volume of hydrogen consumed.

However, hydrogenation continues to be necessary for hardening or raising the melting point of an oil or fat. As long as consumers are not willing to give up confections and baked goods, and continue to buy solid margarines and shortening, hydrogenation remains the most economical way to produce these goods. Palm oil (zero *trans* fat) can be used to make solid fats (margarine, shortening, more stable cooking oils), but this can raise the saturated fat content higher than lightly hydrogenated soybean oil. Also, zero IV stock (zero *trans*) can be added to make a pourable liquid shortening, but this adds saturated fat, while adding very little to oxidative stability. A better pourable shortening would be lightly hydrogenated soybean oil (maybe 2 g *trans* fat per serving size), a reasonable level of saturated fat (less than adding zero IV stock or palm oil). This oil will have good oxidative stability and no winterization because the object was to have a cloudy pourable liquid shortening. In addition, natural oils that go into industrial applications such as castor oil or tall oil will continue to be hydrogenated.

Hydrogen Supply

The materials needed for hydrogenation are the oil feedstock, a catalyst, and hydrogen. Hydrogen can be generated on-site or purchased as needed in liquid or gaseous form.

On-Site Steam Methane Reforming

Historically, most manufacturers who hydrogenate edible oils have also opted to generate their own hydrogen using on-site steam methane reformers (SMRs). SMR technology emerged in the 1950s as the most cost-effective approach for generating hydrogen on a large scale. The large plant technology was mature by 1980 following improvements to the energy efficiency of the reformer. Today, SMR plants can be constructed for centralized hydrogen production with capacities of over 100 MM SCFD (100,000 Nm^3/h).

More recent developments have transpired for small SMR plants. In the early 2000s, there was a national incentive in the U.S. to develop an on-site hydrogen generator for refueling buses or cars powered with hydrogen fuel cells. After a decade of development and testing, the U.S. DOE-sponsored Las Vegas Hydrogen Refueling Station, led by Air Products, demonstrated the first reliable, long-term performance of an economical, unattended, small-scale on-site hydrogen generator. Although the fuel cell refueling industry may take a decade or more to develop, the small on-site generator has filled the market niche to supply the smaller volume demand required for edible oil processing, float glass production, or steel processing.

For any SMR, regardless of technology or vintage, the basic chemistry consists of three reactions:

1) $CH_4 + H_2O \rightarrow CO + 3H_2$

2) $CO + H_2O \rightarrow CO_2 + H_2$ *(water gas shift)*

3) $CH_4 + 2H_2O \rightarrow CO_2 + 4H_2$

Parameters that affect the performance of a reformer are the steam-to-carbon ratio, reaction temperature, reaction pressure, coke formation, catalysts, and catalyst support, all of which can alter over time or in the absence of maintenance. The process requirements for hydrogen generation exceed those of edible oil manufacture. For example, in a large SMR, the reaction pressure is 150–450 psig, and the reaction temperature is 800–950°C, compared with 60 psig and 205°C for edible oil hydrogenation. Thus, making the hydrogen is a significant addition to the edible oil plant's overall manufacturing profile and requirements.

Because the SMR process yields H_2 at about a 75% concentration, further hydrogen purification is needed. This can be done by absorption (scrubbing), pressure swing adsorption (PSA), membranes, or cryogenic condensation. Prior to 1980, the prevalent clean-up method consisted of wet scrubbing with an amine (monoethanolamine, or MEA, for example). After scrubbing, the final product was 95–98% H_2, but also contained CH_4, N_2, CO_2, and CO, and was saturated with water. This type of clean-up is still in use today in older hydrogen plants. However, in units built since the early 1980s, scrubbers have been replaced by PSA units, which use packed beds to adsorb the impurities. When one bed is adsorbing, the other is regenerating, and the beds switch (swing) back and forth, hence the name. PSA units typically have 4–16 beds. PSAs can yield a very pure hydrogen product (99.999%) and have proven to be very reliable.

Delivered Liquid Hydrogen

Cryogenic distillation produces a very pure (>99.995%) hydrogen product from an SMR. This method is used by merchant hydrogen suppliers who generate the hydrogen at a central location and transport the pure hydrogen in liquid form to the use location (see Fig. 9.1). The liquid is stored on the user's premises in a cryogenic tank and passed through vaporizers when consumed. Typically, the cryogenic tank and vaporizers are owned and maintained by the hydrogen supplier. The tank level is monitored remotely, and a tanker is dispatched according to level and usage patterns. Cryogenic liquid storage vessels are designed to minimize the transfer of heat into the tank; however, some heat leak is inevitable. Hydrogen gas boil-off will be retained in the tank increasing its pressure, used by the customer through the economizer circuit to the houseline, or vented to the atmosphere. Therefore, although the hydrogen supply from a liquid tank is extremely flexible, long-term storage without consumption will result in losses.

Fig. 9.1. Hydrogen tanker filling up at a central location.

Opportunities for Conservation within the Hydrogen Supply System

The majority of end-user owned SMR plants in the fats and oils industry were installed during the heyday of partially hydrogenated oils, some 30–40 years ago. From the standpoint of environmental impact, they have several drawbacks. Today, these plants are often considerably oversized for the current hydrogen requirements and, due to their age, have become maintenance intensive and can produce poor quality hydrogen. Opportunities to improve sustainability and economics within the hydrogen supply system include correcting any mismatch of hydrogen supply versus demand and improving hydrogen purity, both of which can be accomplished by attention to supply mode.

Supply/Demand Mismatch

An old SMR cannot be easily turned on and off. Cold start-up can easily take more than 12–24 hours. Thus, the plant is left to run at its lowest turndown production volume, whether the hydrogen is being consumed or not. In some cases, the hydrogen is compressed and sent to surge tanks to be held until needed. In other cases it is vented. This practice wastes energy, fuel, hydrogen, and plant real estate for the surge tanks.

Another consideration is the safety of retaining large volumes of gaseous hydrogen, a highly flammable gas, in surge tanks on-site.

Low-Grade Hydrogen

Another drawback of older reformers is impure hydrogen. As we have seen, while MEA clean-up removes bulk CO_2, it leaves impurities such as CO and CH_4 in the hydrogen stream, which can poison the hydrogenation catalyst. Even reformers with PSA clean-up can yield impure hydrogen if the PSA is not well maintained and operating properly. Lower purity hydrogen reduces the overall reaction rate, reduces the hydrogenation catalyst activity and lifetime, and causes the build-up of inerts and hydrogen in the headspace of the reactor, necessitating venting. In addition, hydrogen purity studies have also shown that the highest purity hydrogen produces the lowest amount of *trans* isomer at all degrees of conversion. These results of impure hydrogen all increase the cost of edible oil production.

Time for a New Hydrogen Supply Mode

Those who have older hydrogen plants have an opportunity to improve their sustainability in terms of economics, environment, energy, and operations.

Although alternative technologies exist, the SMR is still the most serviceable and economical hydrogen generation method for the edible oil industry. Owing to the demand for hydrogen for fuel cell cars and buses, compact SMR units are now available in small sizes that match the lower hydrogen demand. The units are modular and standardized. They are shop fabricated and containerized, minimizing field work to install. There are several units on the market that utilize different reformer technologies such as autothermal (ATR), partial oxidation (POX), catalytic partial oxidation (CPOX), and SMR. For example, the Air Products PRISM® hydrogen generator (see Fig. 9.2) uses process intensification to miniaturize the SMR and associated equipment. This includes heat integration, a proprietary burner and related advanced controls, an advanced shift catalyst, and specially made adsorbents within the PSA system. The reformer tubes are comparatively short and contained in a single integral tube bundle. The radiant section and convection section are integrated to create a compact and thermally efficient reformer compared to a large SMR. The PSA section is also smaller because it utilizes a rotary valve PSA which has a rapid cycle time. These small plants are highly efficient and may be load following to 50% turndown. Cold start-up time is also greatly shortened—from a day down to less than 1–3 hours—because the small plant uses an advanced control system as well as novel heat integration. All this means unattended operation, a smaller plant footprint, and less venting since the system can automatically adjust quickly to match the edible-oil plant demand. It reduces the need for hydrogen surge capacity and the space and risk that it entails. In addition, the compact reformer uses 30% less natural gas than its predecessors.

While hydrogen generated from natural gas cannot be considered truly green, the new SMRs are indeed an improvement over the older units of the 1980s. In fact, the efficiencies achieved by today's SMR plants have reduced CO_2 emissions by >30% when compared with emissions from older plants. At this writing, the impending availability of shale gas portends a long period of relatively low natural gas pricing, which will certainly improve the economics of the process. In addition, the PRISM on-site unit is capable of operating on biogas with minor modifications to the plant.

Fig. 9.2. Smaller hydrogen generators such as Air Products' PRISM® system use SMR technology.

For instance, if sufficient anaerobic digester gas were available, it could be used as a green feedstock to produce carbon-neutral hydrogen.

Another supply option that can make sense from a green perspective is to purchase liquid hydrogen, especially if the hydrogen is liquefied in a centralized plant fed by "green power." Although the industry history is to self-generate hydrogen, lower overall hydrogen demand, variable use patterns, and the desire to simplify operations can tilt the make-versus-buy decision more toward the buy side. Liquid hydrogen is made via large centralized SMRs and liquefied cryogenically. Supplied as needed, 100% flexible, and extremely pure, liquid hydrogen is a viable option for those whose plants are in localities where it is available.

Other supply modes do exist, but are not optimal. Commercial alternatives include partial oxidation (POX) systems and electrolysis. A Goldilocks situation, this is a case of too big (POX) and too small (electrolysis). POX systems are found in large refinery and petrochemical applications. Electrolysis is a niche, but its inefficiency, energy use, and expense make it unsuitable for use in edible oils. Electrolyzers are now predominantly suited for locations where reliable natural gas supply is unavailable. Truly green new technologies for generation of hydrogen include evolution from algae or specialized enzymatic processes; however, these methods are not commercially proven at this writing.

Hydrogenation

A fat or oil molecule consists of three long-chain fatty acids attached to a glycerol stem (see an example in Fig. 9.3).

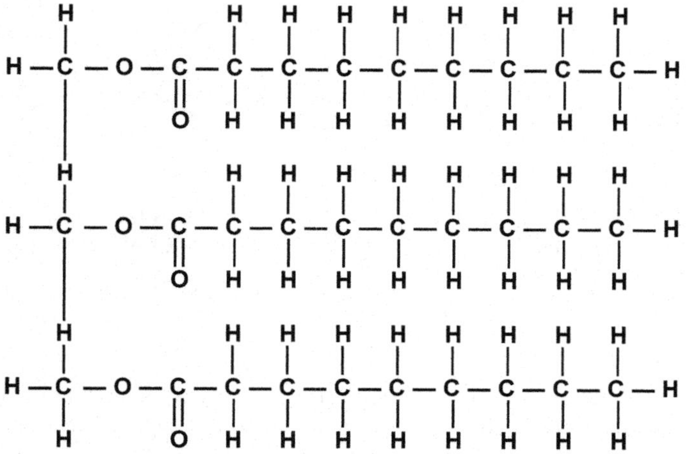

Fig. 9.3. Example of a fully saturated triglyceride molecule (Easterbrook/Farr).

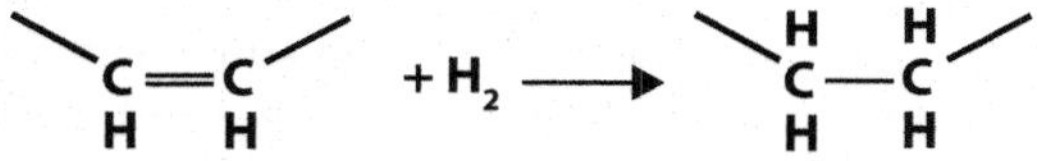

Fig. 9.4. Hydrogenation of a double bond.

Hydrogenation of unsaturated oil reduces the number of double bonds in the fatty acid chains (see Fig. 9.4) and also changes the configuration of the remaining double bonds, converting a portion of the *cis* configuration double bonds to the *trans* configuration geometric isomers (see Fig. 9.5). Additionally, the double bonds may be relocated along the carbon backbone of the fatty acid to create positional isomers. Pressure, temperature, and catalyst structure all play a role in the final isomer structure of the fatty-acid chains and the corresponding properties of the product. Hydrogenation also raises the melting point and increases the oxidative stability. Although *trans* labeling requirements have greatly reduced demand for partially hydrogenated oils, hydrogenation is still required for such products as baking and frying fats and margarines.

It is an industry standard to use iodine value (IV) to identify the degree of initial and desired final saturation. IV represents the amount of iodine absorbed by free bonds on the saturated molecule. One pound of oil requires 0.015 ft^3 of hydrogen to reduce IV by one point. How much total hydrogen is required will depend on how many double bonds the starting material contains and how many double bonds are desired to remain at the end.

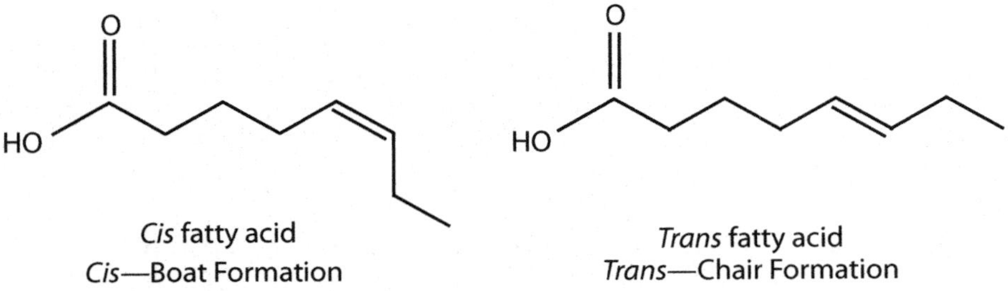

Fig. 9.5. Geometric *cis* and *trans* isomers (showing the carbon backbone structure).

Current Practice

The feed to the hydrogenation reactor (converter) usually consists of refined, bleached oil. This oil must be preheated and combined with a heterogeneous catalyst—usually nickel—and hydrogen. As the oil and hydrogen molecules contact the surface of the catalyst particle, the double bond is broken. Then the saturated molecule must move away, making room for the next combination of unsaturated oil and hydrogen to contact the catalyst surface. The normal progress of a batch or semi-batch hydrogenation reaction can be monitored by the hydrogen uptake. For full saturation, it often assumes the shape of a bell curve, increasing as the temperature increases and decreasing as the concentration of double bonds decreases. At first, hydrogen uptake is fast, but tails off with lowering IV, particularly at IV of around 85. With this exothermic reaction and rapid IV drop, a tremendous amount of heat is developed. The surface area of the cooling coil and volume of cooling water must be such that the temperature can be controlled at set point.

Hydrogenation is an exothermic reaction, which means that after the reaction initiates, it generates its own heat. This heat is removed from a batch converter via cooling coils. The endpoint of the batch is determined by the IV test and the hydrogen is shut off when the desired IV is obtained. Then the batch is cooled prior to filtration. The filtered oil is then sent for the final step, deodorization, before going to storage or transport under nitrogen blanket.

Hydrogenation enables the manufacturer to control end-product qualities, such as solid fat index (SFI), mouth feel, IV, melting point, and shelf life. The key variables are starting oil quality and consistency; catalyst activity, selectivity, and dosing; reaction temperature; hydrogen purity; hydrogen uptake rate; and mixing/mass transfer/agitation.

The edible oil hydrogenation process has undergone comparatively little change over the years, and yet there are many opportunities for green practices to be adopted. The places where edible oil hydrogenation may be made more sustainable within the framework of typical existing operations are as follows:

- Increasing batch turnover
- Reducing batch time
- Increasing catalyst activity

- Increasing catalyst life
- Reducing hydrogen loss
- Capturing and re-using waste heat.

Opportunities for Conservation Within the Hydrogenation Plant

Improved Hydrogen Quality

Hydrogen that carries impurities will function, so some plants continue to live with it. However, the impurities are poisons that will reduce catalyst activity and its selectivity (see Table 9-A). This means the reaction will take longer, and the preferential conversion of the highly unsaturated bonds will be impeded. Economically, the plant is not getting the full value of its catalyst. In addition, poisoned catalyst will allow unreacted hydrogen to reach the headspace along with the impurities. This results in venting and hydrogen loss. The use of pure hydrogen (99.999%) via modern on-site generation or hauled in liquid or bulk hydrogen is the greenest alternative

Converter Enhancements

The continuous stirred tank reactor (CSTR) (see Fig. 9.6) continues to be the workhorse of the edible oil industry, although other effective alternatives do exist and have caught on in other industries, such as basic and specialty chemicals and pharmaceuticals. Continuous processing has several advantages in terms of smaller equipment, better consistency, and higher hydrogenation rates. The use of supercritical fluids, such as CO_2 in continuous oil hydrogenation, has also been demonstrated. Adopting alternative converter systems would require process development to achieve the desired results with the new process. However, opportunities for green processing within the existing converter system include mixing, mass transfer, capture of heat of reaction, avoidance of venting (dead-end hydrogenation), and single purposing of the reactor.

As shown in Fig. 9.6, the CSTR is basically a vessel with heating and cooling coils for heat transfer, baffles for mass transfer/mixing, a triple-bladed agitator, and a sparge ring for introduction of hydrogen. The CSTR should be run to achieve as close to stoichiometric hydrogen consumption as possible. As we have seen, the hydrogen molecule must simultaneously encounter a catalyst particle and a double bond in order to react. If it does not, it will rise to the headspace of the reactor where it will need to be re-introduced to the batch or vented. The entire batch must be well-mixed with no dead zones. Dead zones allow side reactions to occur, introducing undesirable components. Achieving good mixing is dependent upon the agitator type, the placement and type of blades, and the gas distribution. Hydrogen is introduced through

Table 9-A. Impurities Remaining After MEA Clean-up Can Reduce Catalyst Activity and Selectivity.

Reformer with MEA Clean-up	H_2	O_2	N_2	CH_4	CO	CO_2	H_2O
Product Composition (%)	97.92	1.8	1,040	18,200	1,260	236	38

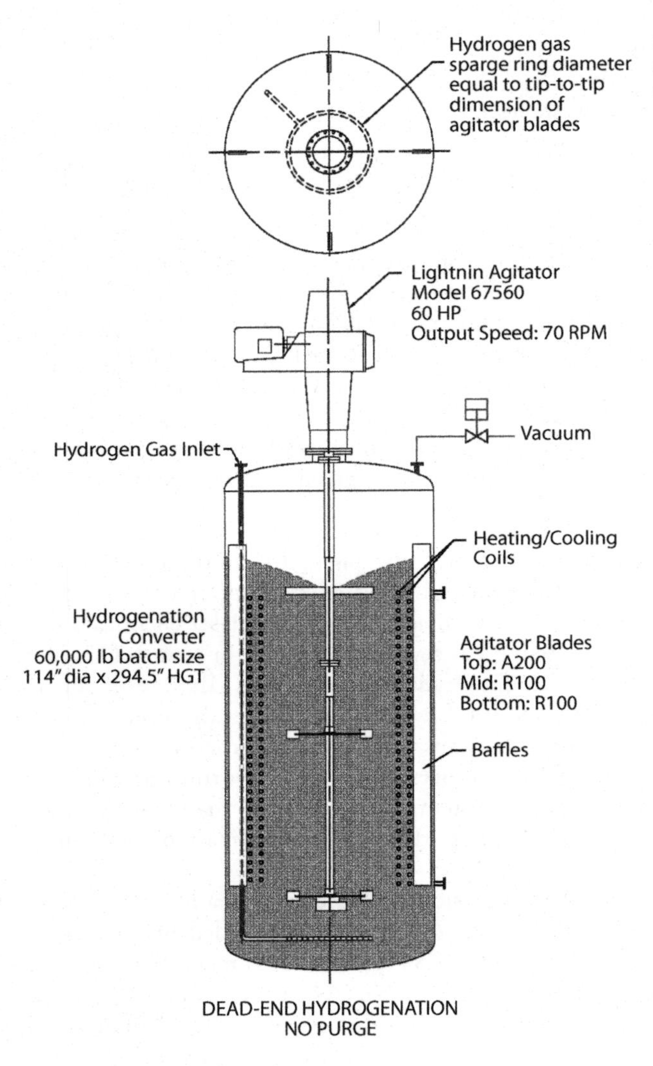

Fig. 9.6. Converter schematic. (Schematic courtesy Walter E. Farr and Owensboro Grain.)

a sparge ring at the bottom of the reactor vessel. Important parameters include the position of the sparge ring relative to the bottom of the vessel and relative to the bottom impeller, and the diameter of the hydrogen distributor relative to the diameter of the impeller blade. Correctly setting these parameters enables the hydrogen to react before reaching the headspace. The agitator is usually a standard three impeller design. The top impeller is downward pumping (axial flow) to reintroduce the headspace gas into the batch. The bottom and middle impellers are usually radial flow. A variation, the draft inducing impeller, was installed in a few plants, but has gone out of favor, after the specially designed Lightnin Agitator proved to be just as effective as the draft tube, with fewer maintenance problems. Hydrogenating with absolutely no purging or venting of headspace gases is known in the industry as dead-end hydrogenation. Besides the savings of not venting hydrogen gas, the practice also allows for improved process control by measuring hydrogen input to the reaction because there is no error associated with purge gas. This then prevents any overshoot of end point, and reduces rework and blending to meet the actual specification of the base stock. Until the mid-1970s, all hydrogenation systems had purging or venting on the converters. Typically, there was a small line with a fixed-valve setting, and the purge continued at this fixed setting throughout the reaction. Before the early 1960s, this purging was necessary due to several factors:

- Soap and phosphorus impurities existed in the feedstock.
- Highly active nickel catalysts had not yet been developed.
- Most on-site hydrogen plants were not producing high-purity hydrogen.
- An agitator had not yet been developed that would facilitate dead-end hydrogenation.

For dead-end hydrogenation to work properly and with high efficiency, all four of the above problems had to be corrected. From 1968 to 1972, Walter E. Farr, working at a major processor in the Midwest, had addressed each of the problems in a systematic way. First, quality systems were developed to measure completeness of refining, that is, to confirm phosphorus and other trace metals (including soap) were very low. The previous few years were spent determining and developing the best catalysts. At first, this plant had the old style SMR with MEA purification: its best purity possible was 98.5%. A few years later, a new SMR with PSA purification was installed, producing 99.95% H_2 purity. With this, no purge at all was needed. Plants that can use liquid hydrogen only will have the advantage of ultra-pure hydrogen, 99.995%.

In the early 1970s, Farr worked with Dr. Oldshue of Lightnin Mixing Co. to develop a new agitator design that included three axial flow turbines, all pumping downward. The top blade was located near the surface of the oil, creating a vortex drawing headspace hydrogen back into the oil. This was the major contributor in making dead-end hydrogenation with zero purge possible. This design has been incorporated in almost every hydrogenation system built since that time.

Other process improvements have also been put forth by Farr. For example, in some existing plants, the oil is fed to the converter cold and heated in place prior to starting the reaction. After the reaction, the batch is then cooled in the converter. This greatly reduces the number of batches per day that can be processed. It is more efficient to run sequential batches through the converter with heating and cooling taking place in a separate vessel. This also allows the hydrogen plant (SMR) to run continuously as it is designed to do.

Heat Recovery

Hydrogenation is an exothermic reaction; it generates heat which is often wasted, but can be recovered and reused in other places within the process. Heat is required to raise the temperature of the oil prior to the addition of hydrogen. Currently, many plants use steam to heat up the cold oil and cooling water to take away the heat of reaction. Instead, an oil-to-oil heat exchanger can be added to allow the hot hydrogenated oil to raise the temperature of the cold oil for the next batch. The cooling water in the hydrogenation reactor can be circulated through a flash tank in which steam is generated for plant use. The increasing cost of energy and water makes these practices more viable today than ever.

Hydrogen Safety

In this chapter, much has been said about the advisability of consuming all the hydrogen that is generated. Besides green motives, safety is another reason why this is important. Hydrogen has a wide flammability range of 4–74% in air. It takes very little energy to achieve ignition. Sparks from electrical equipment or static electricity, open flames, or extremely hot objects must be carefully eliminated. Hydrogen is colorless, odorless, and burns almost invisibly. ASME and ANSI codes should be followed for vessels and piping according to the temperatures and pressures in use.

It is essential for plant personnel to maintain a high degree of caution when working around hydrogen. Purge hydrogen piping with nitrogen prior to maintenance or start-up. Open and close valves slowly. Make sure regulators and relief valves vent outside the building and well away from possible sources of ignition. As hydrogen is not detectable by human senses, it is advisable to have stationary monitors and a handheld monitor available for leak checks.

Conclusions

As we have seen, hydrogen is a reactant that cannot be generated without sophisticated process technology involving the consumption of natural gas, water, and energy, as well as high process temperatures and pressures. While hydrogen's value in use justifies its generation, it would be unsustainable, as well as uneconomical, to expend the resources of natural gas, water, and energy only to waste the product. This is what happens when hydrogen is vented into the atmosphere either through hydrogen supply and demand mismatches, the presence of impurities, or through poor hydrogenation practice. Technology has advanced such that compact and efficient small SMR-based H_2 generators are now on the market. The resulting economics can often allow edible oil producers to shut down older, over-sized hydrogen plants and install a smaller, more efficient hydrogen generator. The make-versus-buy decision should also be re-evaluated since lower volume and higher variability in hydrogen demand can make the purchase of liquid hydrogen economically attractive. The hydrogenation plant also affords opportunities to conserve hydrogen and energy through such measures as heat recovery and improved hydrogen uptake.

Bibliography

Air Products Safetygram #4, Gaseous Hydrogen. Air Products and Chemicals, Inc.: Allentown, Pennsylvania, USA, 2010.

Air Products & Chemicals, Allentown, 1998 "Impact of Mixing and Hydrogen Purity in Batch Oil Hydrogenation Processes." Kerri Gaumer Freidl, Reinaldo M. Machado and Michelle L. Achenbach.

Bailey's Industrial Oil & Fat Products, Sixth Edition, Fereidoon Shahidi, Ed., 2005, John Wiley & Sons, Inc., Volume 5, Edible Oil and Fat Products: Processing: Hydrogenation: Processing Technologies, Walter E. Farr.

Chemical Economics Handbook. SRI Consulting, 2010.

Farr, W.E. Hydrogenation. Proceedings of the World Conference on Oilseed Processing Utilization; Wilson, R.F., Ed., AOCS Press: Urbana, Illinois, 2001.

http://faculty.clintoncc.suny.edu/faculty/michael.gregory/files/bio%20101/bio%20101%20lectures/biochemistry/biochemi.htm (Fig. 9.4, Fat Molecule)

http://www.natural-health-information-centre.com/trans-fats.html (Fig 10.6 *Cis* & *Trans* Configuration).

Naqvi, S.N. SRI PEP Review 2007-3: Hydrogen: A Technical Review, March 2007.

10

Dry Condensing Vacuum Systems for Deodorizers for Substantial Energy Savings

Dr. Ir. M. Kellens
Desmet Ballestra , Minervastraat 1, 1930 Zaventem, Belgium

Introduction

The final quality of refined edible oils and fats is determined to a large extent by the final processing step of deodorization. During this step, fatty acids and other volatile components are removed from the oil by establishing an intense contact between the oil and a stripping gas, usually steam, at an elevated temperature and low absolute pressure. Due to the thermal treatment, some minor components may be altered or destroyed, like oxidation products or arytenoids, resulting in a heat-bleaching effect. Other thermochemical reactions are unwanted, like transisomerization and polymerization. It is thus essential to choose and control the proper operating conditions to obtain an oil with the right organoleptic, nutritional, and functional quality.

The pressure is an important parameter as it directly affects the stripping efficiency of a deodorizer. In the last decade, a lot of attention has been given to the reduction of the deodorizing pressure to minimize the unwanted side reactions by reducing deodorizing time and/or lowering deodorizing temperature, but above all to reduce overall process unit operating cost.

Types of Vacuum Production Systems Used in Oil Deodorizing

Deodorizers operate at low pressure to remove unwanted components while at the same time preserving the nutritional value and improving the organoleptic quality of the oil. The required pressure in a deodorizer, usually between 2 and 4 mbar, is commonly generated by vacuum systems consisting of a combination of steam-jet ejectors (boosters), vapor condensers which are cooled using cooling tower water and mechanical (liquid ring) vacuum pumps (Fig. 10.1). Liquid ring pumps are used in the final stage of the vacuum system to remove the noncondensable gases. Due to high vapor volumes to be

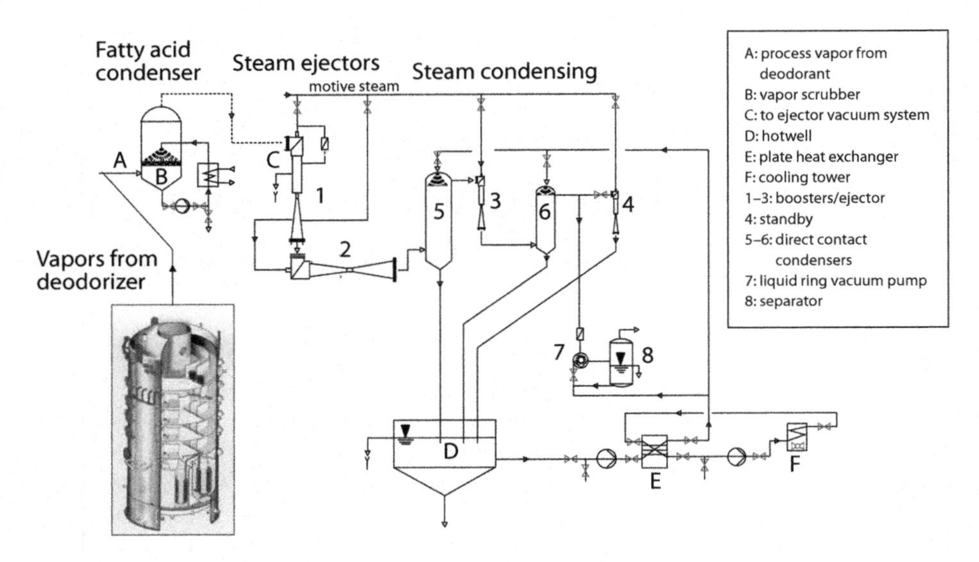

Fig. 10.1. Multistage ejector vacuum system with liquid ring vacuum pump (Desmet-Körting).

removed, motive steam consumption in such steam-jet ejectors is quite high and may account for up to 75% of the steam consumed in a deodorizer.

One way to reduce the motive steam consumption in a steam-jet ejector system with barometric condensers is to lower the temperature of the barometric condense water (Table 10-A). This is done by means of a chiller. The benefit of the lower motive steam consumption, however, must be weighed against the extra required chilling capacity and consequently electrical energy to cool the barometric condenser water. Another benefit from using a lower barometric condenser water temperature is a better condensation of volatile odoriferous material, which in turn reduces the odor emission problem. Together with the condensed steam and highly volatile material, a small amount of fatty matter is usually found back in the condenser water, ±1% of the stripping steam. This fatty matter may decant partially and separate from the

Table 10-A. Effect of Barometric Condenser Water Temperature on Motive Steam Consumption.

Pressure		kg motive steam per kg stripping steam	
Booster	**Deodorizer**	**30°C (1)**	**10°C (2)**
2.5	3 mbar	4.5	1.6
1.5	2 mbar	6.2	2.5

Note: (1) Barometric condenser water inlet temperature: 24°C; outlet temperature: 30°C . (2) Barometric condenser water inlet temperature: 5°C; outlet temperature: 10°C.

water. The waste water is usually sent to a water-effluent treatment plant where it is mixed with other effluent streams from the refinery.

In both cases, all noncondensed vapor from the fatty acids scrubber, mainly stripping steam, is going through the steam ejectors before being condensed in the barometric condenser. The total effluent produced is the sum of the stripping and of the motive steam and the noncondensed fatty matter.

In the search to even further lower effluent and operating cost, a new vacuum production principle was introduced in edible oil deodorizing in the 1990s: so-called dry condensing.

The triple point of water is at 0.01°C and 6.11 mbar. At that temperature and pressure, it is possible for ice, water, and steam to coexist. If the pressure and temperature are below that triple point, steam can go directly into the ice phase without passing through the liquid state.

By lowering the temperature of the steam condenser to below the freezing point of the steam at low pressure, it is possible to condense the main part of the stripping steam before it reaches the steam ejectors. This has a direct impact on the amount of noncondensed vapor to be removed by the ejectors and hence on the amount of motive steam needed for that, so a much lower overall deodorizer effluent (Fig. 10.2). In the example presented, the total amount of steam (sparge and motive steam) used in a dry condensing system is only a fraction of the steam used in a conventional barometric condensing system:

Another and maybe more important result is that this dry condensing (DC) principle allows operating deodorizers at much lower pressures, under more cost-efficient conditions (down to 1.0–2.0 mbar in the deodorizer vessel).

Working Principle of Dry Condensing (DC)

The dry condensing system is becoming more and more the standard in new oil deodorizing plants. In this system, the sparge steam is condensed on surface condensers working alternately at very low temperature (around –30°C, depending upon the required suction pressure). The remaining noncondensables are removed either by mechanical pumps or roots blowers in series with a liquid ring pump or by a vacuum steam ejector system. The dry condensing system reduces the motive steam consumption but requires extra electrical energy.

Due to the relatively high capital cost, the return of investment (ROI) for a dry condensing system may take several years and depends largely upon the ratio between the cost of steam and electricity. In Europe, with higher fuel costs, the production cost of steam is higher, which reduces the ROI of a dry condensing system as compared to a classical vacuum system. As an additional benefit, much lower quantities of wastewater are produced by dry condensing which significantly reduces the cost of effluent treatment and in turn the ROI.

The pressure in the deodorizer itself is always slightly higher (0.5–1 mbar) than at the suction side of the vacuum unit, due to pressure losses caused by the oil

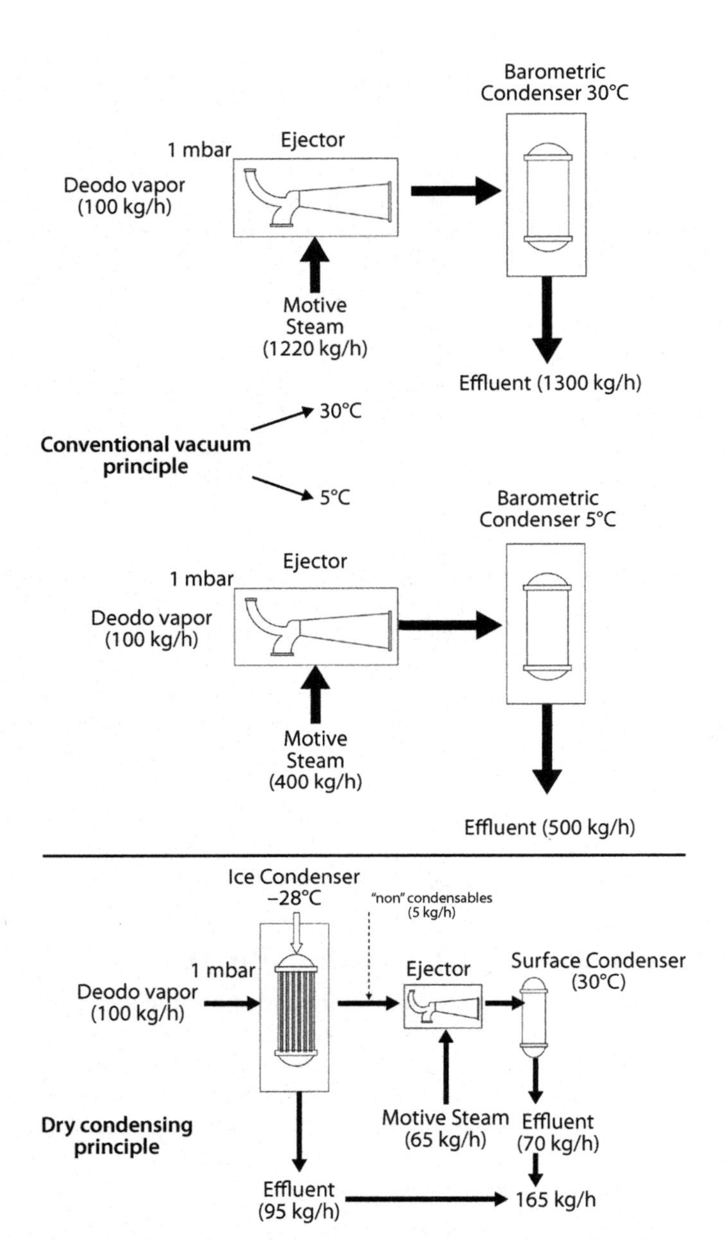

Total steam consumption (basis 100 kg/h sparge steam, 1 mbar suction):

Steam condensing system	motive steam	total steam
Cooling tower water barometric condenser	1200 kg/h	1300 kg/h
Chilled water barometric condenser	400 kg/h	500 kg/h
Ammonia dry condenser	65 kg/h	165 kg/h

Fig. 10.2. Schematic presentation of the main vacuum production principles used in oil deodorizing.

Table 10-B. Effect of Suction Pressure on Condensation Point of Stripping Steam.

Pressure at condenser side	Condensation point of steam
0.5 mbar	−27.3°C
1 mbar	−20.3°C
2 mbar	−12.9°C
3 mbar	−8.4°C

demisters, the fatty-matter scrubbers and other equipment. Consequently, to reach an effective deodorizing pressure of 2 mbar, a pressure of around 1 mbar at the suction side is required.

To obtain efficient steam condensation at this low pressure, special stripping steam condensers operating at extremely low temperatures are required.

In Table 10-B, the effect of the suction pressure on the steam condensation point is given. One could conclude that a condensation temperature of −20°C should be sufficient to achieve a 1-mbar suction pressure, but in reality a temperature closer to −30°C is required. Reason is the presence of other non-condensable gases (air) which lowers the condensation temperature of the steam. It is the ratio of steam–to-air at the condenser outlet which determines the overall condenser temperature and not the ratio of steam-to-air at the inlet (see Fig. 10.7 later in this chapter).

Available Dry Condensing (DC) Systems

The in-the-market available DC systems either consist of condensers with horizontally-oriented tubes ("hairpins") or vertically-oriented straight tubes.

Horizontal DC Systems

Dry condensing equipment with horizontally-oriented freeze condensers is, for example, supplied by GEA-Atlas and Körting. A schematic diagram and lay-out of the equipment is shown in Fig. 10.3.

The freeze condensers consist of hairpin tubes mounted in a shell. The tubes are internally cooled by evaporating a refrigerant (for example, ammonia). This refrigerant is supplied from a vertical duct from the separator to the bottom of the bundle. The refrigerant partially evaporates in the tube bundle and flows back to the separator. The driving force for the refrigerant circulation is gravity (the mass density difference between the downward liquid refrigerant flow and upward two-phase flow). Such cooling systems are usually identified as gravity systems.

During the freezing process, an ice layer of frozen sparge steam and some entrained fatty components accumulate on the tubes in the freeze condenser. As the layer reaches a certain maximum thickness, the freezing process is stopped by closing the valve in the entrance duct and opening the other in the entrance duct of the second, clean freeze condenser.

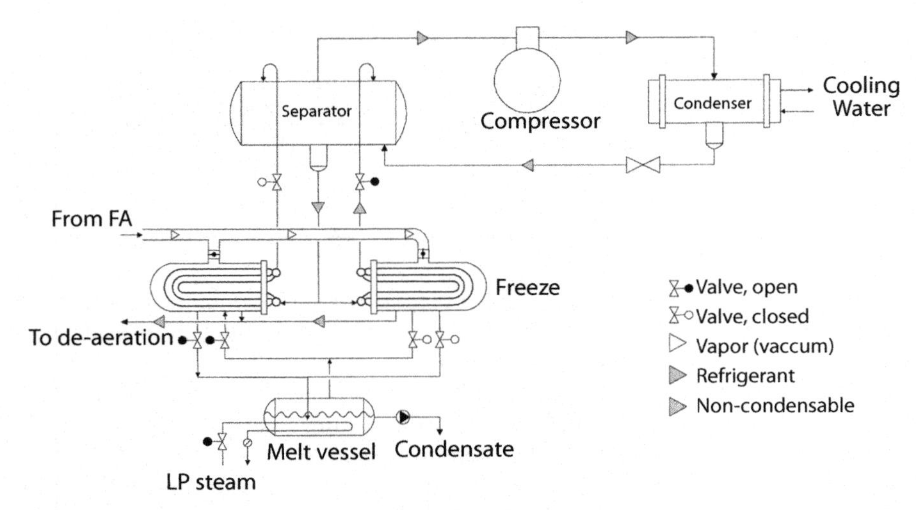

Fig. 10.3. Schematic diagram of a DC system with horizontally-oriented freeze-condensers (For example, GEA-Atlas and Korting).

The ice layer on the tubes of the first freeze condenser is now removed as follows. First, the valve in the refrigerant return duct is closed. Secondly, the shell of the freeze condenser is connected to a so-called melt vessel with relatively warm water by opening the two valves in the connecting ducts between this vessel and the freeze condenser. The warm water now partially evaporates (flashing) and condenses on the cold ice layer, which now starts to melt. The melted ice and fatty components are drained by gravity into the melt vessel. After some time, the tubes are de-iced and clean again.

The two valves to the melt vessel are closed, and the valve in the refrigerant duct is opened again. The refrigerant "falls" into the warm tubes and cools the bundle to the normal operating temperature again (from about +60°C to –30°C). Subsequently, the freeze-condenser remains on stand-by until the operating condenser should be de-iced again.

Advantages of the "horizontal" DC system are the relatively simple and compact construction of the freeze condensers. On the other hand, the mass of refrigerant in the gravity system is typically high. Other points to be noted are as follows:

- High peak loads on the refrigeration plant during cooling of a freeze condenser after de-icing and cleaning.
- Static pressure of the refrigerant in the gravity system, along with two-phase flow pressure drop during refrigerant evaporation in the tubes reduces efficiency of the refrigeration plant.
- Complete removal of fatty deposits from the horizontal tubes during cleaning might be cumbersome.

Vertical DC Systems

DC equipment with vertically-oriented freeze condensers is presently supplied by the Graham Corporation. A schematic diagram of this system is shown in Fig. 10.4.

The working principle of the vertical system is essentially the same as for horizontal systems described previously. The design of the freeze condenser is further optimized by a variable tube pitch, in order to minimize shell-side pressure losses in the condenser. The tubes, however, are mounted with both sides in tube sheets. In case of temperature differences between individual tubes, this leads to thermal stresses on the tubes. The tubes, therefore, must be able to withstand these forces. On the other hand, the condenser has to be designed for high-refrigerant pressures (occurring, for example, during stand-still), which necessitate a stiff construction of the tube sheets (high thickness) in order to prevent unallowable buckling forces on the central tubes.

An advantage of this vertical system, as compared to horizontal systems, is the more efficient removal of fatty matters from the tubes, which drain freely by gravity. A disadvantage of the vertical orientation is again the static pressure of the refrigerant column in the vertical tubes, which causes evaporating temperatures being higher at the bottom (for ammonia order 3K) than at the top. In order to guarantee sufficient condensation of stripping steam over the entire height of the tubes, this

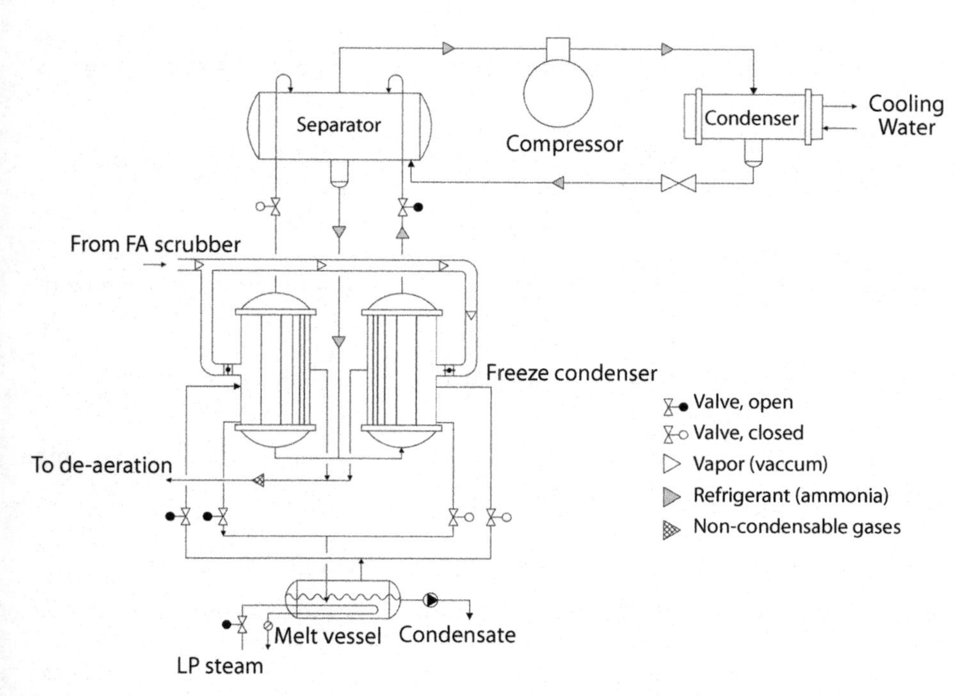

Fig. 10.4. Schematic diagram of a DC system with vertically oriented freeze-condensers (For example, Graham).

evaporating temperature-increase needs to be compensated for by a similar reduction of the refrigerant temperature in the separator, reducing the energy efficiency of the refrigeration plant.

To overcome this problem, the system can be foreseen with refrigerant level control valves at the bottom of the freeze condensers. These control valves increase the boiling level of the refrigerant in the tubes step by step, each time higher pressures or exit temperatures in the non-condensable outlet duct are measured. When the refrigerant level has reached a certain threshold, the freeze condenser is switched to the defrost mode.

To deal with certain disadvantages inherent to the classical horizontal and vertical DC systems, another vertical DC system was developed by Desmet Ballestra (SUBLIMAX). Typical characteristics of this DC system are the stress-free movement of each vertical tube during freezing and heating, because of the absence of an upper tube sheet and the downward flow of the ammonia refrigerant in a falling film along the inner wall of the tubes, creating a high heat transfer (see Fig. 10.5). Sparge steam entering the top of the sublimator flows downward through a series of internal baffling plates to ensure maximum contact between the vapor and cooling tubes. The ice layer on the tube walls is formed from top to bottom.

As a result, the following new features are incorporated in the design:

- Refrigerant evaporation from thin films ("falling films") inside the tubes, with high heat transfer coefficients and constant evaporating temperatures along the entire tube length.
- Efficient cleaning of the outer surface of the tubes is guaranteed.
- Self-draining equipment for refrigerant (inside tubes) and water (outside).
- Large tube pitches and tube diameter with long freezing times.
- No upper tube sheet: no expansion forces on the tubes.
- Location of refrigeration plant can be "freely" chosen.
- Peaks in the cooling capacity are minimized by controlled cooling of the equipment from high temperatures (for cleaning) to low operating temperatures.
- Low energy consumption due to high evaporation temperatures.
- Low system refrigerant mass.

Optionally, the vertical DC system can be foreseen further with energy-saving measures such as hot gas defrosting (by waste heat of the refrigerant compressor).

Condensation Process Description in a DC System

Water vapor from the deodorizer respectively condenses and freezes on the cold tube surfaces in the freeze condenser. The condensation process starts as soon as this vapor/air mixture is cooled below its condensation temperature. In case of a vertical ice condenser, the ice layer formation starts at the top and continues to grow further along the tube downward. Once it reaches the bottom of the tubes and starts to affect

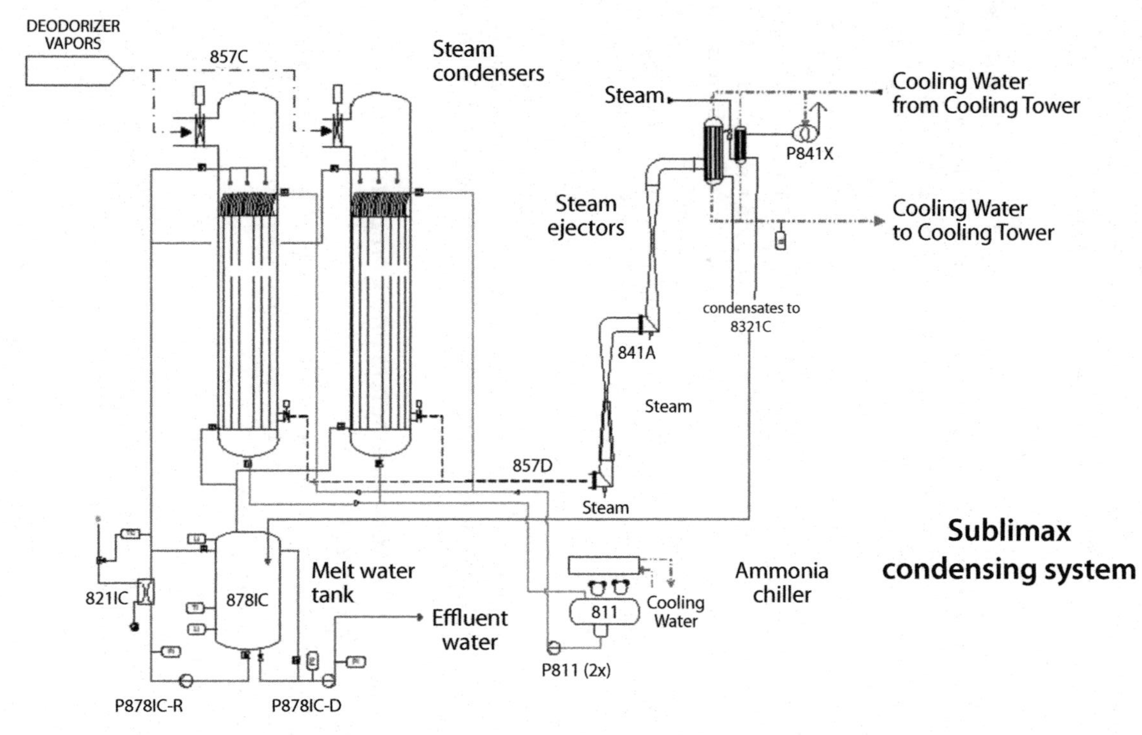

Fig. 10.5. Schematic diagram of the SUBLIMAX falling film DC system (Desmet Ballestra).

negatively the heat exchange, the condenser needs to be put in de-icing mode to melt the ice layer and clean the tubes. This explains why at least two dry condensers are needed to ensure continuous and consistent low pressure in a deodorizer operation.

Condensation temperatures depend on the vapor/air mass ratio, as well as on the prevailing pressure in the freeze condenser, as shown in Fig. 10.6 for 0.5 mbar, 1 mbar, and 1.5 mbar absolute pressure, respectively.

Fig. 10.6 should be read as follows.

If the desired pressure in the deodorizer is 1.5 mbar and the flow resistance of the ductwork and valves from deodorizer to freeze condenser amounts up to 0.5 mbar, then the required pressure in the freeze condenser is 1 mbar.

If, for example the amount of sparge steam is taken to be 150 kg/h and the air leakage rate into the system 5 kg/h, then the mass ratio of vapor to air is 30.

Fig.10.7 shows that the condensation temperature at the inlet of the freeze condenser is for this example –20.5 °C.

If the de-aeration system is selected for a capacity of 10 kg/h (5 kg vapor and 5 kg air) at 1 mbar pressure, then the required condensation temperature at the outlet is –25 °C. At higher temperatures, the mass ratio of water and air increases progressively, as does the required capacity of the de-aeration system.

The maximum allowable condensation temperature is, hence, mainly determined by the conditions at the outlet of the freeze condenser and the required suction pressure.

The temperatures read from Fig. 10.7, however, are not equal to the required refrigerant evaporation temperature in the tubes, since this temperature is lower due to heat transport resistances of the vapor diffusion processes to the cold surface of the ice layer, the tube wall, and heat transfer to the refrigerant, respectively. These influences are accounted for in the next section.

Freeze Condenser Design Parameters and Computation of Refrigerant Temperature

The operating conditions and energy consumption of a dry condensing plant are mainly determined by the design of the freeze condenser. Some commonly-applied starting points for this design are:

- Maximum pressure at the vapor inlet of the freeze condenser, typically 1 to 1.5 mbar.
- Maximum pressure drop over the freeze condenser, typically 0.1 to 0.3 mbar.
- Mass flow rates of sparge steam, noncondensable gases and fatty acids.

Parameters to be optimized for these starting points are:

- Surface area and orientation of the tube bundle.
- Tube diameter, length and pitch.
- Location and size of baffles in the shell (for enhancement of condensation).
- Diameter of inlet and outlet ducts.

Fig. 10.6. 3-D representation of a SUBLIMAX dry condenser and ice formation.

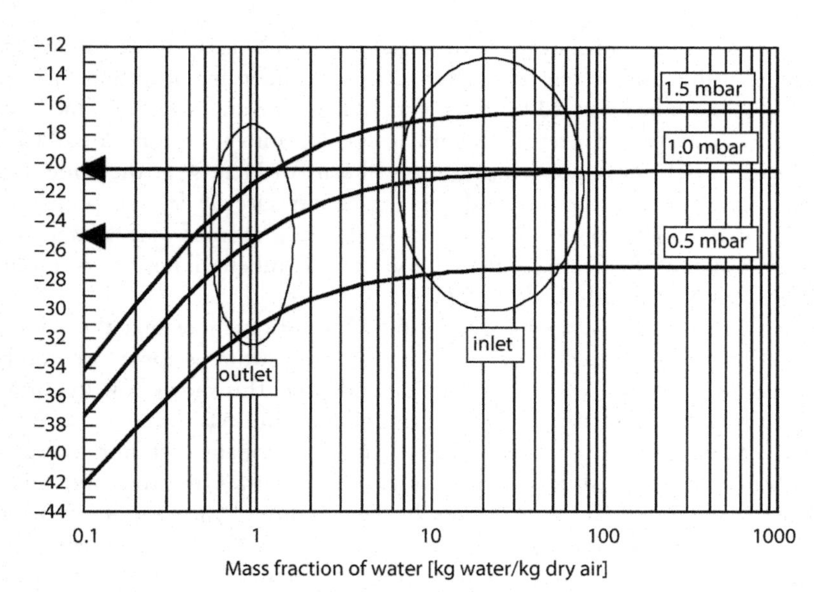

Fig. 10.7. Condensation temperatures of water in air for 0.5, 1, and 1.5 mbar absolute pressure, as a function of the water/air mass ratio.

The design process itself consists of the computation of heat and mass transport rates in the freeze condenser and corresponding pressure losses, and has to take into account practical aspects such as construction codes and cost price. This optimization process is not further discussed here.

An important design parameter is the surface area of the tube bundle. This parameter namely determines to a large extent the required evaporation temperature of the refrigerant and therefore the energy efficiency of the DC system. This is now further demonstrated for a practical case, based on the following:

- Pressure at vapor inlet of freeze condenser: 1.3 mbar.
- Maximum pressure drop over the freeze condenser: 0.2 mbar.
- Mass flow rate of sparge steam: 150 kg/h; air: 3 kg/h; fatty acids: 5 kg/h.

Selected parameters:

- Bundle area: 50 m^2.
- Tube diameter: 70 mm; length: 6000 mm; wall thickness: 2 mm.

Fig. 10.8 shows the computed ice layer thickness distribution and evolution over time from inlet to outlet of the freeze condenser, for a given temperature of the inner tube wall of –27.5°C.

Ice is mainly deposited near the inlet of the condenser. This inhomogeneity of the ice layer thickness can be understood from Fig. 10.7, which shows an initial pronounced decrease of the mass ratio of water and air (and thus high condensation rates) at the inlet of the condenser, which reduces toward the outlet.

Corresponding temperatures of condensation in the vapor, at the surface of the ice layer and at the inside of the tubes after 120 min. condensing operation, are shown in Fig. 10.9. As can be observed from the data in Fig. 10.8 and Fig. 10.9, once the ice layer reaches a certain thickness (8–10 mm at top, 2–3 mm at bottom of the tubes in this particular case), the vapor temperature will start to increase at the outlet of the tube. This is the signal to stop the condensing cycle and start the de-icing program.

Fig. 10.10 shows curves of the history of the mass ratio of water and air at the de-aeration outlet of the freeze condenser, for –27.5°C tube wall temperature and for an increased wall temperature of –26.5°C.

For the higher tube temperature of –26.5°C, a distinct increase of the mass ratio is visible after 100 minutes of freezing. This tube temperature is apparently too high if one wants to extend the freezing time of a condenser. This shows how important a correct refrigerant temperature is, but at the same time, one has to take into consideration that a too-low freezing temperature will disproportionally cost more energy and will not bring the expected benefit in cycle time extension. Once the tubes are covered entirely with an ice layer, the heat exchange will drop; hence, vapor temperature at the outlet will start to rise, indicating a higher steam-to-air ratio, and hence, more duty for the final vacuum group to handle the extra amount of noncondensables.

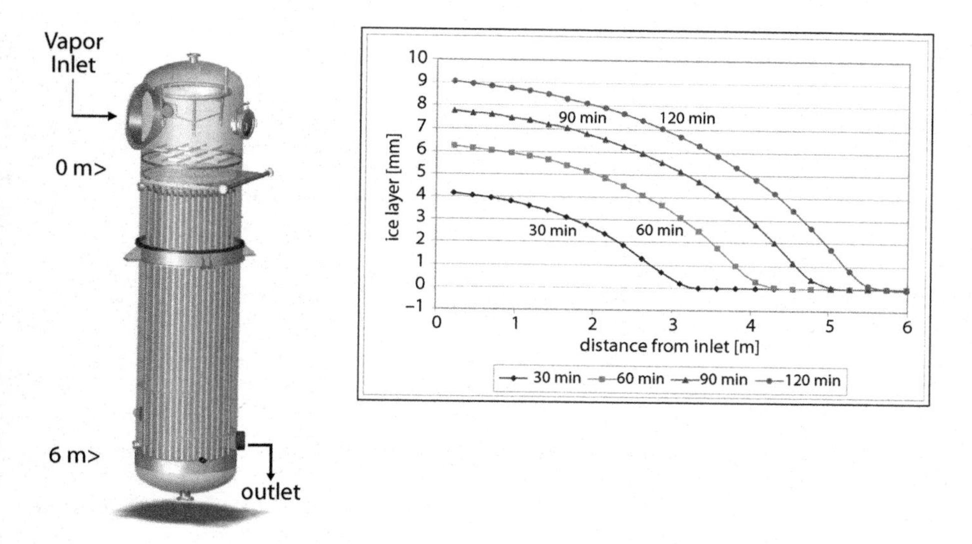

Fig. 10.8. Computed ice layer thickness distribution in the freeze condenser tubes, from inlet to outlet, after 30, 60, 90, and 120 minutes of freezing. Temperature of the inner tube wall: –27.5°C.

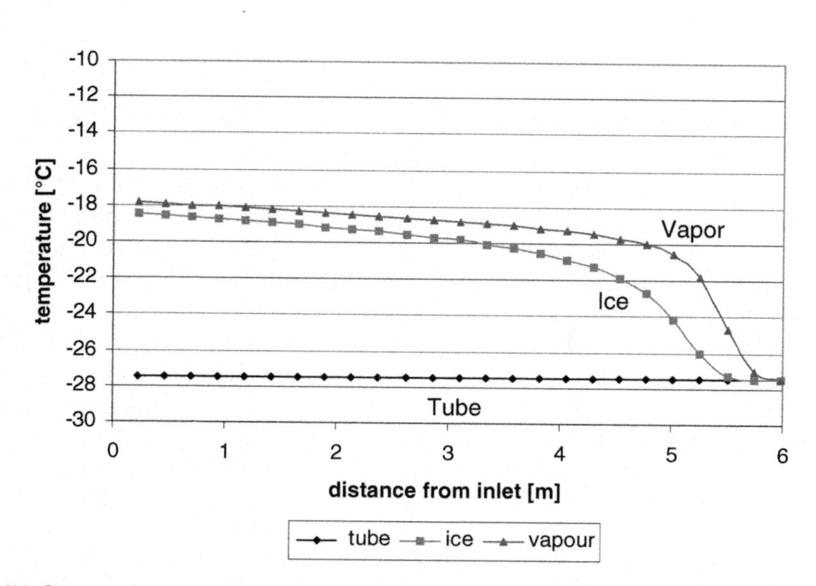

Fig. 10.9. Computed temperature distributions at the end of a freezing period (120 min.), in the vapor, at the surface of the ice layer and at the inside of the tube wall, from inlet to outlet of the freeze condenser.

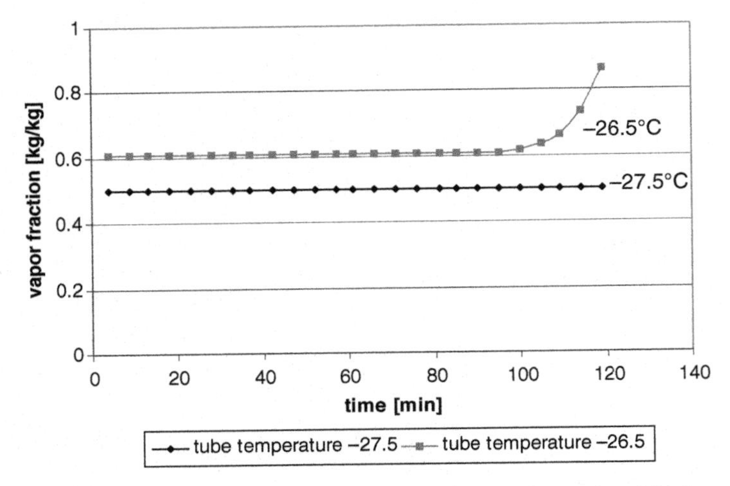

Fig. 10.10. History of the mass ratio of water and air at the de-aeration outlet of the freeze condenser, for –27.5 °C tube wall temperature and for an increased wall temperature of –26.5 °C.

The evaporation temperature of the refrigerant (t_r) can be computed from the wall temperature of the tubes (t_w) and the heat flux through the tube wall ($\dot{q}$) by

$$t_r = t_w - \frac{\dot{q}}{\alpha} \qquad \text{Eq. 10.1}$$

Here α denotes the perimeter averaged convective heat transfer coefficient for the refrigerant flow through each tube. Values for α depend for the most part on the prevailing refrigerant flow pattern inside the tubes and the percentage of refrigerant vapor. These α values have been computed for, respectively, a horizontal-gravity ammonia feed system (as in Fig. 10.3) with stratified flow pattern in the tubes, a vertical-gravity feed ammonia system (see Fig. 10.4) and for the SUBLIMAX system (Fig. 10.5) with falling film evaporation in the tubes.

Furthermore, in the present example, t_w is –27.5°C, and $\dot{q}$ is 2550 W/m^2. Computed refrigerant temperatures using (1) are listed in Table 10-C.

Temperature t_0 denotes the refrigerant temperature in the separator. This temperature is computed from t_r as follows:

The length of the vertical ammonia duct from separator to freeze condenser inlet (see Fig. 10.3 and Fig. 10.4) is taken at 4 meters, corresponding to 0.26 bar static pressure. Typically 35% of this pressure difference is attributed to flow resistances in the ammonia feed ductwork to the tube bundle. The remaining 65% is consumed in the boiling two-phase flow inside the tube bundle and in the return duct to the separator. The boiling pressure of the ammonia is then 0.17 bar higher than the pressure in the separator, corresponding to a 3.5 K higher boiling temperature.

Table 10-C. Computed average heat transfer coefficients (α) in the boiling ammonia in the ice condenser tubes, boiling temperature (t_r) and ammonia temperature in the separator (t_0).

DC System	α [W/(m².K)]	t_r [°C]	t_0 [°C]
Horizontal	2000	−28.8	−32.3
Vertical	2500	−28.5	−32.0
Vertical falling film	3400	−28.3	−28.3

The temperature difference between t_r and t_0 for the falling film system is negligible, because of the open connection between separator and tubes via a self-draining return duct and low refrigerant vapor velocities.

The energy consumption of the compressor depends, among others, on the refrigerant temperature t_0 in the separator, as is computed in the next section.

Energy Consumption

The energy consumption of DC systems mainly consists of steam for the boosters of the de-aeration group and electricity for the compressor of the refrigeration plant. The heat demand of the melt vessel is usually supplied from waste heat sources.

If waste heat is not available, then the DC system can be equipped with a hot gas defrosting system that uses waste heat from the refrigerant compressor for de-icing of the freeze condenser. This option further reduces the energy demand of the melt vessel.

For the aforementioned example of 5 kg/h non-condensables, the steam consumption of the de-aeration group is in the order of 200 kg/h, for each of the DC systems (vertical or horizontal).

The electricity consumption of the compressor depends on both the average cooling demand (about 150 kW_{th}, including losses, for the example of 150 kg/h sparge steam) and process conditions, such as the mass flow rate of sparge steam, cycle times, and the evaporation temperature of the ammonia (t_0 in Table 10-C).

The electricity consumption of the compressor is computed by division of the average cooling demand by the so-called coefficient of performance (COP) of the compressor.

A COP-curve of an ammonia screw compressor is shown in Fig. 10.11 for two condensation temperatures.

The electricity consumption of the compressor of each DC system is then for a 20°C condensation temperature and the evaporation temperatures from Table 10-C:

Horizontal:	52.9 kW_e
Vertical:	52.4 kW_e
Vertical falling film:	46.9 kW_e

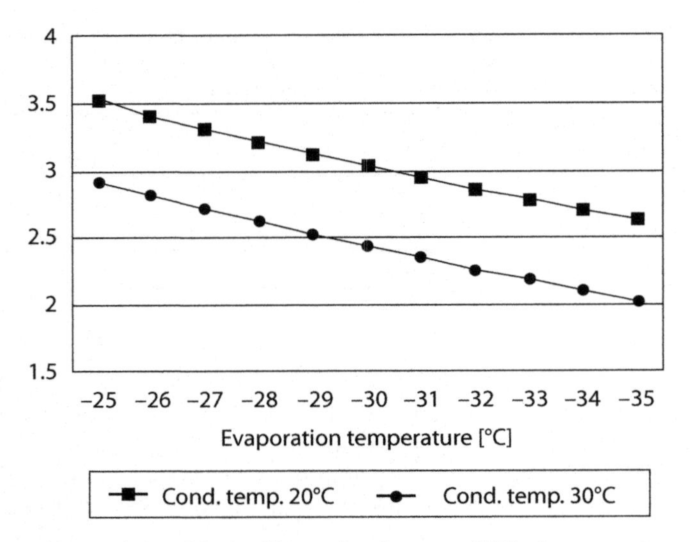

Fig. 10.11. Typical dependency of the coefficient of performance (COP) of an ammonia screw compressor on the evaporation temperature, for 20°C and 30°C condensation temperatures.

The higher evaporation temperature in a vertical falling film condenser hence reduces the energy consumption of the compressor. In addition, the capacity of the ammonia compressor increases typically by about a 4% per degree increase of the evaporation temperature, so in this example the compressor for the vertical falling film DC system (SUBLIMAX) may be chosen about 15–20% smaller.

Another aspect is the occurrence of high peak loads on the refrigeration plant for the horizontal and vertical systems, each time when the temperature of a freeze condenser is brought back to its normal operating temperature after de-icing and cleaning at about +60°C. These peak loads add to the normal operating cooling load (here 150 kW_{th}), and therefore, increase the size of the compressor. With the vertical falling film system, peak loads can be minimized by short ammonia injections during a relatively long (available) cooling period, further minimizing the size of the compressor.

Lay-out and Safety Aspects

The position of an ice condenser in an oil refinery is best as close as possible to the deodorizing vessel, to minimize pressure losses between the deodorizer fatty acid scrubbing system and the steam DC system.

The presence of a rather large amount of ammonia in most of the DC systems, and especially in the ammonia compressor, requires some special precautions and safety conditions.

Care has been taken over the years in the design of the DC systems to minimize the amount of ammonia in the overall DC system. The vertical falling film DC system (SUBLIMAX) is a typical example. Less than 1000 kg of ammonia is required in the whole DC system, with only a fraction of this ammonia (<10%) circulating in the ice condensers.

The ammonia compressors are usually placed in a separate dedicated room, also for the noise. The condensers are normally installed in the same building as the deodorizer vessel.

An example of a DC system installed in an edible-oil refinery is given in Fig. 10.12.

Economics of Dry Condensing

Due to the relatively high capital cost for a dry condensing system, the return of investment (ROI) may take several years. How long depends on the ratio between the cost of steam and electricity (Fig 10.13). In Europe, with typically high fuel cost as compared to electricity, the production cost of steam is accordingly high, which reduces the ROI of a dry condensing system as compared to a classical vacuum system. But more important than the ROI is the much lower environmental pressure as much lower quantities of effluents and emissions are generated by dry condensing (Table 10-D). Especially in populated areas, the environmental impact of a refinery is becoming more and more a matter of concern. With dry condensing, the amount of waste water and smell can be reduced substantially.

Conclusions

A dry condensing vacuum system for deodorizers is today considered state-of-the-art. The larger capital cost is more than compensated for by the much lower operating cost; thus a reasonable payback time is guaranteed. But more important is the possibility to operate the deodorization process at a much lower pressure, offering new

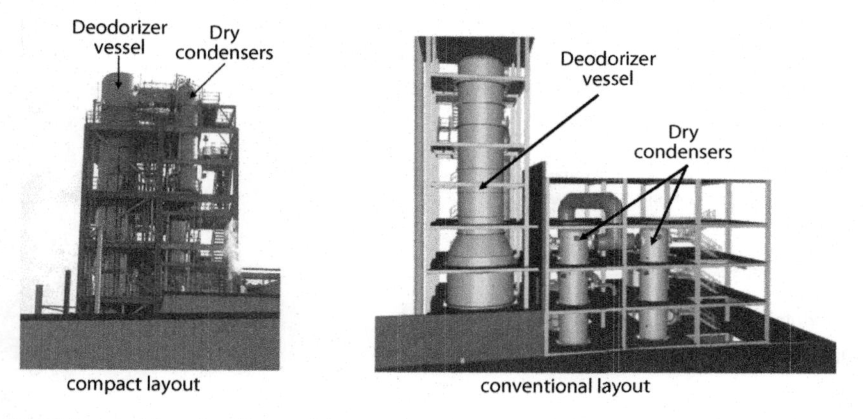

Fig. 10.12. Deodorizing plant layout with vertical ice condenser system (SUBLIMAX, Desmet Ballestra).

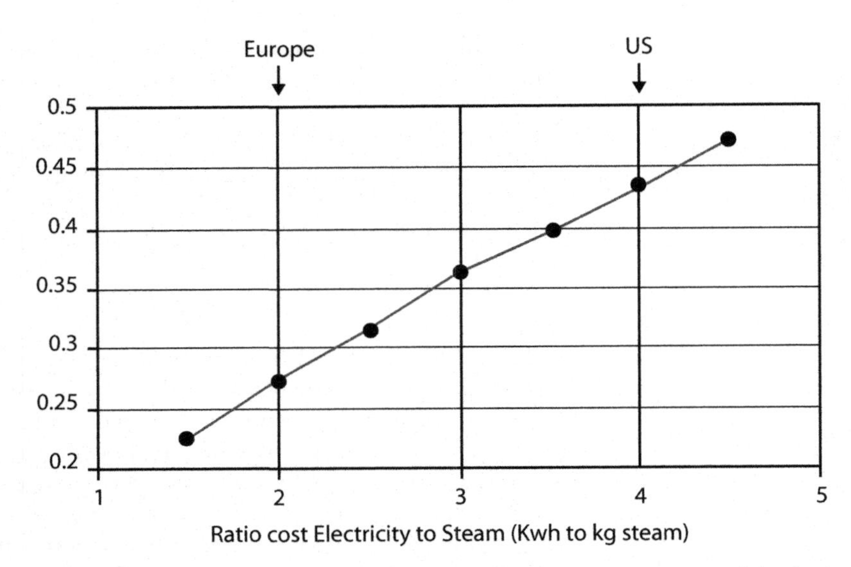

Fig. 10.13. Relative cost comparison between vacuum plant with steam ejectors and dry condensing (Desmet).

Table 10-D. Cost Comparison between Classical Vacuum Plant with Steam Ejectors and a Dry-Condensing Plant (Europe, 2010).

(1): excl. effluent treatment cost (2): incl. effluent treatment cost (1.8 us$/m^3)

Vacuum system	Capital cost	Running cost	Effluent	Pay back
US$	US$(*)	m^3/year	years	
Conventional plant	150.000	333.750 (1)	20.500	/
With ejectors		381.500 (2)		
Dry-condensing plant	900.000	129.750(1)	2.100	3.7 (1)
		133.500(2)	3.0 (2)	

Note: Basis for Comparison: Continuous deodorization of oils.

Deodorizer capacity (500 tpd)	21 tph
Stripping steam (12kg/ton)	250 kg/h
Deodorizer pressure	2 mbar
Suction pressure at de-aeration	1 mbar
Vapor temperature	70°C
Air (300 ppm)	7 kg/h
FFA passing scrubber	3 kg/h
Cost steam	30 US$/ton
Cost electricity	0.09 US$/kWh

Costing based on 8,000 h per year continuous operation.

perspectives to a more efficient removal of unwanted minor-components from the oil under milder refining conditions, while at the same time reducing the environmental impact by a much lower effluent production and lower odor emission.

Current dry condensing systems make use of liquid ammonia to get the condensation temperature below −25°C. The application of other refrigerants may further push the operating pressure down to below 1 mbar in an economical way. Liquid carbon dioxide is a possible alternative as it would allow for lower condensation temperature (down to −58°C), still yielding efficient compressor COPs.

The development of especially the vertical falling film condenser has also created another opportunity, which until now was considered practically impossible. A modern vertical-dry condenser, when properly designed and correctly integrated in the process, can receive the total "dirty" vapor coming from a deodorizer, without the need to pass first through a fatty acid condenser. This total condensing concept (SUBLICON) would make the need for an intermediate fatty matter scrubber superfluous. This would open other windows of opportunity to go back to a milder refining of oils, especially those which contain sensitive components like omega-3 fatty acids or other essential neutraceutical products.

Acknowledgments

I want to thank Jelle Nijdam and Pieter Jellema of Solutherm, partner of Desmet in dry condensing technological developments (SUBLIMAX and SUBLICON) , for their valuable contributions to this chapter.

Bibliography

Carlson, K. Deodorization, in Baileys Industrial Oil and Fat Products, Fifth ed., Vol. 4, Wiley Interscience (1996).

De Greyt, W.; Kellens, M. Deodorization, in Baileys Industrial Oil and Fat Products, Sixth ed. Vol. 5, Wiley Interscience (2005).

Desmet Ballestra, internal information, Belgium.

Gea-Atlas, technical documentation, Denmark.

Kellens M.; De Greyt, W. Deodorization, in "Introduction to Fats and Oils Technology" Chapter 13, 2nd ed., AOCS Press: Champaign, IL (2001).

Korting, technical documentation, Germany.

Lines, J.R. Freeze condensing vacuum systems for deodorization, inform, 12 (3), AOCS Press, Champaign IL (2001).

Solutherm, technical documentation, the Netherlands.

11

Enzymatic Interesterification

Christopher Loren Gene Dayton
Bunge Global Innovation, White Plains, New York, USA

Introduction

Interesterification is a process that allows the modification of the melting characteristics of a blend of oils without changing the fatty acid composition (FAC) of the original blend. Utilizing interesterification allows the manufacturer to produce a functional product with less saturated fat and with essentially no *trans* fatty acid. The traditional method for interesterification was to use a strong chemical to rearrange the distribution of the fatty acids in a random fashion. Commercial enzymes are now available that allow the same rearrangement without the use of harsh chemicals. This chapter will review the chemical interesterification (CIE) process and then compare it to the enzymatic interesterification (EIE) process. It will not review the reaction mechanisms or the products created via the interesterification process.

Structure

Solid or semi-solid triacylglycerols at room temperature are fats, while liquid triacylglycerols are oils. The structure of triacylglycerols consists of three fatty acids of various chain lengths and degrees of saturation on a glycerol backbone (see Fig. 11.1). Triacylglycerols are prochiral compounds and use the Stereospecific Numbering (*sn*) System to identify the positions of the placement of the fatty acids in TAGs, diacylglycerols (DAGs), and monoacylglycerols (MAGs) (Fig. 11.2). The *sn*-2 position in naturally-occurring triacylglycerols is almost always an unsaturated fatty acid. The major exception is mother's milk, where 70% of the palmitic acid is in the *sn*-2 position of glycerol (Kennedy et al., 1999).

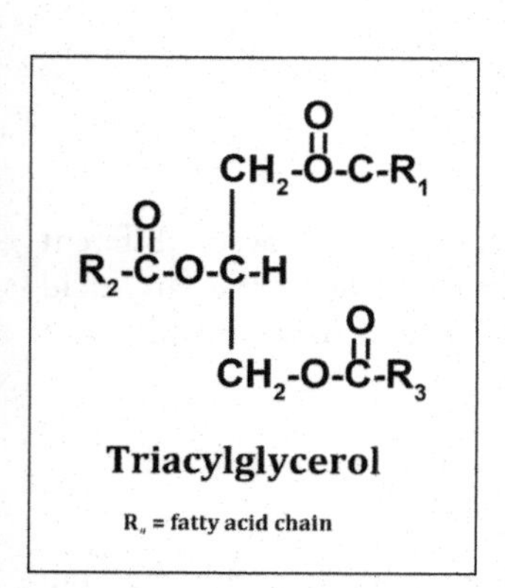

Fig. 11.1. Triacylglycerol structure (Dayton, original).

Interesterification

Interesterification encompasses four different chemical reactions: alcoholysis, acidolysis, glycerolysis, and transesterification.

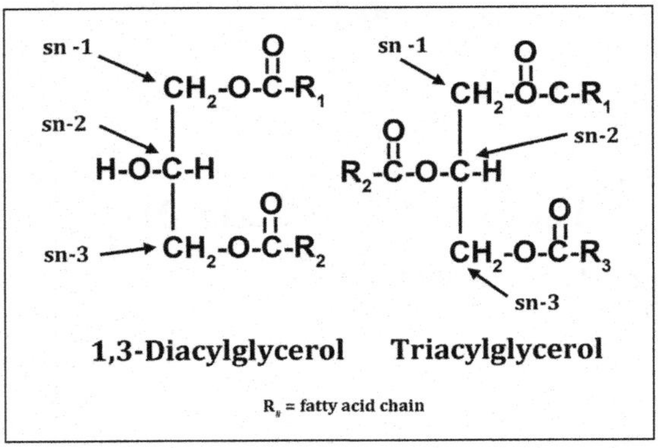

1,3-Diacylglycerol Triacylglycerol

$R_{\#}$ = fatty acid chain

Fig. 11.2. Stereospecific numbering (Dayton, original).

Alcoholysis—monohydric alcohols to produce methyl esters or polyhydric alcohols from Monoacylglycerols (MAGs)

$$RCOOR_1 + R_2OH \rightarrow RCOOR_2 + R_1OH$$

Acidolysis—acid interchange

$$R_1COOR_2 + R_3COOH \rightarrow R_3COOR_1 + R_1COOH$$

Glycerolysis—production of mono- and Diacylglycerols (DAGs)

$$\text{Triacylglyceol} + \text{glycerol} \rightarrow \text{DAG} + \text{MAG}$$

Transesterification—rearrangement of fatty acids, either intramolecular (on the triacylglycerol) or intermolecular (between different triacylglycerols)

$$R_1COOCHHCH(OOCR_2)CR_3 + R_4COOCHHCH(OOCR_5)CR_6 \rightarrow R_1COOCHHCH(OOCR_5)CR_6 + R_4COOCHHCH(OOCR_2)CR_3$$

$R_{\#}$ = Fatty acids of different chain lengths and/or degrees of saturation
C = Carbon
H = Hydrogen
O = Oxygen

This chapter will focus on the rearrangement of fatty acids between different TAGs in order to transform the melting profile without changing the fatty acid composition of the initial blended oil or impart *trans* fatty acids to the blend.

Chemical Interesterification

Catalysts

"Interesterification of fats can occur at high temperatures as 300°C and higher, but requires an inordinately long time and is usually accompanied by decomposition and polymerization" (Sreenivasan, 1978). Chemical reaction rates may be increased substantially, and reaction temperatures can be lowered by employing a catalyst. Catalysts aid in the reaction, but are not consumed by the reaction itself. A very large number of catalysts has been used to produce chemically interesterified (CIE) products including metal salts [acetates, carbonates, chlorides, nitrates, and oxides of tin (Sn), zinc (Zn), iron (Fe), cobalt (Co), and lead (Pb)]; alkali hydroxides [sodium hydroxide (NaOH), potassium hydroxide (KOH), lithium hydroxide (LiOH), alkali hydroxide plus glycerol]; metal soaps (sodium stearate) or metal soaps containing an alkali or alkaline earth metal and an amphoteric metal (lithium-aluminum stearate, and sodium-titanium stearate); alkali metals [sodium (Na), potassium (K), Na-K alloys]; metal alkylates [sodium methoxide ($NaOCH_3$), sodium ethoxide ($NaOC_2H_5$)]; metal hydrides [sodium hydride (NaH)]; and metal amides [sodium amide ($NaNH_2$)]. Sreenivasan compiled a list of CIE catalysts and their reaction conditions (Table 11-A). Industrially, sodium methoxide is the catalyst of choice due to its relative ease of use and commercial availability. All of these catalysts are potentially dangerous chemicals with unique handling, storage, usage, and disposal procedures that must be followed to ensure a safe working environment and food-grade finished products.

Chemical Catalyst Consumption

Impurities present in the feedstock will lead to the destruction of the catalyst and unwanted side reactions reducing the desired yield and the need for additional purification steps. The major impurities causing competitive reactions with sodium-methoxide catalyst in an industrial CIE process are water, free fatty acids (FFAs), and peroxides. An example using these impurities producing unwanted side reactions

Table 11-A. CIE Reaction Conditions (Sreenivasan, 1978).

Catalyst	Percentage	Temperature (°C)	Time (minutes)
Metal Salts	0.1–2	120–260	30–360 Under vacuum
Alkali Hydroxides	0.5–2	250	90 Under vacuum
Alkali Hydroxides plus Glycerol	0.05–0.1 + 0.1–0.2	60–160	30–45 Under vacuum
Metal Soaps	0.2–1	250	60 Under vacuum
Alkali Metals	0.1–1	25–270	3–120
Metal Alkylates	0.2–2	50–120	5–120
Metal Hydrides	0.2–2	170	3–120
Metal Amide	0.1–1.2	80–120	10–60

transforming the catalyst into unwanted products is modeled below with an oil that meets the typical commercial specifications for a fully refined, bleached, and deodorized (RBD) oil:

> 1000 kilograms (kg) blend of oil containing 100 parts per million (ppm) moisture, 0.05% FFAs, and a 1.0 milliequivalents/kilogram (meq/kg) Peroxide value (PV) will inactivate the following amounts of sodium methoxide catalysts.

Water

The 100 ppm of water (H_2O) present in the feed material may be expressed as 5.56 moles of H_2O. The chemical equation for the reaction of water with sodium methoxide yields methanol (CH_3OH) and sodium hydroxide (NaOH) (see Fig. 11.3). The 5.56 moles of water will consume 5.56 moles of sodium methoxide, converting it into 5.56 moles of methanol and 5.56 moles of sodium hydroxide. The very small amount of water present in the starting material will consume 300 grams of the sodium methoxide catalyst producing 178.1 grams of CH_3OH and 222.4 grams of NaOH.

Free Fatty Acids (FFAs)

The 0.05% FFA is equal to 500 grams of fatty acids present in 1,000 kilograms of the example feed material. Fig. 11.4 shows the reaction between fatty acids and sodium methoxide yielding fatty acid methyl ester (FAME) and sodium hydroxide. The amount of sodium methoxide consumed due to the side reactions with FFAs is dependent on the carbon chain length of the fatty acids. Table 11-B demonstrates the differences in grams of sodium methoxide due to the differing chain lengths.

The 293.2 grams of sodium hydroxide generated from the destruction of the catalyst with both H_2O and FFAs will produce additional sodium soaps from saponifying neutral oil, thereby increasing the losses. Sodium soaps are very strong emulsifiers and will cause additional neutral oil entrapment due to their emulsification properties, as was discussed earlier in Chapter 6, in the separation process when water or dilute acid is utilized to neutralize the catalyst via a water-washing step.

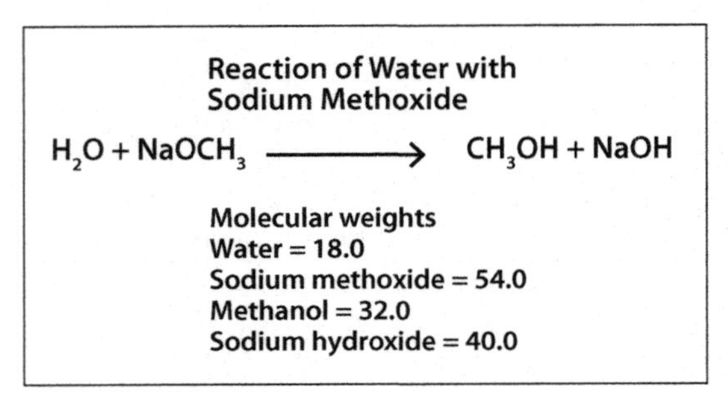

Fig. 11.3. Hydrolysis of fat (Dayton, original).

Peroxides

The AOCS Official Method Cd 8b-90 for PV measures all substances that oxidize potassium iodide under the test condition and is expressed in terms of milliequivalents of oxidized potassium iodide per kilogram

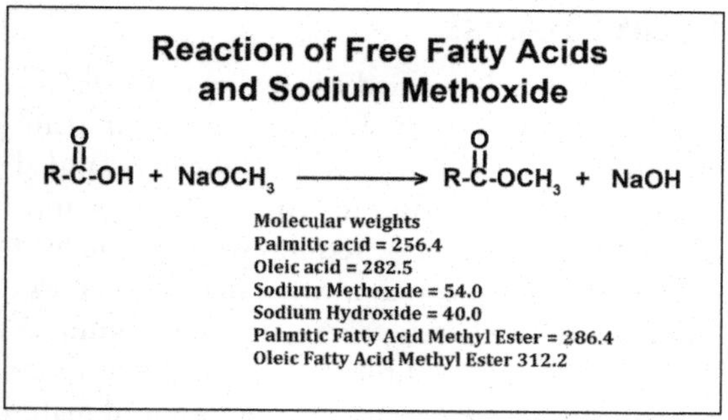

Reaction of Free Fatty Acids and Sodium Methoxide

$$R\text{-}\overset{O}{\overset{||}{C}}\text{-}OH + NaOCH_3 \longrightarrow R\text{-}\overset{O}{\overset{||}{C}}\text{-}OCH_3 + NaOH$$

Molecular weights
Palmitic acid = 256.4
Oleic acid = 282.5
Sodium Methoxide = 54.0
Sodium Hydroxide = 40.0
Palmitic Fatty Acid Methyl Ester = 286.4
Oleic Fatty Acid Methyl Ester 312.2

Fig. 11.4. Reaction of sodium methoxide and fatty acids.

(AOCS, 2011). The number of milliequivalents is equal to the molar concentration in the present example of 1,000 kg of oil. A PV of 1 is equivalent to 1 mole of oxidizing compounds that would consume 1 mole of sodium methoxide, equaling 54 grams of lost catalyst.

Combining the catalyst consumed by the water, FFA, and PV in the example above leads to at least 450 grams of lost catalyst and the formation of unwanted products (Table 11-C). The feedstock in the example met the industry standards for a finished product, and any additional impurities would only increase the required amount of catalyst and neutral oil consumed in the CIE reaction with sodium methoxide. If 0.5 kg of catalyst were to be used in the reaction, the reaction would not have run to completion and would lead to variability between batches.

Table 11-B. Mass of Sodium Methoxide Consumed Based on the Fatty-Acid Chain Length.

	Palm (palmitic acid)	Soybean (oleic acid)
0.05 % Free Fatty Acids		
Fatty Acids (g)	500	500
Fatty Acids (moles)	1.95	1.77
Sodium Methoxide Consumed		
Catalyst (moles)	1.95	1.77
Catalyst (g)	105.3	95.6

Table 11-C. Sodium Methoxide Consumed Due to Impurities.

	Palm	Soybean
Sodium Methoxide Consumed (g)		
Water	300	300
Free Fatty Acids	105	96
Peroxide Value	54	54
Total	459	450

Chemical Interesterification Process

The feed material for the CIE process is physically refined oil (palm type oils) or refined and bleached (RB) oil for caustic refined oils. If the FFA level is high in palm type oils from storage or poorly refined oil, a little sodium hydroxide may be added to reduce the FFA to an acceptable level of 0.1% in order to reduce the consumption of catalyst prior to the oil being dried. The presence of sodium soaps in the feed material will not interfere with the interesterification reaction when sodium methoxide is utilized. The CIE reaction is predominantly a batch process due to the number and variety of products produced and the preferred equipment. The material to be interesterified is blended together, followed by a drying step. The reaction feed material is dried at 120 to 150°C under a vacuum of 24 mbar. The oil is then cooled to 70 to 100°C, where 0.1% of sodium methoxide powder is added and mixed for 30 to 60 minutes. In a number of references, it was claimed that the reaction is continued until a "distinctive brown color denoting randomization" was detected (Lanining, 1985; Lampert, 2000). A method was developed and patented for monitoring the CIE reaction via the dark brown color development (Liu & Lamert, 2001). However, this color change is not a measurement of the chemical randomization of the fat, but of the destruction of the tocopherols in the feed material by the catalyst. Most of the catalysts described in the literature for CIE processes are very strong bases and are not selective for randomizing the fatty acids on the triacylglycerol. The chemical catalysts will react with all compounds present in the feed material including so called "unsaponifiable" material (tocopherols, tocotreinols, and sterols).

The chemical interesterification reaction is stopped by the addition of water, dilute solution of acid, and/or an acid activated silica like Trysil® 300. If water is used to neutralize the sodium methoxide, it is separated from the oil via high speed centrifugation, producing "soapy" water which may be acidulated to recover the fatty acids generated from the saponification reaction. Alternatively, a dilute acid (citric or phosphoric) or silica is added prior to the vacuum bleaching process, where 0.5 to 2% bleaching earth is added to remove the soaps and color bodies generated from the basic catalyst. The CIE oil is either deodorized directly or blended with another base stock and deodorized to complete the purification process.

Chemical Interesterification Oil Losses

Industrially, for an oil with the level of contaminates listed in the previous example using RBD quality, 0.1% of sodium methoxide would be utilized to modify the fat blend, and 0.5% of adsorbents would be used to remove soaps and the dark brown color. The following are the losses based solely on the stoichiometric values for the catalyst and adsorbents utilized (see Table 11-D).

1,000 kg of feed material requires 1,000 grams of catalyst.
1,000 grams of sodium methoxide is 18.51 moles of catalyst.
18.51 moles of sodium soap will be formed.
18.51 moles of FAMEs will be formed.
10 kg of adsorbents remove 3 kg of neutral oil.

Table 11-D. Oil Losses Based on the Formation of Soaps, FAMEs, and Adsorbents.

	Palm	Soybean
Sodium soap		
moles	18.51	18.51
mass (kg)	5.15	5.63
Fatty acid methyl esters		
moles	18.51	18.51
mass (kg)	5.30	5.78
Adsorbents (kg)	3.00	3.00

The sodium soaps may be recovered in the water washing step via acidulation and conversion into fatty acids, or they may be converted into fatty acids in the oil by the addition of excess citric acid in the catalyst neutralization step and recovered in the deodorization process as deodorizer distillate. If the fatty acids are to be recovered directly from the oil in the deodorization process, the vacuum system must be designed to account for the increased vapor to be condensed and recovered in comparison to the chemical refining process. It was estimated by DeGreyt that an additional loss of 30% will be generated based on the amount of FFA and FAME to be removed in the deodorization process (DeGreyt, 2004). The total losses produced in the CIE process will be at least 1.6 to 1.8% for the given example for the finished product. The losses from a typical processing facility are significantly higher due to the levels of impurities in the starting material. Losses as high as 5% have been observed (DeGreyt, 2004). The total stoichiometric losses generated in the CIE process for palm and soybean oils are shown in Table 11-E for the RBD quality oil above.

Stability of the Chemical Interesterified Products

The CIE process is a robust method for modifying the melting properties of triacylglycerols without changing the fatty acid composition, but the stability of the finished product is depressed. Many researchers (Lau et al., 1982; Lo & Handel, 1983; Tautorus & McCurdy, 1990) showed that the oxidative stability of the CIE fat was dramatically reduced compared to the parent oils. It was shown by Zalewski et al. that

Table 11-E. Total Losses Generated from Chemical Interesterification.

	Palm (kg)	Soybean (kg)
Catalyst	10.45	11.41
Adsorbents	3.00	3.00
Deodorization	3.14	3.42
Total Losses	16.59	17.83

the decrease in stability in CIE oils could not be attributed to the fatty acid randomization because they were able to restore the stability with various antioxidants in lard (Zalewski & Gaddis, 1967). Park et al., (1983) were able to clearly demonstrate that the loss in oxidative stability was predominately due to the destruction of the tocopherols in CIE soybean oil. When they supplemented the oils with the lost tocopherols (alpha, gamma, and delta), they nearly achieved the same oxidative stability (Park et al., 1983). This author believes that the loss of tocopherols and the generation of pro-oxidants cannot be compensated by the addition of antioxidants for the loss in stability of CIE oils and the products generated from these oils in comparison to their starting materials or enzymatically interesterified fats.

Enzymes

Enzymes are nature's catalysts. They are proteins that have a remarkable ability to catalyze very specific chemical reactions of biological importance. They are very specific in both the substances they react with (substrates) and the reaction they catalyze. In an enzyme catalyzed reaction, none of the substrate is diverted into a nonproductive side reaction, so no waste is generated.

An enzyme's catalytic activity and selectivity are determined by its 3-dimensional (3-D) structure. The 3-D structures are determined by roughly 20 different naturally-occurring amino acids linked together in a chain-like fashion forming a macromolecule. These macromolecules fold into 3-D structures (Copeland, 2000). The individual amino acid's side chain enables the chemical reactivities that produce the enzyme's ability to distinguish, orientate, and bind the substrate to the active site. Once the substance has been engaged with the active site of the enzyme, the chemical reaction takes place followed by the release of the chemical products.

The International Union of Biochemistry and Molecular Biology has developed a nomenclature for naming enzymes, Enzyme Commission Number (EC number). Each enzyme is given a four-sequence number to classify it based on the chemical reaction it catalyzes.

EC 3—Hydrolases are enzymes that breakdown molecules using water. The chemical equation is as follows:

$$A - B + H_2O \rightarrow A - OH + B - H$$

EC 3.1—Hydrolases that react on ester bonds.

EC 3.1.1—Hydrolases that react on the ester bond located on the carboxylic acid.

Lipases (EC 3.1.1.3) are enzymes that in the presence of water cleave the fatty acids present on triacylglycerols. Lipases use an amino acid catalytic triad serine-histidine-aspartic acid/glycine (Ser-His-Asp/Glu) for ester bond hydrolysis (Brumlik, et al., 1996; Shu et al., 2007). In this triad, histidine acts as a general base forming a hydrogen bond with a catalytic serine residue which then allows the formation of the first tetrahedral transition state. At low pH conditions, i.e., pH 3 and lower, histidine will be protonated. If histidine is already protonated, the hydrogen bond formation with catalytic serine residue will not occur, and reaction will not proceed, thereby reducing or eliminating enzymatic activity (Kerovuo, 2008).

Most enzyme reactions are reversible, but the reaction conditions typically must be dramatically changed for a reverse reaction to occur. The formation of triacylglycerols is made using the same lipase by removing the excess water generated during the reaction (Yamane, 1988). Three different methods for utilizing lipases in esterification reactions (building a triacylglycerol) have been demonstrated by controlling the amount of water, thereby limiting any unwanted hydrolysis. The first method utilizes a water-soluble organic solvent (methanol, ethanol, ethylene glycol, glycerol, etc.) for the substrate. The second method employs a water insoluble organic solvent (heptanes, octane, and isooctane) in order to remove the water from the substrate via vacuum or absorbent for drying. Finally, the third method immobilizes the lipase on a carrier where the only water present is contained in the active site and unavailable for hydrolysis (Cowan, 2011). If water is limited to only the active site of the protein, the hydrolysis reaction is able to occur, as well as the reverse reaction re-attaching the fatty acid to another available site.

Denaturation

A protein's macromolecular structure is functionally active only within a very narrow range of environmental conditions such as temperature, ionic strength, and relative acidity. Exposure to conditions outside its active environmental conditions will cause a protein to lose its secondary and tertiary structures (Garrett & Grisham, 2010). The loss of its structural order is called denaturation and is accompanied by loss of functionality. Enzymes may become denatured by precipitation of the protein, temperature, ionic strength, metal salts, pH, organic solvents, and removal of water at the active site. An everyday example of a protein undergoing denaturation is an egg that has been cooked. It is no longer clear and fluid, but opaque and "rubbery." When an enzyme has undergone denaturation, it no longer is able to catalyze a biological reaction.

Lipases

Lipases' natural function is to cleave fatty acids within a biological matrix such as digestion or the decomposition in the environment. Enzymes obtained from the environment or animal sources are known as "wild-type." They have a very narrow temperature range where they are active, typically 5 to 40°C, and require a water environment since the reaction they catalyze is hydrolysis. Industrially it is important to have an enzyme that is able to catalyze the required reaction at a temperature where the fat is liquid (see Table 11-F) and where the protein is dissolved in the water or is immobilized on a carrier and water is present in the active site. The rearrangement of the fatty acids between triacylglycerols in the interesterification reaction occurs in a very low water environment to prevent the removal of the fatty acids.

Table 11-F. Melting Temperatures of Pure Fatty Acids and Triacylglycerols (Bailey, 1950).

Carbon Chain	Fatty Acid (°C)	Triacylglycerol (°C)
Caprylin (C8:0)	16.7, 16.5[a]	8.3
Caprin (C10:0)	31.6	31.5
Laurin (C12:0)	44.2	46.4
Myristin (C14:0)	54.4	57.0
Palmitin (C16:0)	62.9, 61.3[a]	65.5
Stearin (C18:0)	69.6	73.1
Arachidin (C20:0)	75.4	75–78[b]
Behenin (C22:0)	80.0	83.0[c]
Lignocerin (C24:0)	84.2	86[d]

[a]Nu-check Prep, 2007-2008 product catalog
[b]http://www.chemicalbook.com/ProductChemicalPropertiesCB1391710_EN.htm
[c]http://www.chemicalbook.com/ProductChemicalPropertiesCB7406890_EN.htm
[d]http://lipidbank.jp/wiki/LBGTGccc:24000SC01:24000SC01:24000SC01:01

Enzymatic Interesterification

Enzymatic interesterification is a relatively new process developed in the early 1980s by Unilever, Novozymes, Fuji, and others working in lipase research (Holm & Cowan, 2008). The research predominately focused on finding a path to produce cocoa butter equivalents (CBEs) utilizing less expensive starting ingredients (Macrae, 1985). Three different types of lipases may be applied for interesterification: (1) non-specific lipases, (2) *sn*1,3 specific lipases, and (3) fatty acid specific lipases. Non-specific lipases are lipases that cleave fatty acids at all three positions of the triacylglycerols. An interesterification reaction utilizing a non-specific lipase will generate the same triacylglycerol profile as a CIE processed oil. A reaction using an *sn*1,3 lipase may generate triacylglycerols with a symmetrical configuration like 1,3-dipalmitoyl-2-oleoylglycerol (POP), 1,3-distearoyl-2-oleoylglycerol (SOS), 1-palmitoyl-2-oleoyl-3-stearoylglycerol (POS) for cocoa butter substitutes. Finally, a reaction using a fatty acid specific lipase from *Geotrichum candidum* that selectively cleaves the unsaturated fatty acids from triacylglycerols in the Δ (delta) 9 position may produce an oil with increased unsaturation. (Macrae, 1983; Charton & Macrae, 1992). The author knows of no commercial process or products produced utilizing a fatty acid specific enzyme from *Geotrichum candidum.*

Enzymatic Interesterification Process

Unilever (formerly Lever Brothers Company) disclosed a process for utilizing a number of different lipases for interesterification of lipids (Coleman & Macrae, 1981). They disclosed a method for immobilizing lipases and reusing them in a batch slurry reaction system. The immobilized lipases required the addition of up to 10% water for their activity to be restored. The "re-hydrated" immobilized enzymes were mixed with blended lipids that were already dissolved in a petroleum ether (PE) or hexane solvent. The mixture was stirred at 40°C for 16 hours to complete the reaction (Macrae, 1983). Two major drawbacks to the initial process were the use of large amounts of water for the hydration/activation of the enzyme, which lead to unwanted hydrolysis of triacylglycerols, and the use of organic solvents to reduce the viscosity of the lipids and limit the amount of hydrolysis.

Fuji Oil Company revealed a method for vacuum drying an enzyme preparation in the immobilization process and a method of transesterification where water is limited in the feed material by vacuum heat drying to reduce the unwanted formation of diacylglycerols (Matsuo et al., 1983a; 1983b).

In 1988, Unilever disclosed a continuous interesterification process utilizing an immobilized lipase packed in a column (Macrae & How, 1988). The reactor consisted of a lower chamber containing 1:1 ration of Celite®:H_2O separated from the upper chamber by a glass wool plug. The upper chamber contained the wetted immobilized enzyme. The lipids were dissolved in PE and pumped to the bottom of the column to ensure a water saturated organic phase entering the catalyst portion of the column for reaction. The interesterified product was removed from the top of the column with a residence time of less than 2 hours.

A solvent-free packed bed process for interesterification was first described at the World Conference on Emerging Technologies in the Fats and Oils Industry held in Cannes, France, by Novozymes (formerly Novo Industri A/S) (Hansen & Eigtved, 1986). Novozymes patented the immobilized *Mucor miehe* described in the presentation (Eigtved, 1989). The 1,3-specific lipase was immobilized on a macroporous weak anion exchange resin with a particle size distribution enabling a process where the pressure drop across the enzyme bed would not require the need for solvents (Hansen & Eigtved, 1986). A water-saturated ion exchange resin pre-column was utilized to saturate the oil with water prior to exposure with the enzyme column (Fig. 11.5). The pressure drop across the packed bed at the higher flow rates was 1 bar with little compression of the packed bed at 60°C.

Ajinomoto Co. Inc., in Japanese patent application JP 2203789-A, disclosed the addition of an alkaline substance effects an increase in enzyme activity of *Rhizopus javanicus* and *Rhizopus delemar* (Ajinomoto, 1990; Pedersen et al., 2011).

Archer Daniels Midland (ADM) disclosed a process for deodorizing the initial substrate prior to the enzymatic interesterification process in a packed bed (Lee, 2008). It was demonstrated that the freshly deodorized oil without any peroxides would increase the half-life of the enzyme from 9 days to 20 days. In another filing by ADM, they presented information for improving the productivity of the enzyme by purifying the oil with a pretreatment of either an amino acid on silica or a texturized vegetable protein (Binder et al., 2006). The blended feed material oil was heated to 70°C, and then it was pumped through a purification media column before entering a column at an initial flow rate of about 4 gallons per minute. The flow rate of the

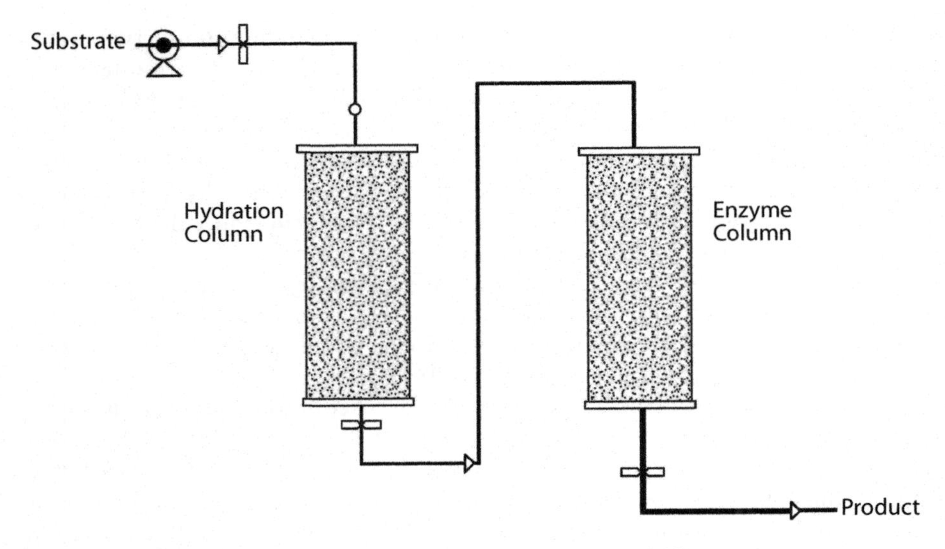

Fig. 11.5. Process flow for the solvent free interesterification process (AOCS).

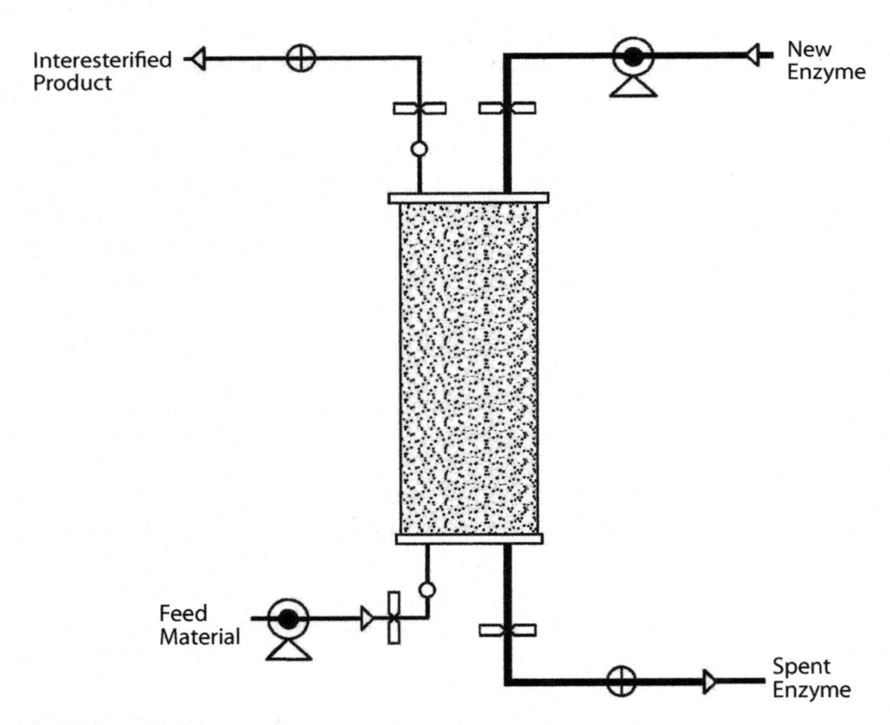

Fig. 11.6. Process flow diagrams based on Peeters' disclosure (U.S. patent application).

feed material was reduced as the enzyme activity decreased to maintain a consistent finished melting product (Binder et al., 2006).

Cargill revealed a process of interesterification where the immobilized enzyme is periodically removed from the inlet of the column, and an equivalent portion of immobilized enzyme is added at the outlet of the packed column (Peeters et al., 2007, 2011). Fig. 11.6 is a depiction of the column from the description provided in the United States patent application.

Bunge disclosed a continuous process for enzymatic treatment for a lipid-containing composition at a substantially constant flow rate (Dayton & dos Santos, 2013a,b). The method is comprised of contacting the lipid feedstock with a pre-treatment processing aid, then causing the feedstock to pass at a substantially constant flow rate through a treatment system comprised of a plurality of enzyme-containing fixed-bed reactors connected to one another in series. Each of the fixed bed reactors is individually serviceable; the flow rate of the feedstock remains substantially constant through the system when one of the fixed bed reactors is taken off line for servicing (see Fig. 11.7). The pretreatment processing aid comprises a silica having an average pore size greater than 150 angstroms, and contains less than 10% volatiles. The processing aid can be placed in one or more of the fixed bed reactors, disposed above the enzyme in the reactor, or it can be in a pre-treatment system which can comprise one or more reactors. The silica can be the commercial product TriSyl® 150IE manufactured by Grace Davidson. The advantages of the Bunge disclosure over the prior

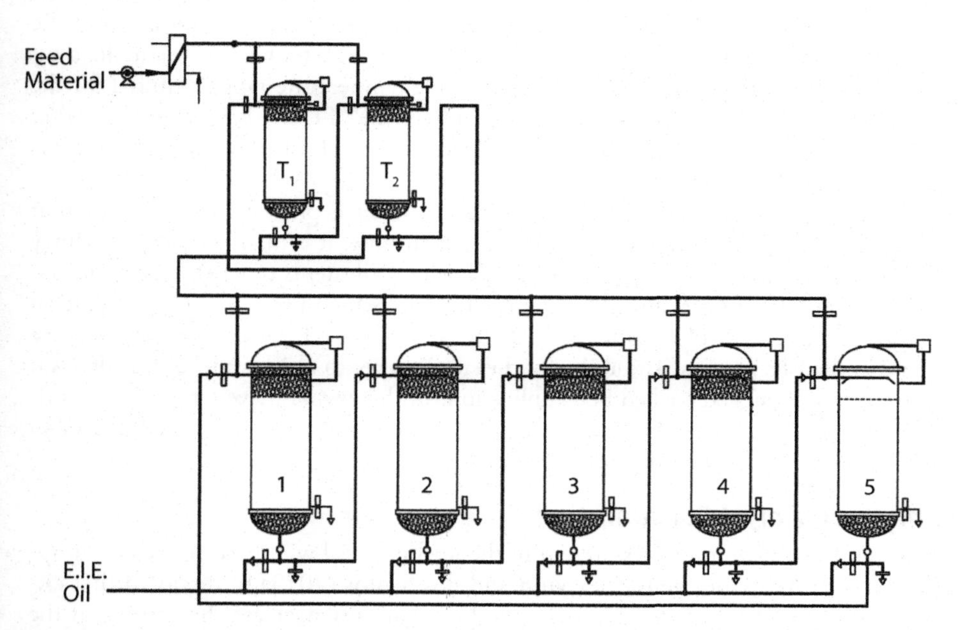

Fig. 11.7. Industrial EIE process schematic (U.S. patent application).

art are twofold. First, the process is continuous without the need to stop the process due to changing and replacing the enzyme, taking a reactor out of service for needed repair, or requirement to manipulate the flow rate in order to achieve identical results due to changing enzyme activity in the reaction beds. The second advantage disclosed is the ability to use a RB feed material instead of a RBD feedstock, thus reducing the processing requirements and producing an oil with greater amounts of tocopherols and sterols that would be lost in the re-deodorization process.

DeSmet Ballestra Engineering revealed an enzymatic process to extend the activity of the enzyme by the introduction or in situ production of soap in the oil prior to exposure to the catalyst (Kellens et al., 2012). The soap is either introduced or produced via a dilute caustic solution to yield a concentration of 5 to 200 ppm, followed a homogenizing the fat and soap mixture, pumping the material through a lipase packed bed, and finally a step to remove the soap from the interesterified material.

Commercial Enzymes

Novozymes has developed two immobilized lipases for food grade interesterification of fats and oils, Lipozyme® RM IM derived from *Rhizomucor miehie/Aspergillus oryzae* and Lipozyme® TL IM from *Thermomyces languinosus/Aspergillus oryzae.* Lipozyme® RM IM is predominately used for high value structured lipids while Lipozyme® TL IM is the workhorse for interesterification. Both enzymes are 1,3-specific lipases, meaning they only cleave and re-attach fatty acids at the *sn*1 and *sn*3 positions of the triacylglycerol (see Fig. 11.2). These enzymes have been evolved to maintain their secondary and tertiary structure over a very wide temperature range required for the industrial application of interesterification. Lipozyme® TL IM is fully functional at up to 75°C where almost all fatty acids and triacylglycerols of abundance and importance are liquid (see Table 11-D).

Lipozyme® RM IM is immobilized protein on a phenolic resin carrier that maintains nearly 100% of the selectivity of the enzyme. It is mainly utilized in slurry batch reactions for the production of cocoa butter equivalents (CBEs). Lipozyme® TL IM is immobilized on a silica (hydrophilic) carrier which promotes acryl migration (Yemc & Cowan, 2011). Consequently, the *sn*-1,3 lipase behaves as a non-selective lipase in the EIE process. The ability of the silica carrier to resist compression allows it to be used in a rugged packed bed application. Both enzymes have temperature optimums of 70°C and are completely denatured at 90°C after 1 to 2 hours. Table 11-G is a summary of the two different enzymes.

Enzyme Catalyst Consumption

Acids and oxidation products present in the feed material will increase the consumption of the enzymatic catalyst. Dayton and dos Santos (2013a,b) disclosed that citric acid present in the feed material would dramatically decrease the activity of the enzyme. Novozymes demonstrated that the citric acid molecule would fit in the

Table 11-G. Comparison of Commercial Enzymes.

	Lipozyme® TL IM	Lipozyme® RM IM
Microbial source	*Thermomyces lanuginosus*	*Rhizomucor miehei*
Expression system	*Aspergillus oryzae*	*Aspergillus oryzae*
Carrier	silica	phenolic resin
Specificity	*sn*-1,3, but acyl migration occurs	*sn*-1,3
Application	interesterification	CBEs
Temperature optimum	70°C	70°C
Denaturation temperature	90°C	90°C
pH optimum	7–9	5–7

active site of the Lipozyme® TL IM lipase inactivating it (Cowan et al., 2008). Priozzi and Halling at the Fourth Italian Conference on Chemical and Processing Engineering held in Florence, Italy, May 2–5, 1999, demonstrated the decrease in relative enzyme activity of Lipozyme® IM 20 *Mucor miehei* lipase immobilized on a weak anion exchange resin) with a lipid source containing various levels of peroxides (see Fig. 11.8). Unlike the consumption of chemical catalyst due to impurities that result in competitive reactions and unwanted byproducts, acids and oxidation products only increase the amount of enzyme required to complete the reaction.

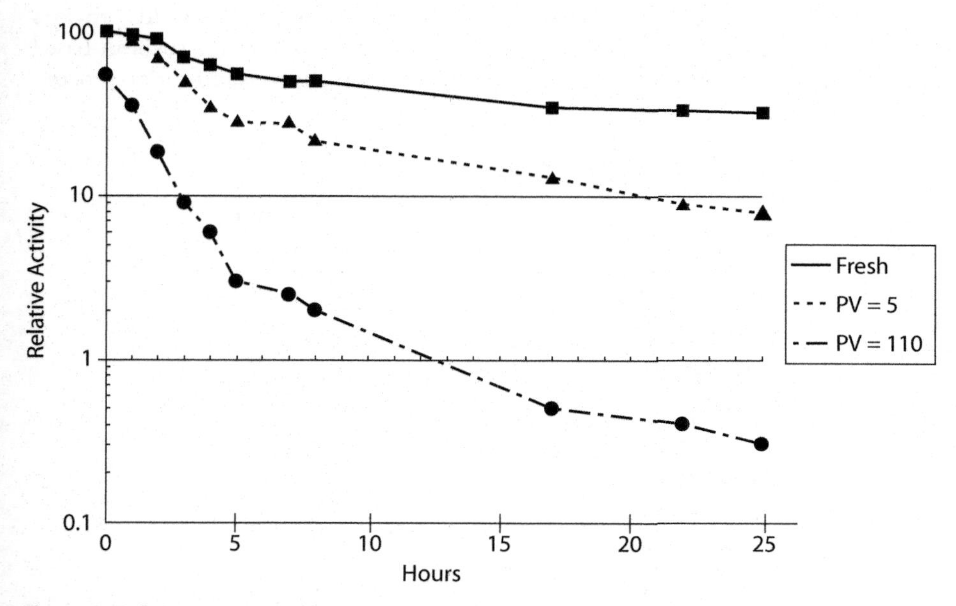

Fig. 11.8. Relative enzyme activity versus time (Diks, 2011).

Other impurities like soaps and phospholipids present in the oil do not increase the consumption of the enzymes, but will affect the separation of the immobilized catalyst from the interesterified product. Table 11-H contains a list of impurities and their suggested limits as suggested by their manufacturer. The feed material for the CIE process is the same as the EIE process.

Yields

Oil losses from the enzymatic process are caused by unwanted hydrolysis due to excess water in the feedstock and adsorption of the material in the immobilized catalyst. The amount of hydrolysis is equal to the amount of water present in the oil and the amount of enzyme used in the reaction as is shown in Fig. 11.1. The amount of fatty acids generated by the hydrolysis of water is calculated by the number of moles of water in the system. If the feedstock containing 0.01% water and 400 grams of enzyme containing 5% water is utilized to produce the reacted fat, then 6.67 moles of fatty acids will be generated. If the fatty acids generated are based on oleic acid, then 1.88 kg of FFA will be generated versus 1.71 kg of FFA based on palmitic acid. The oil losses due to adsorption of the feed material into the enzyme-enzyme carrier are roughly 30% by weight or 120 grams for the 400 grams of enzyme utilized. Utilizing the previous estimate by DeGreyt that an additional loss of 30% will be generated based on the amount of FFA to be removed in the deodorization process, roughly 0.5 kg of additional losses will be incurred (Table 11-I) (DeGreyt, 2004).

The hydrolysis of the oil is minimized by the drying of the oil prior to exposure of the oil to the enzyme.

Comparing the CIE process versus the EIE process for product yields provides a dramatic difference of roughly 14.2 kg for palm and 15.3 kg for soybean-based product per 1,000 kg of starting material for the enzymatic interesterification process.

Table 11-H. Impurities Limits.

	Level
Soap (ppm)	<2
Phosphorus (ppm)	<2
Iron (ppm)	<0.5
Nickel (ppm)	<0.5
Copper (ppm)	<0.5
Citric acid (ppm)	<25
Extracted water pH	6.5–9.0
Anisidine value	<4.5
Peroxide value (meq/kg)	<1.5

Table 11-I. Hydrolysis of Fat.

	Palm (kg)	Soybean (kg)
Hydrolysis	1.77	1.88
Adsorption	0.12	0.12
Deodorization	0.53	0.56
Total	2.42	2.56

Environmental Impact

A Life Cycle Assessment (LCA) addressing the environmental impact of the chemical process versus the enzymatic process was completed by Novozymes for a palm oil based margarine hard stock using international protocols (Nielsen at al., 2007; Wenzel et al., 1997). The environmental load per 1,000 kg of hard stock produced may be found in Table 11-J (Cowan, 2011b)

Conclusions

Advancements in biology have enabled the production of very sophisticated biological catalysts that are able to perform chemistries at "low" temperatures that are specific for both their substrates and the products they produce. These renewable catalysts are produced in simple fermentation batch reactions allowing the production of proteins that are easily denatured and readily disposed of, unlike most chemical catalysts.

It is not enough to be "green" for an industrial process to be implemented over an existing technology, but it must also be economical. The Enzymatic Interesterification technology enables an economic value over the existing CIE process by eliminating various processing steps and dramatically reducing the waste material generated in the process. An additional value typically not included in most calculations is the improved quality of both the oil and finished product.

Table 11-J. Life Cycle Assessment.

Impact potential	Savings by replacing chemical interesterification with enzymatic interesterification per 1,000 kg of palm hard stock
Fossil energy (MJ)	270
Global warming (kg CO_2 equivalents)	22
Acidification (g SO_2 equivalents)	66
Smog formation (g ethylene equivalents)	–4

References

Ajinomoto. Process for modifying oils and fats, Japanese patent publication number 2-203789, published August 1990.

American Oil Chemists' Society Official Method, Peroxide Value Acetic Acid-Isooctane Method Cd 8b-90 reapproved 2009, accessed December 15, 2011.

Bailey, A.E. Chapter 3—Melting and solidification of pure compounds found in Melting and solidification of fats. Interscience Publishers, Inc., New York 1950.

Binder, T.P.; Bloomer, S.; Lee, I.; Solheim, L., Wicklund, L.E. Methods for producing fats and oils, United States Patent Application Publication No. U.S. 2006/0257982 published November 2006.

Brumlik, M.J.; Buckley, J.T. Identification of the catalytic triad of the lipase/acyltransferase from Aeromonas hydrophila. J. Bacteriol., 1996, 2060–2064.

Charton, E.; Macrae, A.R. Substrate specificities of lipases A and B from *Geotrichum candidum* CMICC 335426. Biochimica et Biophysica Acta, 1992, 1123, 59–64.

Coleman, M.H.; Macrae, A.R. Fat process and composition, United States Patent 4,275,081 issued June 1981.

Copeland, R., Chapter 3. Structural components of Enzymes. In Enzymes 2nd ed., Wiley VCH, New York (2000).

Cowan, D.; Hemann, J.; Yee, H.S.; Holm, H.C. "The Influence of lipase type and fat origin on Enzymatic Interesterification." Presented at the 6th EuroFed Lipid Congress, Athens, Greece September 2008.

Cowan, D. "New applications for enzymes in oil processing" presented at the Society of Chemical Industry Meeting of Enzymatic Processing and Modification—Current and Future Trends, University of Ghent, Belgium, June 2011a.

Cowan, W.D. Edible oil processing—modification: Enzymatic Interesterificaton, The AOCS Lipid Library accessed January 10, 2012.

Dayton, C.L.G.; dos Santos, M.A. Continuous process and apparatus for enzymatic treatment of lipids, United States Patent 8,361,763 issued January 2013a.

Dayton, C.L.G.; dos Santos, M.A. Continuous process and apparatus for enzymatic treatment of lipids, United States Patent 8,409,763 issued April 2013b.

De Greyt, W. Chemical versus enzymatic interesterification, presented at IUPAC-AOCS Workshop on Fats, Oils, and Oilseeds Analysis & Production, December 2004, Tunis, Tunisia.

Diks, R. "Processing aspects of enzymatic rearrangement" presented at the Society of Chemical Industry meeting of Enzymatic Processing and Modification—current and future trends, University of Ghent, Belgium, June 2011.

Eigtved, E. Immobilized *Mucor miehe* lipase for interesterification. United States Patent 4,798,793 issued January 1989.

Garret, R.A.; Grisham, C.M. Enzymes—Kinetics and specificity. In Biochemistry 4th ed., Brooks/Cole Boston Mass (2010).

Hansen, T.T.; Eigtved, P. "A new immobilized lipase for interesterification and ester synthesis" in the Proceedings of the world conference on emerging technologies in the fats and oils industry, A.R. Baldwin, Ed., AOCS Press, Urbana, IL, 1986.

Holm, H.C.; Cowan, D. The evolution of enzymatic interesterification in the oils and fats industry, Eur. J. Lipid Sci. Tech. 2008, 110, 679–691.

http://www.chemicalbook.com/ProductChemicalPropertiesCB1391710_EN.htm, Accessed January 10, 2012a

http://www.chemicalbook.com/ProductChemicalPropertiesCB7406890_EN.htm, Accessed January 10, 2012b

Kellens, M.; Gibon, V.; Petrauskaite, V.; Maes, J. Enzyme Interesterification Process. United States Patent Application Publication No. U.S. 2012/0270283 published October 2012.

Kennedy, K.; Fewtrell, M.S.; Morley, R.; Abbott, R.; Quinlan, P.T.; Wells, J.C.K.; Bindels, J.G.; Lucas, A. Double-blind, randomized trial of synthetic triacylglycerol in formula-fed term infants: effects on stool biochemistry, stool characteristics, and bone mineralization. Am. J. Nutr. 1999, 70(5) 920–927.

Kerovuo, J. Declaration of Janne Kerovuo under 37 CFR § 1.132 for 10/556, 816 December 2008.

Lampert, D. Processes and Products of Interesterification, Chapter 12 of Introduction to fats and oils technology, editors O'Brien, Farr and Wan, AOCS Press, Urbana, IL, 2000.

Laning, S.J. Chemical Interesterification of Palm, Palm Kernel and Coconut Oils, JAOCS, 1985, 62(2) 400–407.

Lau, F.Y.; Hammond, E.G.; Ross, P.F. Effect of randomization on the oxidation of corn oil, JAOCS 1982, 59(10) 407–411.

Lee, I. Methods for producing fats and oils, United States Patent No. 7,452,702 issued November 2008.

Liu, L.; Lampert, D.S. Partial interesterification of triacylglycerols, United States Patent No. 6,238,926 issued May 2001.

Lo, Y.C.; Handel, A.P. Physical and chemical properties of randomly interesterified blends of soybean oil and tallow for use in margarine oils. JAOCS, 1983, 60, 815.

Macrae, A.R. Lipase-Catalyzed Interesterification of Oils and Fats. JAOCS, 1983, 60(2).

Macrae, A.R. Microbial lipases as catalysts for the interesterification of oils and fats. In: Biotechnology for the oils and Fats Industry, C. Ratledge. P. Dawson. J. Rattray, Eds. AOCS Press: Champaign, IL, 1985, pp. 189–198.

Macrae, A. R.; How, P. Rearrangement Process, United States Patent 4,719,178 issued January 1988.

Matsuo, T.; Sawamura, N.; Hashimoto, Y.; Hashida, W. Methods for enzymatic transesterification of lipid and enzyme used therein, United States Patent 4,472,503 issued November 1983a.

Matsuo, T.; Sawamura, N.; Hashimoto, Y.; Hashida, W. Methods for enzymatic transesterification of lipid and enzyme used therein, United States Patent 4,416,991 issued November 1983b.

Nielsen, P.M.; Oxenboll, K.O.; Wenzel, H. LCA case study: Cradle-to-gate environmental assessment of enzyme products produced industrially in Denmark by Novozymes A/S. International Journal of LCA, 2007, 12, 432–438.

Nu-Check Prep, Inc., 2007–2008 product catalog. Elysian, Minnesota.

Park, D.K.; Terao, J.; Matsushita, S. Influence of Interesterification on the autoxidative stability of vegetable oils. Agric. Biol. Chem. 1983, 47(1), 121–123.

Pedersen, L.S.; Pearce, S.W.; Holm, H.C.; Husum, T.L.; Nielsen, P.M. Enzymatic Oil Interesterification. United States Patent Application number 2008/024189 published October 2008. Prior Art reference, "Translation of Demandant's Exhibit No. 1 (JP 2-203789)," July 2011.

Peeters, E.H.G.; Kruidenberg, M.B.; Dell, A.J. Enzymatic modification in a continuously regenerated bed column. United States Patent Application Publication No. U.S. 2007/0243591 published October 2007.

Peeters, E.H.G.; Kruidenberg, M.B.; Dell, A.J. Enzymatic modification in a continuously regenerated bed column. United States Patent Application Publication No. U.S. 2011/0227828 published February 2011.

Shu, Z.Y.; Yan, Y.L.; Yang, J.K. *Aspergillus niger* lipase: gene cloning, over-expression in *Escherichia coli* and in vitro refolding. Biotechnol. Lett. 2007, 29, 1875–1879.

Sreenivasan, B. Interesterification of Fats, JAOCS, 1978, 55(11) 796–805.

Tautorus, C.L.; McCurdy, A.R. Effect of randomization on oxidative stability of vegetable oils at two different temperatures, JAOCS, 1990, 67(8) 525–530.

Wenzel H.; Hauschild, M.; Alting, L. Environmental assessment of products. Volume 1: Methodology, tools and case studies in product development. Chapman and Hall. 1997.

Yamane, T. Enzyme technology for the lipids industry: an engineering overview, JAOCS, 1987, 64, 175–181.

Yemc, B.; Cowan, D.; Lipozyme RM IM/Lipozyme TL IM private communication.

Zalewski, S.; Gaddis, A.M. Effect of transesterification of lard on stability, Antioxidant-synergist efficiency, and rancidity development, JAOCS, 1967, 44, 576–580.

12

CLA Production by Photo-isomerization of Linoleic Acid in Linoleic Acid Rich Oils

Vishal Jain[1] and Andrew Proctor[2]

[1]Mars Global Chocolate, Elizabethtown, Pennsylvania, USA and

[2]University of Arkansas Department of Food Science, Fayetteville, Arkansas, USA

Introduction

Conjugated linoleic acid (CLA) is a mixture of positional and geometric isomers of linoleic acid (*cis*-9, *cis*-12 octadecadienoic acid) with the two conjugated double bonds at various carbon positions in the fatty acid chain. Each double bond can be *cis*- or *trans*- and the conjugated double bonds appear between carbon 7 and 14. About 28 isomers of CLA have been identified by Adolf et al. (1999).

Conjugated linoleic acid came to prominence when Pariza and his group reported anticarginogenic health benefits of CLA found in beef. Since then, several researchers have looked into the health benefits of natural and synthetically-prepared CLA. The earliest reported research about CLA was by Booth et al. (1935) showing the fatty acids obtained from butter fat by short time saponification had absorption at 230 nm. However, similar major components of butter treated in a similar manner did not exhibit absorption at 230 nm. Later, Moore et al. (1937) found that the absorption at 230 nm was due to the presence of the conjugated double bonds. Studies of milk fat fractions by Hilditch and Jasperson (1941) suggested that the conjugated unsaturation was associated with the C18 polyunsaturated fatty acids.

Preliminary studies by Ip et al. (1994) showed that approximately 3 g CLA/day are required to obtain the health benefit effects in humans, although no RDA value exists for CLA. However, CLA intake from natural sources is around 10% of the suggested intake, and any increase of beef or dairy products would significantly increase the levels of saturated fats in the diet. This finding led to synthesis of CLA with simpler, cheaper, feasible means.

Health Benefits of CLA

The most common CLAs obtained in nature are the *cis*-9, *trans*-11 octadecadienoic acid and the *trans*-10, *cis*-12-octadecadienoic acid, and most studies report health benefits for the two isomers. Ha and Pariza (1987) discovered that CLA fatty acids from grilled beef inhibited chemically-induced mutagenicity and the genesis of tumors in the skin of mice. This work led to further studies to examine the anticarcinogenic effects. Ip et al. showed that carcinogen-induced carcinogenesis in rats can be prevented by CLA by suppressing the proliferative activity in the rat mammary glands (Ip et al., 1991; 1994). A commercial blend of CLA with 2 isomers (*cis*-9, *trans*-11 CLA and *trans*-10, *cis*-12 CLA) taken at a ratio of 0.5–1.5% of energy intake inhibits the initiation and promotion of chemically induced skin, mammary, and colon cancer in different animal models (Haugen et al., 2003). Studies by Lee and Kritchevsky (1994) showed that rabbits fed with hypercholesterolemic diet with CLA exhibited less atherosclerosis.

Mice supplemented with only 0.5% of CLA had up to 60% lower body fat and up to 14% increased lean body mass compared to controls (Park et al., 1997). The CLA-fed animals showed greater activity of enzymes involved in the delivery of fatty acids to the muscle cells and the utilization of fat for energy, while the enzymes facilitating fat deposition were inhibited. A recent study confirmed that feeding male mice a CLA-enriched diet for six weeks resulted in 43% to 88% lower body fat, especially in regard to abdominal fat (Haugen et al., 2003). CLA appears to directly affect adipocytes and skeletal muscle cells, which are principal sites of fat storage and fat combustion, respectively.

A study using diabetic rats indicated that part of CLA's effectiveness in preventing obesity might lie in its ability to act as a potent insulin sensitizer, thus lowering the insulin resistance and consequently the insulin levels (Risérus et al., 2002) Since elevated insulin is the chief pro-obesity agent, it is important to keep insulin within the normal range. By activating certain enzymes and enhancing glucose transport into the cells, CLA acts to lower blood sugar levels and normalize insulin levels.

Although human studies have shown insignificant body weight loss with CLA enhanced diet, there was a significant body fat reduction (Thom et al., 2001) and a significant reduction in the waist size (Risérus et al., 2001). Larsen et al. (2003) found that *trans*-10, *cis*-12 CLA may produce liver hypertrophy and insulin resistance by a redistribution mechanism of fat deposition resembling lipodystrophy. They also found evidence that CLA decreased the fat content of both human and bovine milk.

Clinical trials showed that 1.8 g CLA/day significantly reduced body fat in healthy exercising men and decreased mean BMI (Ha et al., 1987; Ip et al., 1991). Risérus and others showed that CLA decreases abdominal fat in obese men with metabolic syndrome (Ip et al., 1994). Recently, the *trans-*, *trans*-CLA isomers were shown to have health benefits, in particular, *trans*-9, *trans*-11CLA isomer inhibits bovine aortic endothelial cell proliferation by apoptotic pathway and proliferation of human leukemic cell lines (Haugen et al., 2003). Storey et al. showed that *trans*, *trans*-CLA isomers decreased the ultraviolet radiation (UVR)-induced secretion of interleukin (IL-8), prostaglandin E_2 (PGE_2), and tumor necrosis factor (TNF-α)

in human skin cells responsible for UVR-induced inflammation and carcinogenesis (Lee et al., 1994).

CLA Synthesis

CLA Biosynthesis in Nature

Conjugated linoleic acid is formed as an intermediate during bio-hydrogenation of linoleic acid (LA) to stearic acid in the rumen by *Butyrivibrio fibrisolvens* and other rumen bacteria (Kepler et al., 1966). CLA is also formed by conversion of *trans*-vaccenic acid (TVA) (t-11 $C_{18:2}$) by Δ^9 desaturase in the mammary gland (Corl et al., 2003). Though little information is currently available regarding the biochemical mechanism that regulates the metabolism of the different CLA isomers in the ruminant animals, research suggests that it is the biohydrogenation that needs to be controlled to obtain high CLA content in ruminants (Lawson et al., 2001). Fig. 12.1 shows the biosynthesis mechanism for CLA. Kepler and Tove (1967) showed that *cis*-9, *trans*-11 CLA was produced from LA which was subsequently hydrogenated to TVA. However, the first step in the conversion of LA is isomerization of LA when the double bond is moved from the *cis*-12 to *cis*-11 position forming *cis*-9, *trans*-11 CLA, which is done by a bacterial membrane bound enzyme linoleic isomerase (EC 5.3.1.5). This is followed by hydrogenation of CLA at *cis*-9 position to form TVA. It was later proposed that Δ^9 desaturase of monoenes, such as TVA, from oleic acid to *cis*-/*trans*-CLA could be synthesized exogenously which can occur in non-ruminants. The majority of CLA is in

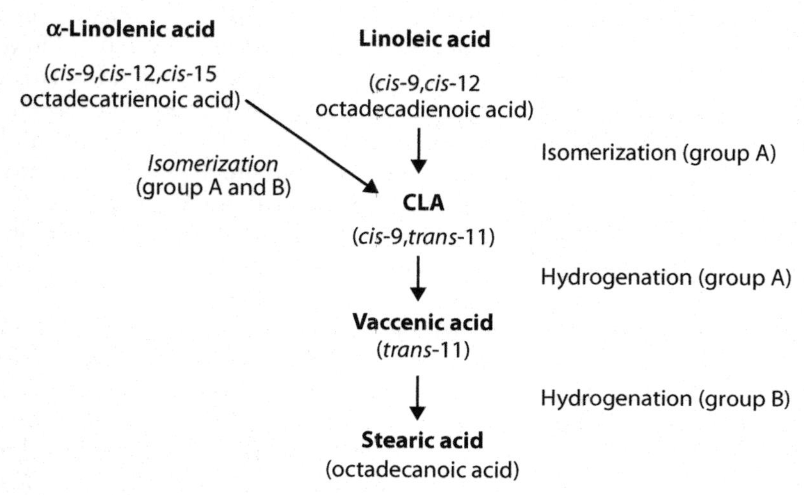

Fig. 12.1. Biohydrogenation of CLA in ruminants.

the rumen in the form of *cis*-9, *trans*-11 isomer, and the transfer efficiency of CLA to milk fat is affected by the presence of different isomers of CLA (Lawson et al., 2001).

The factors that affect the CLA content in milk, meat, and other food products from ruminants are diet, animal breed, and post-harvest related factors (Khanal et al., 2004). While animal-to-animal variation is of great significance, post-harvest related factors appear to be of minor importance. Animal diet is the primary factor and could be manipulated to a great extent for enhancing the concentrations of CLA in food products. Milk fat from cows showed significant correlations to the daily intake of unsaturated fatty acids in diet and the elevation of grazing of cows to CLA obtained in milk fat (Collomb et al., 2004). To study the effect of environmental factors on the CLA production in ruminants, ruminal bacteria was added to commercial preparation of soybean oil and a mixture of soluble carbohydrates. When pH was lowered to 5.0, no CLA isomers were formed, signifying that pH has a great influence on the CLA production by ruminal bacteria (Martin & Jenkins, 2001).

CLA Biochemical Synthesis

CLA Production by Fermentation

Lee et al. (2003) listed the bacteria investigated for CLA production including *Clostridium, Lactobaccillus, Propionibacterium, Lactococcus,* and *Streptococcus.* Many of these have been used in food fermentations, and most of them produced *cis*-9, *trans*-11 CLA as the major LA isomer. More specifically, *L. reteri, L. planetarium, P. freudenreichii* ssp. *freudenreichii,* and *M. elsdenni,* produced *cis*-9, *trans*-11 CLA and *trans*-10, *cis*-12 CLA.

Recently, free linoleic acid obtained by oat lipase activity was isomerized to CLA by fermentation with *Propionibacterium freudenreichii* ssp. *shermanii,* producing CLA (Vahvaselka et al., 2004). The conversion rate was less than 50%. However, 80% of the CLA was the *cis*-9, *trans*-11 CLA isomer. Ricinoleic acid in free fatty acid form has also been successfully converted to CLA using *Lactobaccilus planetarium* (Ando et al., 2003). Most of the CLA was the *trans*-9, *trans*-11 CLA isomer. In a later investigation, an *in vitro* study used a mixed rumen culture to examine the effect of monoensin (an antibiotic) and fish oil on CLA production (Wang et al., 2005). It was found that monoensin reduced the *cis*-9, *trans*-11 CLA relative to fish oil, but increased the *cis*-10, *cis*-12 CLA relative to fish oil and the control.

Immobilized Cells. Recent CLA fermentation reports describe the use of immobilized bacteria to increase CLA yields. Lee et al. (2003) bound *Lactobacillus reuteri* to silica gel to convert LA to CLA by fermentation. The immobilized cells produced 8.6 times more CLA than free cells. Lin et al. (2005) immobilized *Lactobacillus delbrueckii* ssp. *bulgaricus* and *Lactobacillus acidophilus* to either chitisan or polyacrylamide and incubated with LA. Both immobilization methods produced higher yields of CLA than by free cells, but polyacryamide was the most effective. *Lactobacillus delbrueckii* ssp. *bulgaricus* immobilized on polyacryamide produced most CLA.

Producing CLA by current fermentation technology requires LA as a free fatty acid substrate with conventional fermentation media and extensive purification, with cell immobilization to optimize yields. This would be more expensive and tedious than converting soy oil LA to CLA in a single-step process for food use.

Bacterial Enzymes. Production of TAG containing CLA was achieved by esterifying CLA to glycerol with the aid of a *Candida antartica* lipase (Arcos et al., 1998). Enzyme extracts of *Lactobacillus acidophilus* and *Propionibacterium freudenreichii* ssp. *shermanii,* containing linoleic acid isomerase, were successfully used to isomerize LA to CLA *in vitro* (Lin et al., 2003). These studies were on a laboratory scale and used LA only in mg quantities.

CLA Isomerization

Chemical Isomerization Using Methyl Esters. The CLA methyl cis-9, *trans*-11-octadecadienoate has been prepared on a large scale from methyl ricinoleate (Bordeaux et al., 1997). Methyl ricinoleate was purified from castor esters by a partition method. It was converted to the mesylate, which was reacted with a base (1,8-diazabicyclo-undec-7-ene) to give a product that contained 66% of the desired ester. Two urea crystallizations produced a product containing 83% methyl *cis*-9, *trans*-11-octadecadienoate. When methyl linolenate was heated in a potassium hydroxide-ethylene glycol solution at 200°C for 7 h, about 80% conversion (conjugation) was attained. However, the reported method also involves cyclization and other side reactions (Scholfield et al., 1959).

Chemical Isomerization Using Vegetable Oil. In an attempt to overcome the high cost of purification, Kim et al. (2003) produced a high purity CLA from safflower oil. This involved alkali isomerization with sodium hydroxide in propylene glycol in an oil bath, followed by cooling, pH reduction with phosphoric acid, addition of water, and partition of CLA into hexane. The CLA was then crystallized as a urea-inclusion complex by adding to a urea-saturated methanol solution and filtered. The filtrate contained 95% CLA and was further purified by crystallization. Since 1 kg of oil was used, it represented a larger scale preparation than previously described, but the CLA was not part of the oil, and its production required many processing steps. As safflower oil is relatively expensive, Yang and Lui (2004) used the more commonly used soy oil to synthesize CLA. They claimed to be the first to produce CLA from soy oil by a 15-step method involving saponification, alkali isomerization, multiple partition extractions, crystallization, and evaporation processes, similar to those used by Kim et al. (2003). Iwata et al. (1999) subjected fat or oil containing linoleic acid to alkali isomerization reaction in an alkali-propylene glycol solution, and linoleic acid is converted or transformed into CLA.

Hydrogenation. Ju et al. (2003) showed that soy oil hydrogenation could be manipulated to produce CLA by added sulfur. The optimum amount of added sulfur varied with the amount of nickel catalyst used. Results showed that 30 ppm of sulfur with 0.05% nickel produced almost 123 mg CLA/g oil after 1 h of hydrogenation with 0.049 MPa hydrogen pressure at 220°C. This is only 12% CLA, of which *trans-*, *trans-*CLA was the prevalent isomer, and accompanied by the production of 19% *t*-oleic acid.

Heterogeneous Catalysis. Studies to isomerize LA, as a free fatty acid, to CLA over a heterogeneous supported metal catalyst have been favored as a cleaner technology over homogeneous catalysis (Reaney et al., 1999). Bernas et al. (2004) provided a review of CLA production by heterogeneous catalysis and a description of comparative studies of isomerization of LA to *cis*-9, *cis*-11 and *cis*-10, *cis*-12 CLA. The isomerization was done over aluminum hydroxide and carbon supported metal catalysts (Os, Ir, Pt-Rh) under nitrogen at atmospheric pressure. Their study describes the effects of many variables on CLA production in a complex process that requires careful control to optimize the catalysis, for example, choice of metal, support material, purity of raw materials, solvent, pre-adsorption of hydrogen on metal catalyst, agitation rate, and others. However, these studies were not with food-grade materials and used technical grade LA.

Photoprocessing to Synthesize CLA

Photoirradiation of Methyl Esters

Seko et al. (1998), Canaguier et al. (1984, 1986), and Julliard et al. (1987) obtained a 80% yield of CLA methyl esters from LA methyl esters dissolved in petroleum ether, benzene or carbon disulfide after exposure to high intensity light in the presence of an iodine sensitizer. The isomer content was not reported.

Photoirradiation of Vegetable Oil

Overhead Irradiation System. Gangidi and Proctor (2004) produced CLA isomers from soybean oil by photoisomerization of soybean oil linoleic acid and studied the oxidation status of the oil. Refined, bleached, and deodorized soy oil with added iodine concentrations was exposed to 100 W mercury lamp for 0 to 120 h. The lamp was placed 45 cm above the beaker containing oil which resulted in insufficient exposure of oil to the light (Fig. 12.2). There was no sign of oxidation as measured by ^{1}H NMR (absence of the aldehyde protons in the oil), and infrared spectroscopy (lack of a hydroperoxide). While this is not conclusive evidence regarding the absence of oxidation, it strongly suggests that radiation did not greatly compromise oxidative quality. However, the system produced low yields of about 0.6% of total oil of *cis*-9, *trans*-11 and *trans*-10, *cis*-12 CLA isomers, each in presence of about 0.33% iodine and required long irradiation times of about 90 hours, due to minimal exposure of oil to light. Yet, the study was a significant advancement in showing that *"in principle"* CLA-soy oil could be inexpensively produced by photoirradiation.

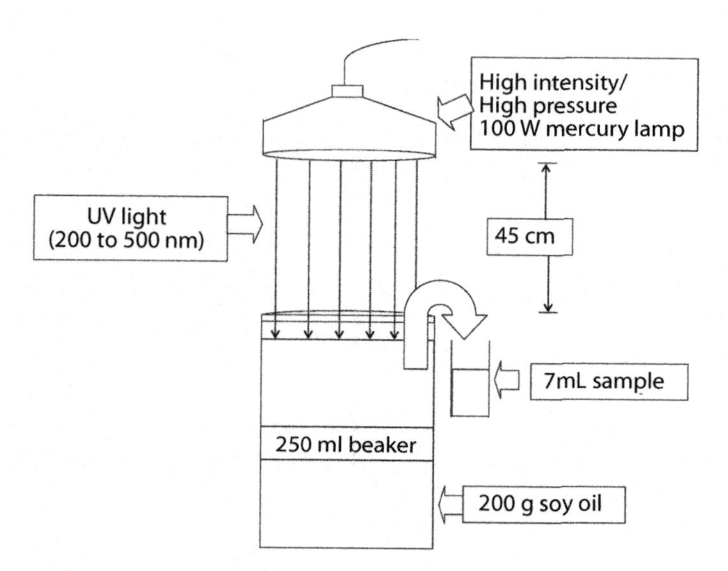

Fig. 12.2. Conventional photochemical reaction setup.

Nevertheless, this study revealed three problems that needed to be addressed:

a) Increasing CLA yields by increased exposure of oil to radiation without increasing the probability of lipid oxidation,
b) Reducing the processing time to increase efficiency and throughput, and
c) Removal of iodine and possible iodo-compounds from the oil.

Central Irradiation System. Jain and Proctor used a customized photocatalysis reaction system (Fig. 12.3) consisting of a borosilicate-glass jacketed reaction vessel. The vessel was flat-bottomed to facilitate magnetic stirring. The reaction vessel accommodated a double-walled, borosilicate glass immersion well (Ace Glass Inc.) with inlet and outlet tubes for cooling. The annular space formed between the reaction vessel, and the immersion well held the oil for irradiation. The immersion well supported a 100-W UV/visible medium-pressure, quartz, mercury-vapor lamp (Ace Glass Inc.). The assembly was placed in a black-walled cabinet to absorb the outgoing radiations from the assembly to ensure a safe operating environment. The reaction vessel and the immersion well were connected to a water supply to maintain the temperature of the oil between 22 and 25°C, necessary for optimum efficiency and maximum lamp life. The customized reaction system facilitated maximum exposure of the oil to light.

Commercial soy oil was deaerated with a sonicator for 30 min. Oil was heated to 70°C, to facilitate iodine solubilization, while flushing with nitrogen to avoid oxidation. Then, 0.25% iodine was added to the oil, and the contents in the beaker were

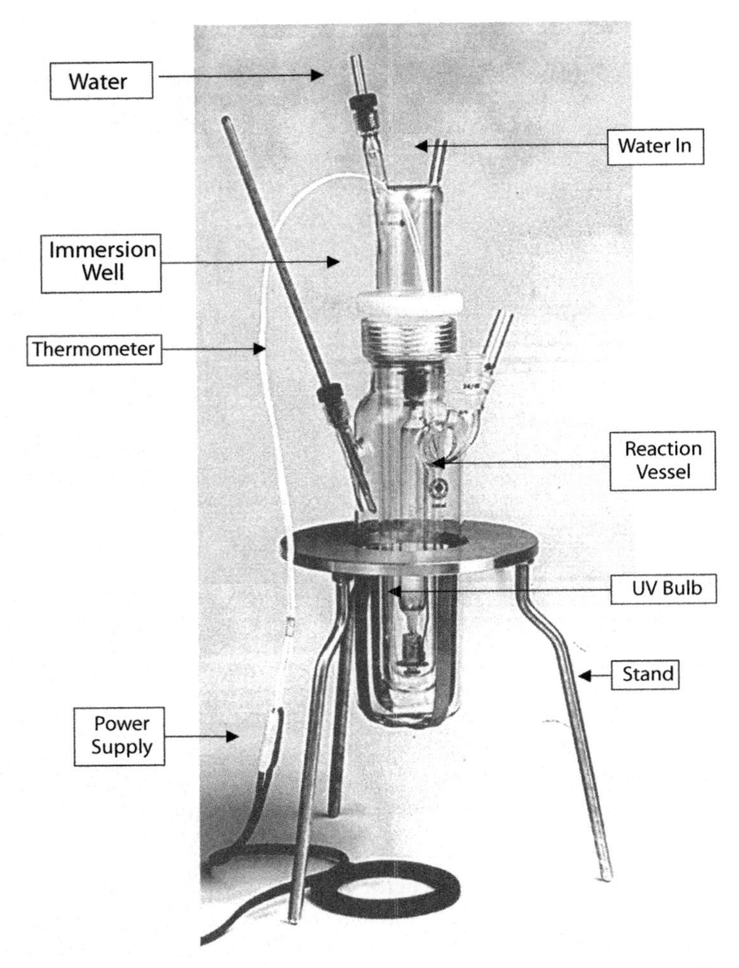

Fig. 12.3. Customized photochemical reaction system. (Picture obtained from Ace Glass Inc., Vineland, New Jersey).

stirred until the iodine completely dissolved (Gangidi & Proctor, 2004). The assembly was placed on a magnetic stirrer, and the oil was stirred continuously during irradiation. The soy oil was irradiated for 240 h, and two 5-mL samples were collected at 24-h intervals in 10-mL amber colored glass vials.

Fig. 12.4 shows the effect of irradiation time on total CLA isomers as a percentage of total oil on a weight-by-weight basis. Irradiating soy oil for 240 h with 0.25% iodine yielded 27.3% CLA isomers. Because soy oil contains 50% linoleic acid (LA), this represents a 54% conversion rate of LA to CLA. A higher CLA yield can be attributed to enhanced oil exposure to light by use of the customized photochemical

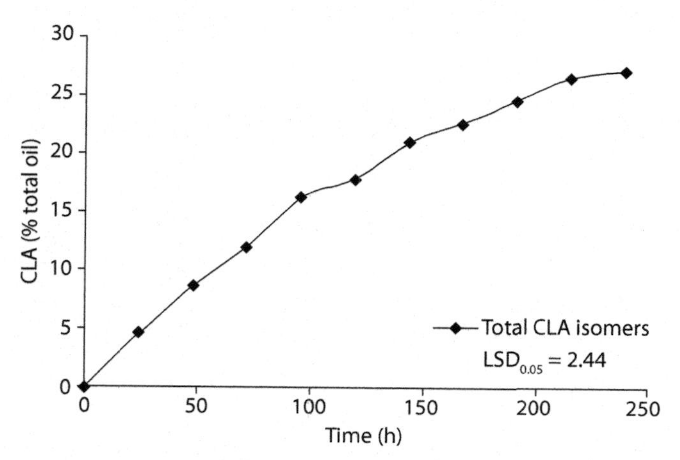

Fig. 12.4. Increase of total CLA isomers (% total oil, w/w) with time of irradiation in soy oil with 0.25% iodine as a photosensitizer.

reaction system as compared to previous studies by Gangidi and Proctor (2004). Fresh soy oil and the 144-h irradiated oil sample with 0.15% iodine showed no measurable hexanal peak at 2.9 min. This indicated that the irradiated oil sample had no significant increase in oxidation levels as compared to control.

Soy oil fatty acid profiles, as by partial GC-FID chromatograms before and after photoirradiation, are shown in Fig. 12.5. The data show that CLA isomers are produced at the expense of linoleic acid but also with some loss of linolenic acid with *trans,trans* being the major CLA isomers.

Pilot Scale Photo-Irradiation System. Jain et al. (2008) designed a customized pilot scale photoirradiation system (Fig. 12.6) to produce CLA-rich soy oil. The innovation of the system was the illuminated laminar flow unit, which allowed maximum exposure of the oil for linoleic acid photoisomerization. The following is a discussion of each unit in the system:

Pre-Irradiation Mixer. The pilot plant unit consisted of a 10-L capacity reaction vessel with a temperature coil and a mechanical stirrer to hold degassed oil for irradiation. In this vessel, oil could be heated under nitrogen blanket and constant stirring to mix resublimed iodine catalyst crystals. Oil with dissolved iodine was pumped in the illuminated laminar flow unit by a peristaltic pump, capable of pumping at various flow rates.

Irradiation in Illuminated Laminar Flow Unit (ILFU). This unit was designed to increase oil-light exposure and penetration relative to laboratory irradiation systems (Gangidi & Proctor, 2004; Jain & Proctor, 2007a). Two borosilicate glass plates were fixed in a 303 stainless steel frame with Teflon-coated grooves in the frame to allow laminar flow. The distance between the plates could be varied by placing the plates in

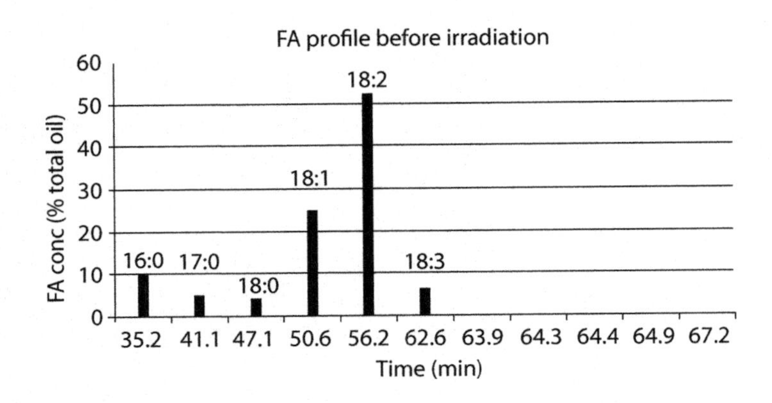

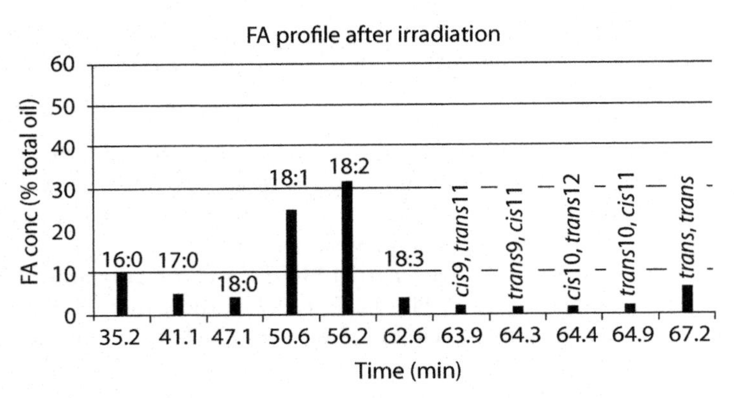

Fig. 12.5. Soy oil fatty acid profiles as shown by partial GC-FID chromatograms before and after photoirradiation.

appropriate grooves within the frame. This would vary the volume of the ILFU from 1 to 7 L. A third plate permanently fixed at a distance of 0.5 cm from the first plate served as a water jacket to provide cooling of oil during irradiation. A thermocouple probe was inserted in the steel frame to monitor the oil temperature. Three 450-W UV/visible lamps (Ace Glass Inc., Vineland, NJ) were placed on the water-jacketed side of the laminar flow unit to ensure maximum light exposure. The laminar flow unit and the UV/visible lamps were placed in a wooden box.

Post Irradiation Adsorption Unit. A wet adsorption process could be conducted in a 10-L slurry tank connected parallel to the reaction vessel and the pump to remove the residual iodine. Water (0.3%) was added to the oil and thoroughly stirred. Adsorbent could be added to the slurry tank from the adsorbent hopper. The slurry was then pumped through a batch filter. Filtered oil was cooled to room temperature and stored in the CLA oil tank.

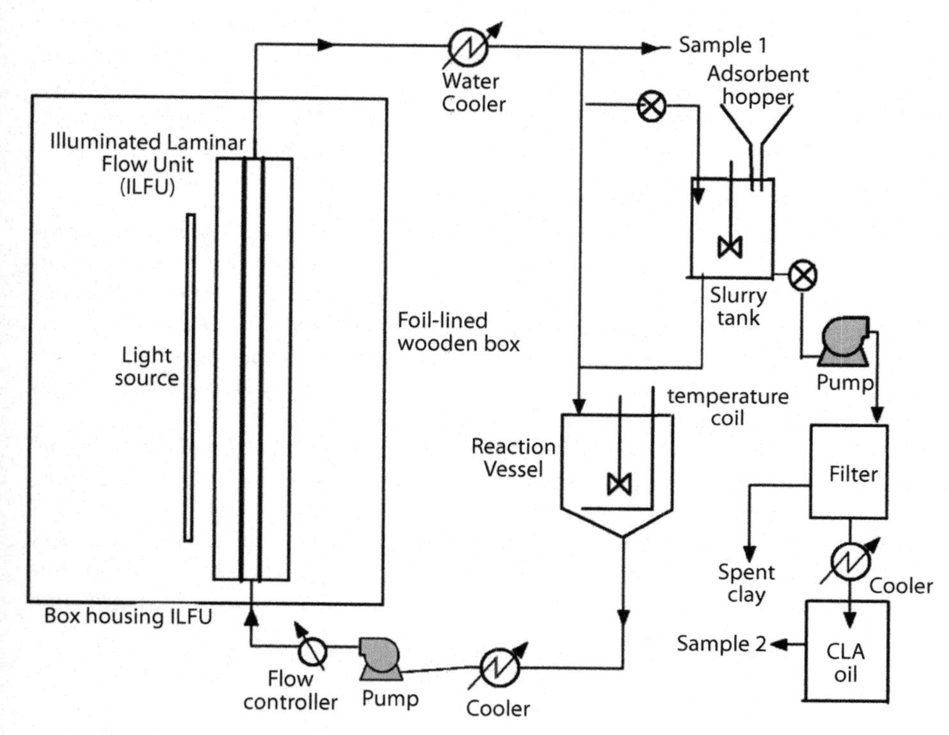

Fig. 12.6. Pilot plant processing unit to produce CLA-rich soy oil by photoisomerization.

Table 12-A. Comparison of CLA Isomers in 144-h-Irradiated Soy Oil Containing 0.15% Iodine with and without Stirring the Contents of the Reaction Vessel.

	CLA isomer concentration [% total oil (w/w)]	
CLA isomer	**Stirring**	**No stirring**
Linoleic acid	30.86a	37.21b
cis-9, *trans*-11 CLA	1.78a	1.53b
trans-10, *cis*-12 CLA	1.72a	1.47b
trans, trans-CLA[a]	17.56a	12.17b
Total CLA	23.79a	17.14b

[a]Constituted by *trans*-8, *trans*-10 CLA, *trans*-9, *trans*-11 CLA, and *trans*-10, *trans*-12 CLA. Values are means, n=3. Means with different letters within a row differ significantly, $p < 0.05$.

This system proved to be a vast improvement over previous photo-processing systems with respect to yields and linoleic acid to CLA conversion rate. The optimized system yielded 22% total CLA and 3.5% *cis-*, *trans-*, and *trans-*, *cis-* CLA. However, almost 80% of CLA isomers identified were the *trans-*, *trans-*CLA isomers—*trans-*9, *trans-*11 CLA, *trans-*10, *trans-*12 CLA, and *trans-*11, *trans-*13 CLA.

Factors Affecting Photo-Processing of Soy Oil

Mixing. As with any reaction mechanism, mixing increased CLA yields in photo-processed soy oil. Mixing causes uniformity in light exposure to oil molecules in presence of iodine, and the entire reaction mixture progresses toward equilibrium together. Table 12-A shows a comparison of the CLA isomers formed in irradiated soy oil with 0.15% iodine concentration with and without stirring the contents in the reaction vessel. No stirring may be one of the reasons for low yields of CLA isomers as experienced in previous studies by Gangidi and Proctor (2004).

However, continuity of irradiation of the reaction mixture also plays an important role in total yields realized. The pilot scale system in a continuous mode produced only about 2.5% total CLA and 0.25% of *cis-*, *trans-* and *trans-*, *cis-*CLA isomers in 12 h as compared to 5.7% total CLA and 0.94% *cis-*, *trans-* and *trans-*, *cis-*CLA isomers in a static mode, all other parameters kept constant. In a continuous system, oil is irradiated in the ILFU, but light exposure is limited to transit time

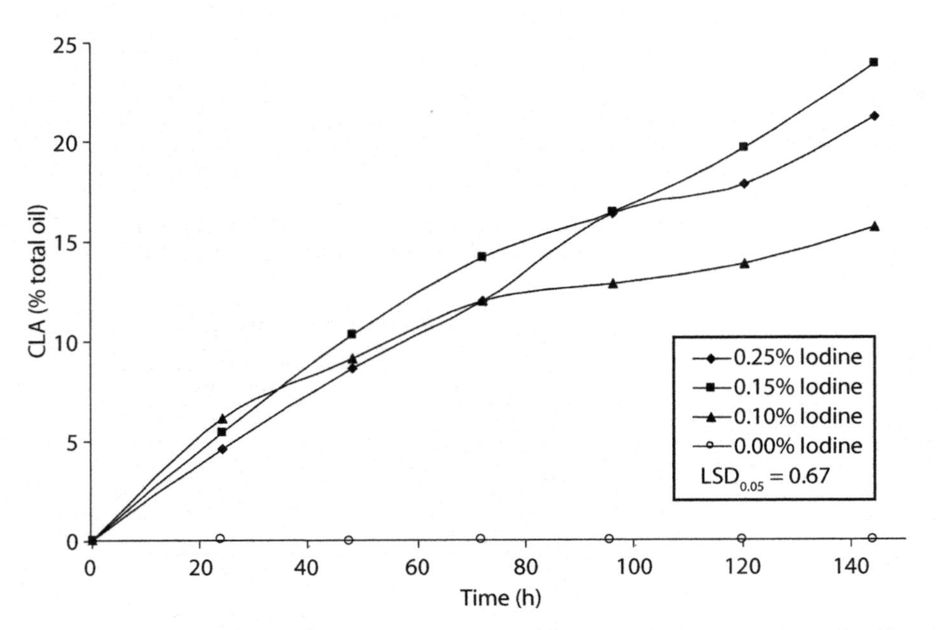

Fig. 12.7. Effect of iodine concentration on total CLA isomers (% total oil, w/w) content in irradiated soy oil.

in ILFU which may not be long enough for optimum iodine radical production and subsequent CLA formation. Furthermore, it was hypothesized that intermediates in CLA formation may revert to linoleic acid in the dark.

Reflective Surfaces. Increasing the availability of UV light for photo-processing increased the CLA yields. Reflective sheets placed in the box holding the ILFU significantly increased the CLA yields in soy oil. Total CLA isomers increased almost three-fold from 5.7 to 16.4%, and *cis-*, *trans-* and *trans-*, *cis-*CLA isomers increased from 1 to 2.4% with the aid of reflective surfaces in the ILFU. The UV light was better utilized by using the reflective surfaces, and thus increased CLA yields.

Iodine Concentration. Optimum amount of iodine available in the reaction mixture plays a significant role to the yields realized and conversion rates of linoleic acid to CLA. Fig. 12.7 shows the effect of different iodine concentrations on the total CLA yield as a percentage of total oil on a weight-by-weight basis with a central irradiation system. Irradiation of soy oil for 144 h with 0.15% iodine produced 23.8% total CLA isomers, which is significantly higher than total CLA isomers produced by irradiating soy oil with 0.10 and 0.25% iodine. Although soy oil with 0.10% iodine started with a higher conversion rate, the isomerization rate subsequently fell below the conversion rates of oil with 0.15 and 0.25% iodine.

Fig. 12.8 shows the total CLA isomer concentrations in irradiated soy oil after 12 h of 0.5-cm-thick oil lamella with 0.1, 0.25, and 0.5% iodine concentrations with a pilot scale system. Soy oil with 0.1% iodine yielded 2.5% total CLA with an increase to 16.5% total CLA with 0.25%. There was a reduction to 14% total CLA

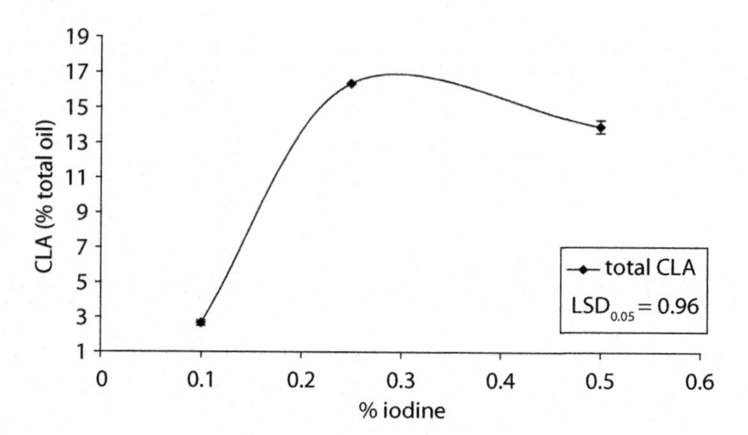

Fig.12.8. Effect of iodine concentration on total CLA yields in soy oil with 0.1, 0.25, and 0.5% iodine obtained by photoisomerization in pilot plant scale system after 12 h. Values are means of duplicates analyses of two replications. Bars represent standard error of the mean.

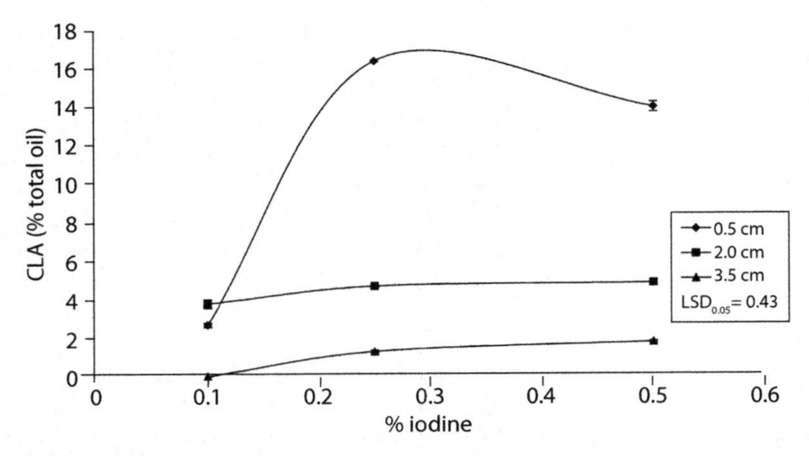

Fig. 12.9. Effect of different levels of oil-layer thickness and iodine concentration on total CLA yields by photoisomerization of soy oil in pilot plant scale system after 12 hr at 35°C. Values are means of duplicates analyses of two replications. Bars represent standard error of the mean.

when the iodine concentration was increased to 0.5%. This reduction is probably due to low light penetration of the oil, as oil is darkened at higher iodine levels. At lower iodine concentrations, iodine was a limiting factor for CLA production. The maxima appear to be around 0.3% iodine.

Similar iodine concentration effects were observed on *cis-*, *trans-*/*trans-*, *cis-*CLA isomer concentrations (*cis-*9, *trans-*11 CLA, *trans-*9, *cis-*11 CLA, *cis-*10, *trans-*12 CLA, and *trans-*10, *cis-*12 CLA). With 0.25% iodine, maximum concentrations of 2.3% of these isomers were formed which is about 15% of the total CLA formed. The concentration of *cis-*9, *trans-*11 CLA and *trans-*10, *cis-*12 CLA produced in photoisomerized soy oil was 70 times the concentration found in beef and milk (Beaulieu et al., 2002; Veth et al., 2004). The other 85% CLA comprises *trans-*, *trans-*CLA isomers, consisting of *trans-*8, *trans-*10 CLA, *trans-*9, *trans-*11 CLA, and *trans-*10, *trans-*12 CLA isomers.

Lamella Thickness. Film thickness affects the conversion rate and the amount of total CLA produced. Fig. 12.9 shows total CLA isomer concentrations in irradiated soy oil after 12 h at 35°C with 0.5-, 2.0-, and 3.5-cm thick oil lamellas and 0.1, 0.25, and 0.5% iodine concentrations in the pilot plant scale system. The largest concentration of total CLA, 16.4%, was obtained with 0.25% iodine and a 0.5-cm oil lamella. Lamella thicknesses of 2 and 3.5 cm produced only 4.5 and 1.5% total, respectively, with iodine concentration having little effect on CLA levels in the thicker lamella. This may be due to lack of full light penetration in thicker oil lamellas.

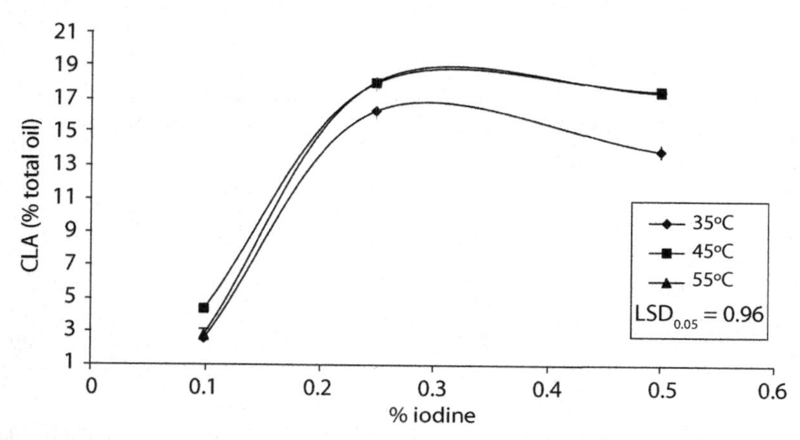

Fig. 12.10. Effect of oil temperature at different iodine concentrations on total CLA yields in soy oil at 0.5 cm oil-layer thickness by photoisomerization in pilot plant scale system. Values are means of duplicates analyses of two replications. Bars represent standard error of the mean.

Total CLA and *cis-, trans-/trans-, cis-*CLA isomers produced by photoisomerization were calculated on a weight basis after 12 h for each lamella thickness. The largest mass of total CLA, about 165 g, was produced with a 2-cm-lamella followed by 150 g by a 0.5 cm lamella. This corresponds to 30 and 22 g *cis-, trans-/trans-, cis-*CLA isomers by 2.0- and 0.5-cm lamella, respectively. The comparison

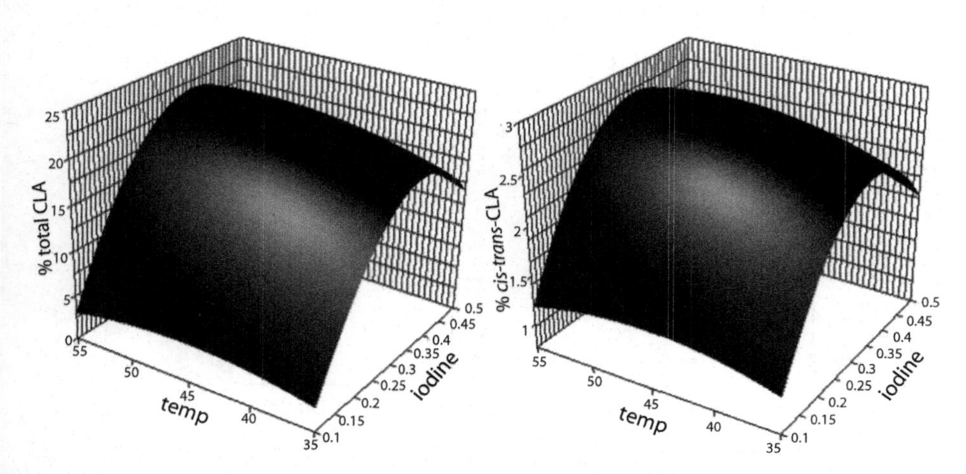

Fig. 12.11. Response surfaces plot of a) total CLA isomers (% total oil) and b) *cis-, trans-/trans-, cis-*CLA isomers (% total oil) in soy oil by photoisomerization with different values of oil temperature and iodine concentration at 0.5 cm oil-layer thickness after irradiating for 12 h in ILFU.

suggests that the overall rate of formation of CLA isomers decreased with an increase in oil-layer thickness as a result of decreased light penetration through the oil. However, the total CLA mass produced increased as more oil was irradiated. Only about 96 g total CLA and 8 g *cis-*, *trans-*/*trans-*, *cis-*CLA isomers were produced by a 3.5-cm lamella after 12 h, suggesting a significant decreased light penetration. Thus, lamella thickness plays an important role in the rate of formation of CLA isomers.

Temperature. Higher temperature affects the reaction kinetics and increases the rate of reaction. Fig. 12.10 shows total CLA isomer concentrations at 35, 45, and 55°C with 0.1, 0.25, and 0.5% iodine after 12-h irradiating a 0.5 cm lamella in a pilot scale system. A temperature of 55°C consistently produced the greatest CLA concentration at each iodine concentration, and was similar to that obtained at 45°C, with 0.25% and 0.5% iodine. Although less CLA was produced at 35°C, the pattern of formation over the range of iodine levels was the same. The increase in temperature could facilitate the increase in non-photochemical reactions, but the photochemical reaction may be a limiting factor above 45°C. In addition, changes in oil physical properties with temperature, such as decrease in density, may facilitate light penetration.

Response Surface Methodology (RSM). Fig. 12.11a shows a response surface plot of total CLA isomers (% total oil) that can be produced in soy oil using different iodine concentrations and oil temperatures and a 0.5-cm-oil lamella. The plot predicts that a maximum of 22% total CLA isomers can be formed in RBD soy oil with 0.35%

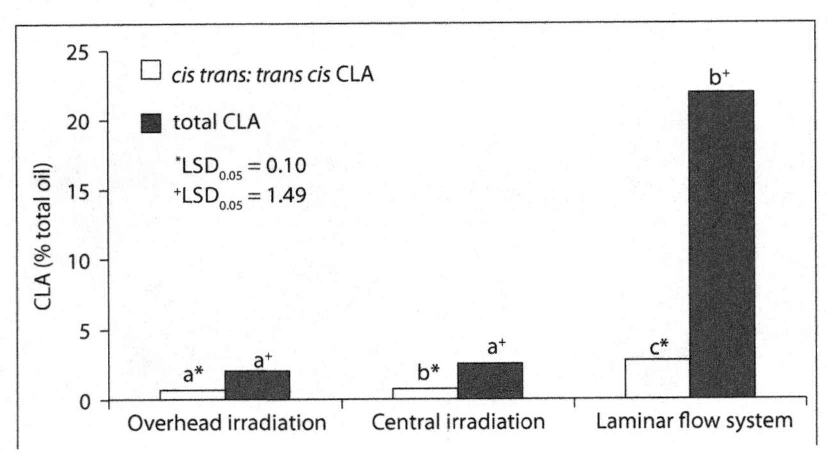

Fig. 12.12. Comparison of optimized CLA yields obtained by photoisomerizing soy oil linoleic acid after 12 h by a) overhead irradiation system (Gangidi and Proctor 2004), b) central irradiation system (Jain and Proctor 2006), and c) pilot scale laminar flow system. Values are means of duplicates analyses of 2 replications. Bars represent standard error of the mean. Similar colored bars with different letters are significantly different ($p < 0.05$).

Table 12-B. Conjugated Linoleic Acid (CLA) Concentration per Serving and per 100 g of CLA-rich Soy Oil Relative to Beef and Milk.

	CLA concentration (g)					
	CLA-rich soy oil[a]		Beef (Beaulieu et al., 2002)		Whole milk (Veth et al., 2004)	
CLA isomer	½ oz	per 100g serving	3 oz	per 100g serving	8 oz	per 100g serving
c-9, *t*-11 CLA	0.12	0.85	0.02	0.02	0.06	0.03
t-9, *c*-11/*c*-10, *t*-12 CLA	0.16	1.11	—[b]	—	—	—
t-10, *c*-12 CLA	0.12	0.85	<0.01	<0.01	<0.01	<0.01
t-, *t*-CLA[c]	2.69	19.19	—	—	—	—
Total CLA	3.09	22.00	0.10	0.12	0.06	0.03

[a]CLA-rich soy oil obtained by optimized conditions in pilot scale photoirradiation system.
[b]Values not reported in literature.
[c]Consists of *trans*-8, *trans*-10 CLA, *trans*-9, *trans*-11 CLA, and *trans*-10, *trans*-12 CLA.

iodine catalyst at 48.1°C. Under similar conditions, 2.81% of *cis*-, *trans*-/*trans*-, *cis*-CLA isomers can be produced in 12 h (Fig. 12.11b).

Fig. 12.12 shows a comparison of the optimized CLA yields obtained by different published photoirradiation methods in photoisomerized soy oil after 12 h. The overhead irradiation system (Gangidi & Proctor, 2004) and the central irradiation system (Jain & Proctor 2006) could yield about 2 and 2.5% of total CLA isomers, respectively. The pilot scale laminar flow system performed nine times better than the previous methods and produced 22% total CLA isomers. The pilot scale system could also yield 3.5 times the *cis*-, *trans*-/*trans*-, *cis*-CLA isomers than the previous methods.

Table 12-B shows a comparison of CLA isomers produced in CLA-rich soy oil obtained by this study relative to that in beef and milk. A half-ounce serving of CLA-rich soy oil can provide 3.1 g of total CLA isomers and 0.24 g of *cis*-9,

Table 12-C. Fatty Acid Analysis of Crude, Alkali-refined, Alkali-refined Bleached, and Alkali-refined Bleached and Deodorized Soy Oil Samples after UV Irradiation in the Pilot Plant System for 12 hr with 0.35% Iodine Catalyst.

Soy oil sample	*cis-,trans-; trans-,cis*-CLA[a] (% total oil)	Total CLA (% total oil)
Crude	0.2d	0.2d
Alkali-refined	1.7c	6.4c
Alkali-refined bleached	2.8b	10.3b
Alkali-refined bleached and deodorized	4.3a	16.3a

[a]Comprised of *cis*-9, *trans*-11 CLA, *trans*-10, *cis*-12 CLA, *trans*-9, *cis*-11 CLA and *cis*-10, *trans*-12 CLA isomers.
Data expressed as means, n = 4. Values within a column with different letters differ significantly, $p < 0.05$.

trans-11 CLA and *trans*-10, *cis*-12 CLA isomers as compared with 0.1 g total CLA in a 3-oz serving of steak filet and 0.06 g total CLA in an 8-oz serving of whole milk. These amounts were calculated based on the total fat content of beef (20%) and milk (4%). Seventeen g of CLA-rich soy oil obtained by pilot scale photoirradiation is equivalent to the amount of fat obtained from a 3-oz. steak filet. However, this oil provides 38 times the total CLA isomers and 15 times the *cis*-9, *trans*-11 CLA and *trans*-10, *cis*-12 CLA isomers concentration than the steak. An ounce of CLA-enriched potato-chip product (Jain & Proctor, 2007) using the irradiated soy oil from the pilot plant method could provide 2.5 g total CLA, 75 times higher than a 1-oz steak filet.

Degree of Processing. Amount of CLA isomers produced in soy oil was greatly affected by the degree of processing of the oil (Jain et al., 2008). Table 12-C shows the CLA content in soy oil samples, processed to various degrees of refining, after irradiation. Only 0.2% total CLA was synthesized in crude soy oil sample comprised of *cis*-9, *trans*-11 CLA, *trans*-10, *cis*-12 CLA, *trans*-9, *cis*-12 CLA and *cis*-10, *trans*-12 CLA isomers (*cis, trans-/trans, cis*-CLA). However, there was a significant increase in CLA yields as degree of refining was increased. Alkali-refined (REF) soy oil sample contained 6.4% total CLA with 1.7% *cis, trans-/trans, cis*-CLA. The highest CLA content was obtained in refined, bleached, deodorized (RBD) soy oil with 16.3% CLA isomers.

Iodine, a catalyst for the photoirradiation reaction, absorbs only in the lower UV range from 200 to 215 nm. The absorbance profiles of soy oil samples suggested that RBD soy oil absorbs the most in the 220–290 nm range, as compared to crude

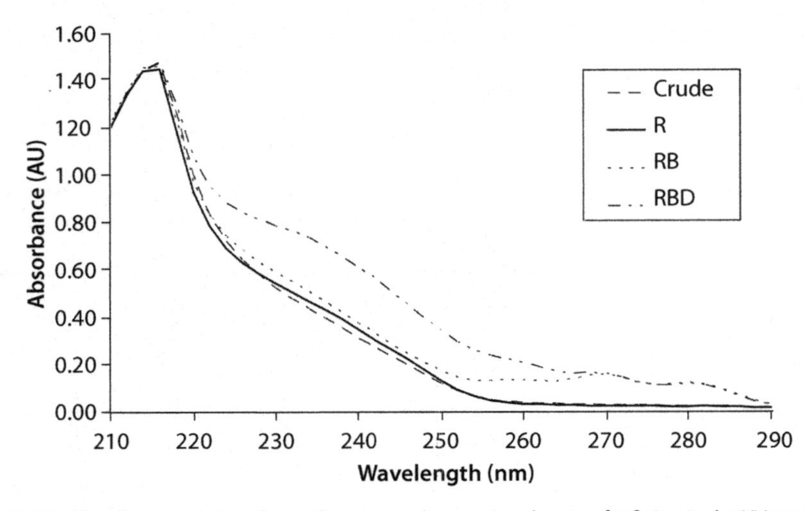

Fig. 12.13. Absorbance spectra of soy oils processed to varying degree of refining in the UV range.

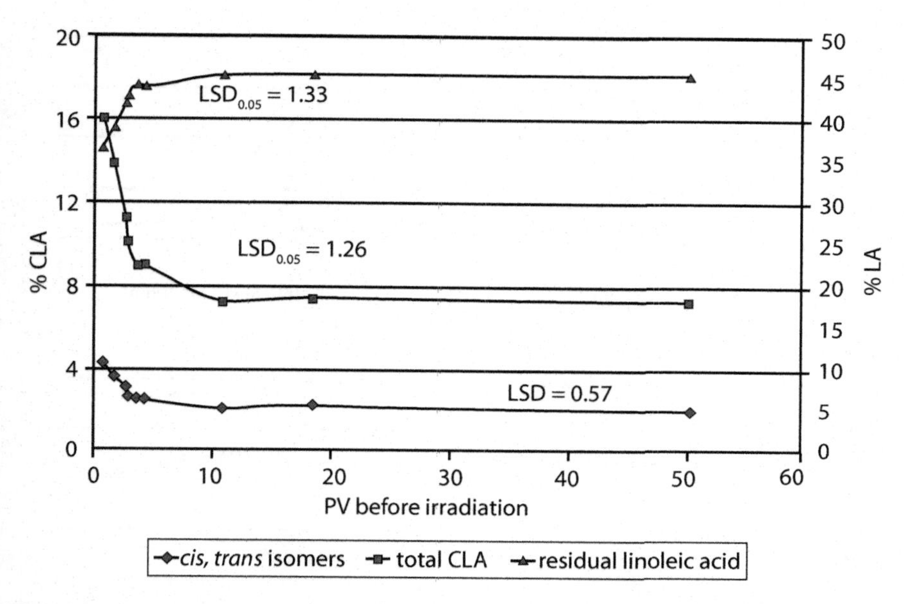

Fig. 12.14. CLA yields and residual LA obtained by photoisomerization of soy oil LA with oils of various peroxide values (PVs).

soy oil (Fig 12.13). The extent of UV radiation absorbed by soy oil samples increases as the degree of refining increases. The extent of UV radiation absorbed by the soy oil samples could be correlated to the amount of CLA isomers synthesized in each sample.

Effect of Minor Components of Vegetable Oil. Soy oils with varying levels of peroxides, tocopherols, phospholipids, free fatty acids (FFA), and lutein were each UV irradiated, with Jain et al. (2008) pilot scale photo-processing system, and the CLA content and oxidative stability were determined (Tokle et al., 2009). Fig. 12.14 shows a significant decrease in CLA yields with increased peroxide levels in oil before irradiation. The control RBD soy oil with 0.8 meq PV produced ~16% total CLA. The CLA yield quickly decreased with slight increase in PV. The yield was about 7% with a PV of 11 meq with no further significant decrease in CLA yield. An increase in PV would mean oxidation of linoleic and linolenic acid to conjugated diene hydroperoxides which are excellent UV absorbers, thus, decreasing the UV light availability by iodine in the reaction mixture.

Soy oil with 1100 ppm of naturally-occurring tocopherols yielded about 14% total CLA. With an increase in tocopherol levels to 1400 ppm, CLA yields increased significantly to 16% (Fig. 12.15A). However, any further increase in tocopherol levels decreased total CLA yields. Tocopherols act as free-radical scavengers, and

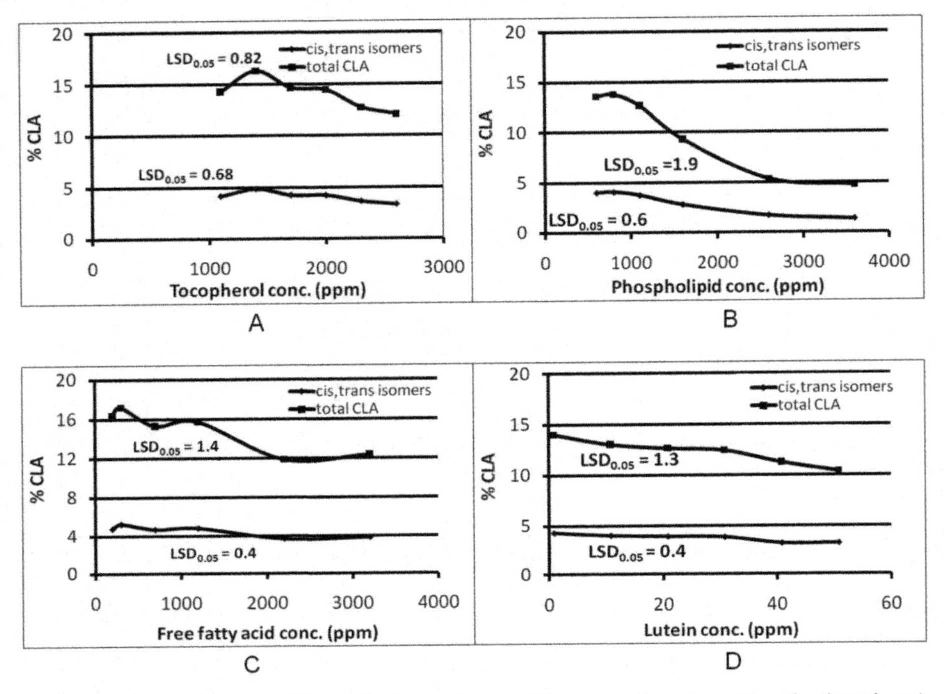

Fig. 12.15. Total and *cis,trans*-CLA yields obtained by photoisomerization of soy LA with oils with various A) tocopherol, B) phospholipid, C) free fatty acid, and D) lutein contents.

interact with free radicals as a result of hydroperoxide propagation, thus increasing the CLA yield. However, above 1400 ppm tocopherols may act as radical scavengers and affect iodine radical formation, thereby decreasing the CLA yields. Phospholipids synergistically act as antioxidants with tocopherols and decreased CLA yields with higher amounts of phospholipids (Fig. 12.15B). Presence of FFA (prooxidants) (Fig. 12.15C), and lutein (UV absorber) (Fig. 12.15D) negatively affected the CLA yields with increasing amounts of these minor components in soy oil.

Choice of Vegetable Oil as a Source for CLA. Conjugated linoleic acid can be obtained naturally from beef and dairy products. Beef fat contains about 25% palmitic acid (C16:0), 12% stearic acid (C18:0), 35% oleic acid (C18:1), and 75 mg cholesterol/100 g fat; while butter fat contains 75% saturated fat that included palmitic (30%), stearic (10%), myristic (C14:0) (15%), and lauric (C12:0) (5%) acids. Developing a source of CLA as a part of unsaturated oil would be an ideal alternative.

High-LA safflower oil has the highest initial LA (72.6%), whereas soy and corn oils have intermediate levels [53.7% and 57.5 (w/w), respectively]. Sunflower oil has less LA, and flax oil has the least. Gammill et al. (2010) photo-processed several

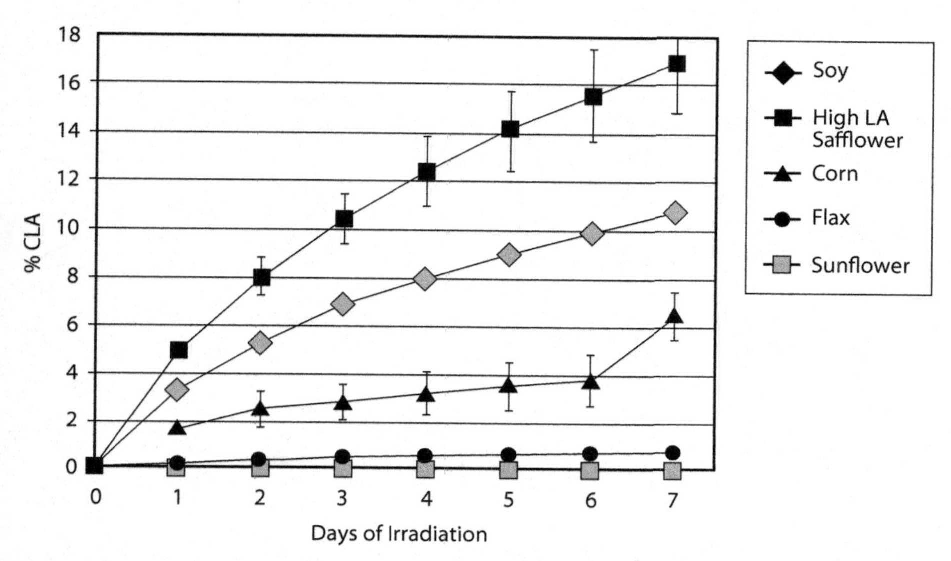

Fig. 12.16. CLA production in high-LA safflower, soy, corn, flax, and sunflower oils by laboratory-scale processing.

vegetable oils with central irradiation system and the pilot-scale system. With both systems, high-LA safflower oil, with the highest amount of linoleic acid, gave the highest total CLA yields (Fig. 12.16). Although sunflower oil had 25% linoleic acid, central irradiation system did not produce any CLA. Also, corn oil, which had more linoleic acid than soy oil, produced far less total CLA with both systems. This showed that levels of linoleic acid alone are not good predictors of CLA production, and confirmed that minor components significantly affect CLA yields.

Soy oil is low in saturated fat (15%) and high in unsaturated fat (61% polyunsaturated, 24% monounsaturated). It contains two essential fatty acids, linoleic and linolenic, which are not produced in the body. The linoleic content in the oil is as high as 53%. Converting a proportion of soy oil LA to CLA would create soy oil with substantially improved health-promoting value, and allow CLA intake in traditional food without increased dietary saturated fat. Soy oil accounts for nearly 80% of vegetable oil produced in the U.S. Thus, it is an ideal type of raw material for CLA production.

CLA Isomerization Chemistry in High Linoleic Acid Oils

Photo-processing of soy oil in the presence of iodine produced a CLA-rich soy oil, of which approximately 75% of the total CLA content was found to be *trans, trans* isomers mainly composed of *trans*-8, *trans*-10 CLA, *trans*-9, *trans*-11 CLA, and *trans*-10, *trans*-12 CLA (Jain et al., 2008). Yetella et al. (2011) proposed the iodine radical

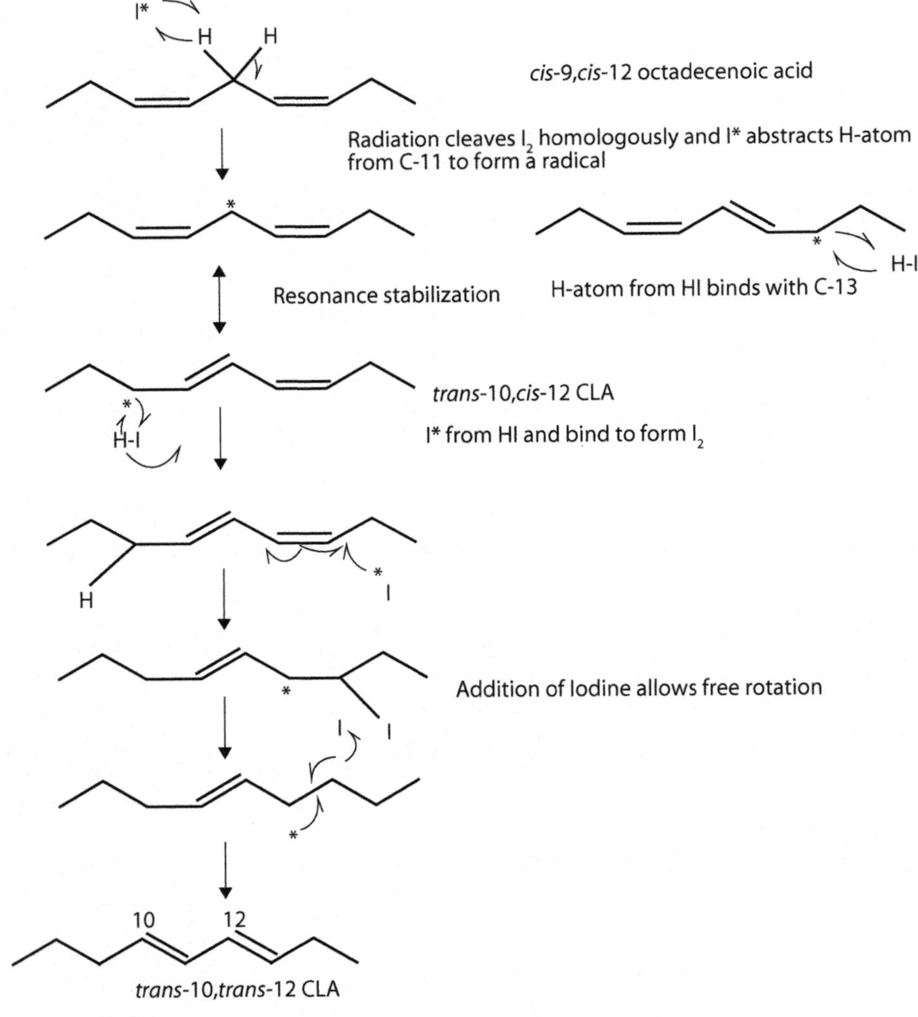

Fig. 12.17. Free radical mechanism for the formation of CLA isomers during photo-isomerization of soy oil.

mechanism (Fig. 12.17). The first step in this reaction is the catalytic abstraction of a hydrogen radical from the allylic Carbon 11 to form an LA radical (McInthosh, 2009). Conventional understanding of radical scavenging antioxidants would predict that increasing antioxidant levels would reduce CLA formation. However, to speculate, if in this case methylene-interrupted diene radicals are formed and isomerized

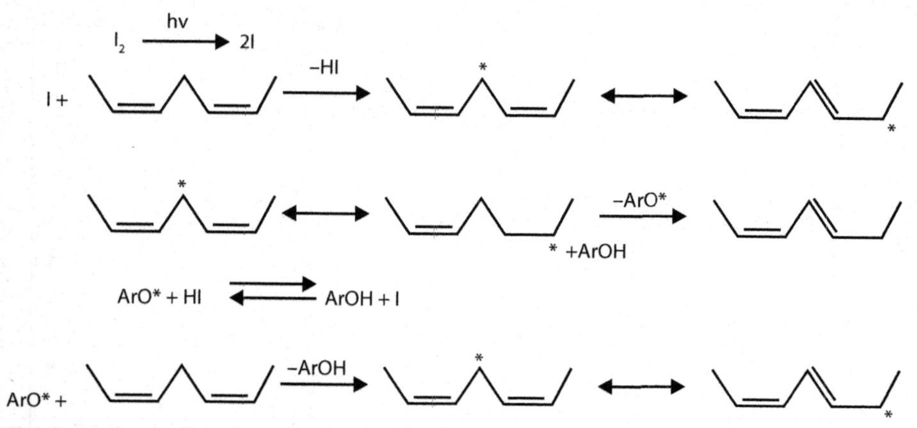

Fig. 12.18. Proposed mechanism of the effect of antioxidants on CLA yield.

faster than the antioxidant-hydrogen radicals donation, then conjugated diene radicals will be stabilized to allow CLA formation (Fig. 12.18).

Jain and Proctor (2007b) proposed a simple reaction mechanism to evaluate the reaction rate order (Fig. 12.19). *Cis*-9, *cis*-12 octadecadienoic acid (linoleic acid) in soy oil is converted to *cis*-9, *trans*-11 CLA and *trans*-10, *cis*-12 CLA isomers. A portion of these is converted to *trans*-9, *cis*-12 CLA and *cis*-10, *trans*-12 CLA, respectively. These isomers would then form the more stable *trans*-8, *trans*-10 CLA, *trans*-9, *trans*-11 CLA, and *trans*-10, *trans*-12 CLA isomers.

Fig. 12.20 shows the fatty acid isomer concentrations during photoirradiation of soy oil with UV/visible light in the presence of iodine. The CLA isomer

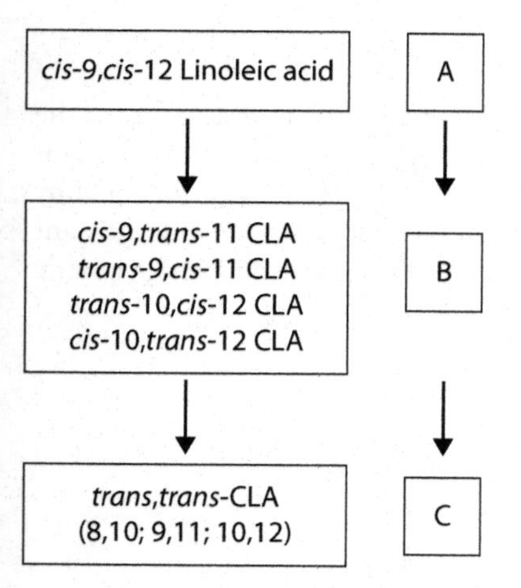

Fig. 12.19. Scheme of proposed soy oil CLA isomer formation by photoirradiation.

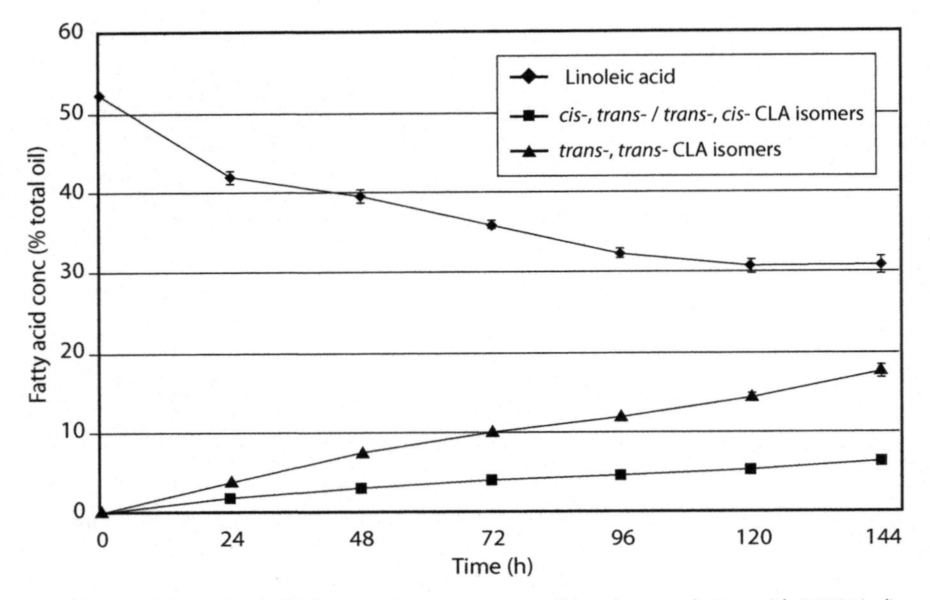

Fig. 12.20. Linoleic acid and CLA isomers formation in soy oil by photoirradiation with 0.15% iodine. Plots are the means, with standard deviation, of duplicate irradiations, with duplicate isomer analysis for each irradiation.

concentrations increased at the expense of A present in the reaction system. There was a ~23 percentage point decrease in the linoleic acid concentration during photoirradiation that corresponds to the sum of CLA isomers concentrations. The *trans-, trans-* CLA isomers were formed about 17%, and the *cis-, trans-* and *trans-, cis-*CLA isomers about 6% of the total oil present in the reaction system.

Kinetics of CLA Isomer Synthesis

Kinetics of the Reaction A to B. Although, zero order and first order kinetic plots for the consumption of A yielded R^2 values of 0.91 and 0.95, respectively, a second order plot gave an R^2 value of 0.99 which best explained the order of the reaction confirmed by residual data analysis (Fig 12.21). Slope of this second order plot, k_1, the rate constant for the disappearance of A, was 9.01×10^{-7} L/mol sec. From a second order kinetic plot, we get the following equation:

$$A = \frac{A_0}{1 + A_0 k_1 t} \quad \text{(Eq. 12.1)}$$

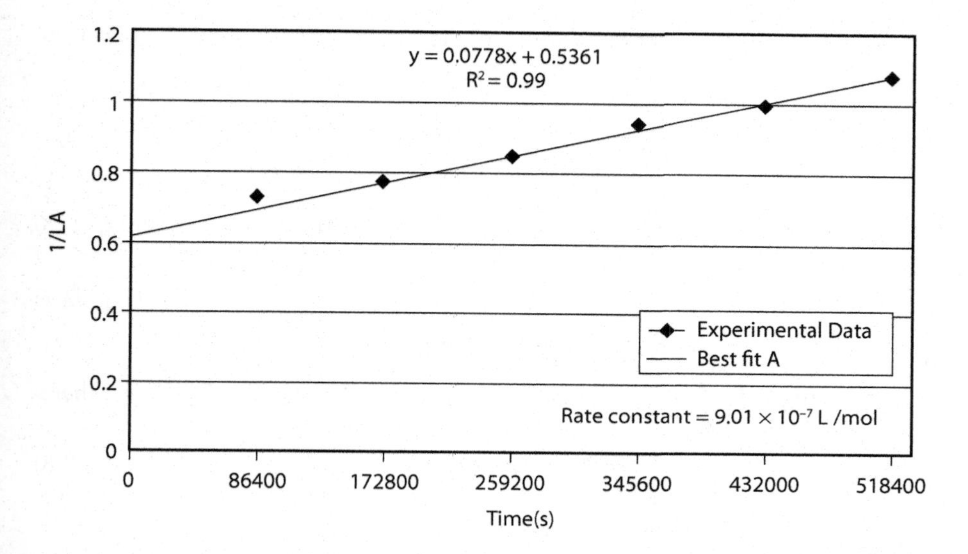

Fig. 12.21. Kinetic plot showing second order kinetics for the conversion of linoleic acid (A) to *cis,trans-/trans,cis*-CLA isomers (B) by photoisomerization of soy oil.

Kinetics of the Reaction B to C with Respect to B. According to the proposed mechanism, intermediate CLA B isomers are formed from A, and they undergo further photoprocessing to form C. Eq. 12.1 can thus be modified to

$$\frac{dB}{dt} = k_1A^2 - k_2B^m \qquad \text{(Eq. 12.2)}$$

integrating the above equation from time t = 0, to t = t,

$$\int_0^t B_t = \int_0^t k_1A^2\,dt - \int_0^t k_2B^m\,dt \qquad \text{(Eq. 12.3)}$$

Substituting 4 in 6 we get,

$$\int_0^t B_t = \int_0^t k_1\left(\frac{A_0}{1+A_0k_1t}\right)^2 dt - \int_0^t k_2B^m\,dt \qquad \text{(Eq. 12.4)}$$

$$B_t = A_0\left(1-\frac{1}{1+A_0k_1t}\right) - \int_0^t k_2B^m\,dt \qquad \text{(Eq. 12.5)}$$

Assuming a zero order reaction for the formation of *trans,trans*-CLA isomers, we would have $B^n = B^0 = 1$. Thus, Eq. 12.4 gives:

$$B_t = A_0\left(1 - \frac{1}{1 + A_0 k_1 t}\right) - k_2 t \qquad \text{(Eq. 12.6)}$$

However, as obtained from the reaction A to B, $k_1 = 9.01 \times 10^{-7} << 1$, so Eq. 12.6 reduces down to:

$$B_t = -k_2 t \qquad \text{(Eq. 12.7)}$$

The plot of B_t vs. t gave the zero order plot for the reaction B to C.

Assuming a first order reaction for the reaction B to C. $B = B_0 e^{-k_2 t}$ and therefore, Eq. 12.7 would give

$$\ln B_t = \ln B_0 + k_2 t \qquad \text{(Eq. 12.8)}$$

The plot of ln B_t vs. t gave the first order plot for the reaction B to C.

Assuming a second order reaction, $B = \frac{B_0}{1 + B_0 k_2 t}$ and Eq. 12.4 gives us

$$\frac{1}{B_t} = \frac{1}{B_0} + k_2 t \qquad \text{(Eq. 12.9)}$$

A plot of $1/B_t$ vs. t gave the second order plot for the reaction B to C.

Residual analysis performed on the different rate orders confirms reaction B to C as a first order reaction. Rate constant for the reaction was obtained from the slope of the first order plot (Fig. 12.22) and equals 2.75×10^{-6} sec^{-1}. Thus, the reaction B to C with respect to B is dependent on the overall rate of formation of B isomers.

Kinetics of the Reaction B to C with Respect to C. Fig. 12.23 shows a zero order plot for the formation of C isomers in the photoirradiation reaction system with an R^2 of 0.99. This thermodynamic predominantly governs formation of C isomers. This would be in congruence with the fact that *trans-, trans-*configuration is a thermodynamically more stable form than *cis-, trans-/trans-, cis-* configuration. Thus, intermediate B isomers form stable *trans-, trans-*CLA isomers, and this conversion is governed by the rate constant for the reaction obtained as 10.66×10^{-7} mol/L sec.

Reaction A to B is a second order reaction (Fig. 12.21), and thus, the rate of consumption of A depends on the amount of A present at a given time. Thus, initially the rate of consumption of A is greater, and it slows with time as observed in Fig. 12.20. Reaction B to C with respect to B isomers is a first order reaction (Fig. 12.22), and depends on the B isomers in the reaction system. Thus, the rate of reaction increases steadily with time and so does the amount of C isomers in the

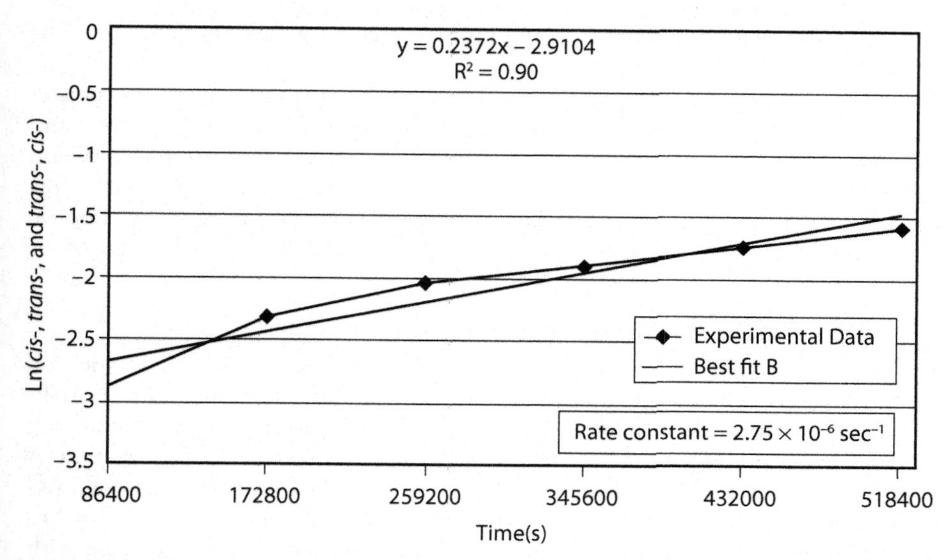

Fig. 12.22. Kinetic plot showing first order kinetics for the conversion of *cis-*, *trans-* and *trans-*, *cis-*CLA isomers (B) to *trans-*, *trans-*CLA isomers (C) with respect to B isomers during photoisomerization of soy oil.

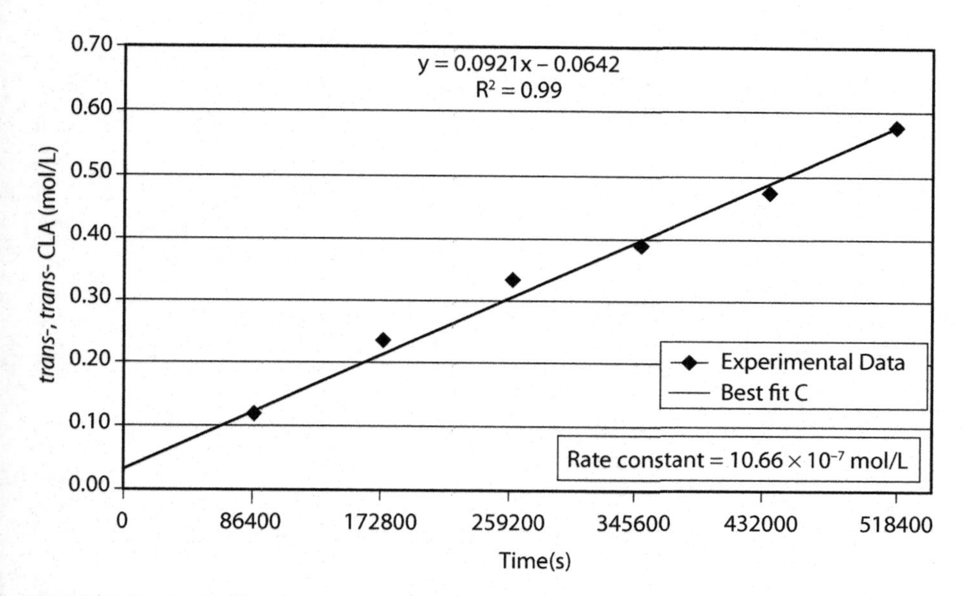

Fig. 12.23. Kinetic plot showing zero order kinetics for the formation of *trans,trans-*CLA isomers (C) with respect to C isomers during photoisomerization of soy oil.

system. The rate-controlling step in the formation of CLA isomers from LA is the consumption of LA to form *cis-, trans-/trans-, cis-*CLA isomers.

The kinetic study of the CLA isomers formation shows that the greater the concentration of linoleic acid, the greater the concentration of *cis,trans-* and *trans,cis-*CLA isomers produced. As the reaction proceeds, the rate of formation of these intermediate isomers slows down. Hence, to maintain a high rate of formation of B isomers, we may need to maintain a large concentration of linoleic acid. In contrast, the formation of the stable *trans,trans-*CLA isomers is governed by thermodynamics.

HPLC-Identified CLA Triacylglycerides

Lall et al. (2009a) identified new triacylglyceride (TAG) fractions in CLA-rich oil by nonaqueous reversed-phase high-performance liquid chromatography (NARP-HPLC). An acetonitrile/dichloromethane (ACN/DCM) gradient and an evaporating light scattering detector/ultraviolet (ELSD/UV) detector were used for detection. A comparison of a CLA-rich soy oil with RBD soy oil showed emergence of new TAG peaks in photo-processed CLA-rich soy oil. The LnLL, LLL, LLO, and LLP (Ln, linolenic; L, linoleic; O, oleic; and P, palmitic) peaks reduced after isomerization with an increase in adjacent peaks that coeluted with LnLnO, LnLO, LnOO, and LnPP

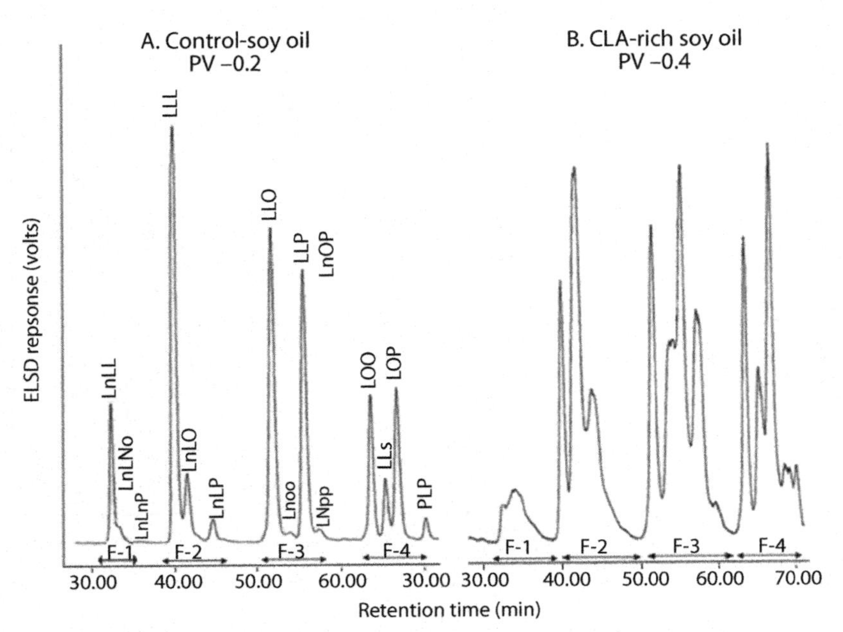

Fig. 12.24. NARP-HPLC fractionation of (A) Control-soy oil (B) 22% CLA-rich soy oil detected by ELSD.

(Fig. 12.24). The newly formed peaks were wider than those of the original oil and absorbed at 233 nm. This suggested coelution of other TAG peaks.

A UV detector response comparison of the two oils showed a significant response at 233 nm for the CLA-rich soy oil TAG fractions suggesting presence of CLA isomers in these fractions. The NARP-HPLC fractions were collected and these fractions of mixed TAGs were further analyzed for their fatty acid profiles. Fig. 12.25 shows the distribution of CLA several isomers in these fractions.

CLA Measurement

Micro-FAME GC FID Analysis

Lall et al. (2009b) developed a micro FAME for analyzing 1 mg of oil sample by limiting the volume of solvents utilized. This method was developed to analyze CLA-rich soy oils, although the method can be easily adapted to pure vegetable oils. Several soy oil samples between 100 to 1 mg were used for method development. Heptadecanoic methyl ester was used as an internal standard, 5% to the weight of sample. Only 1 mL of toluene and 4 mL 0.5M sodium methoxide in methanol was added and mixed for 10–12 min. To inhibit the reaction, 0.2 mL glacial acetic acid was added

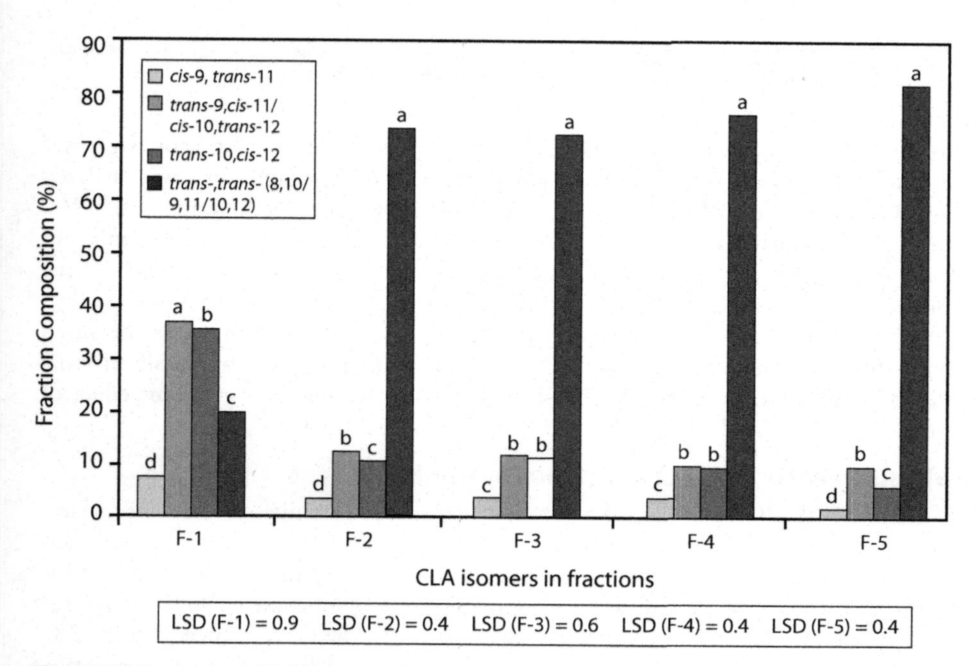

Fig. 12.25. Distribution of % CLA isomers in 34.2% CLA rich oil TAG fractions analyzed by FAME GC. Replication (n) = 3. Means with different letters within each fraction differ significantly, $p < 0.05$.

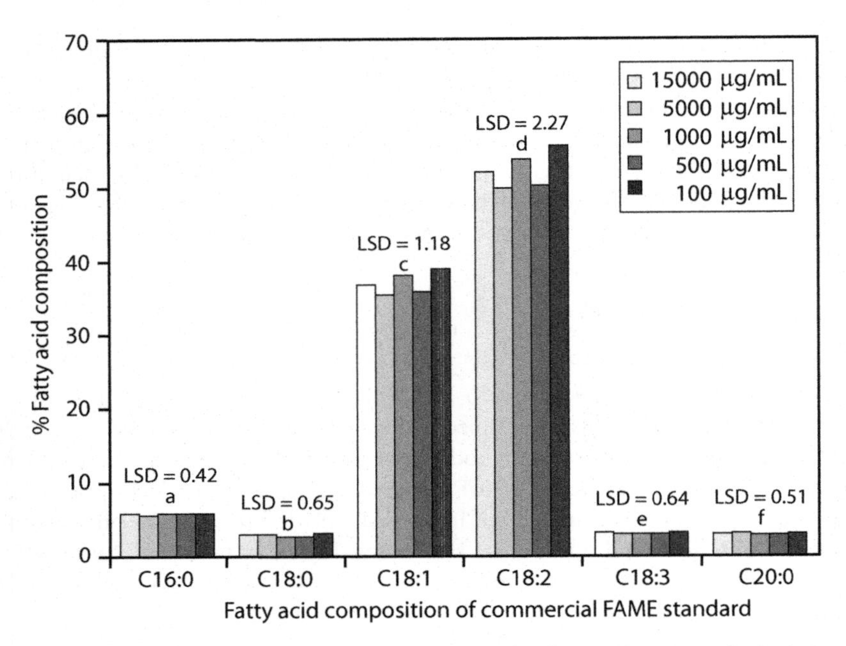

Fig. 12.26. Fatty acid composition of soy oil at various dilutions by novel micro-method relative to a control analysis conducted by a conventional macro method.

followed by 5 mL of distilled water and only 1 mL of hexane. The organic layer was extracted and dried over anhydrous sodium sulfate. Dried organic layer was analyzed with GC FID to quantify the fatty acids. Fig. 12.26 shows a fatty acid comparison of 100 to 1 mg oil samples as measured with the micro-fame method. No statistically-significant difference in the fatty acid quantification and low LSD at 0.05 p value validated the method and was found appropriate for CLA isomer analysis of micro samples.

An orthogonal regression analysis (Fig. 12.27) of 1 mg sample, as measured with the micro-method, compared to 100 mg sample measured by a conventional GC-FID method, produced a R2 value of 0.99 indicating no significant effect of quantity on fatty acid profile obtained with the micro method.

Measurement of CLA in CLA-rich Food by ATR-FTIR

Kadamne et al. (2009) developed a non-destructive, rapid attenuated total reflectance – Fourier transform infrared spectroscopy method to quantify total and geometrical CLA isomers in CLA-rich soy oil, with total CLA content ranging from 0.38 to 25.11% without any sample preparation. An ATR-FTIR spectrum of control and CLA-rich soy oil indicated distinct conjugated diene peaks specific to the CLA isomers (Fig. 12.28). A closer look at the conjugated diene peaks showed that different CLA isomers represented specific parts of the spectrum, and thus, quantification was possible (Fig. 12.29).

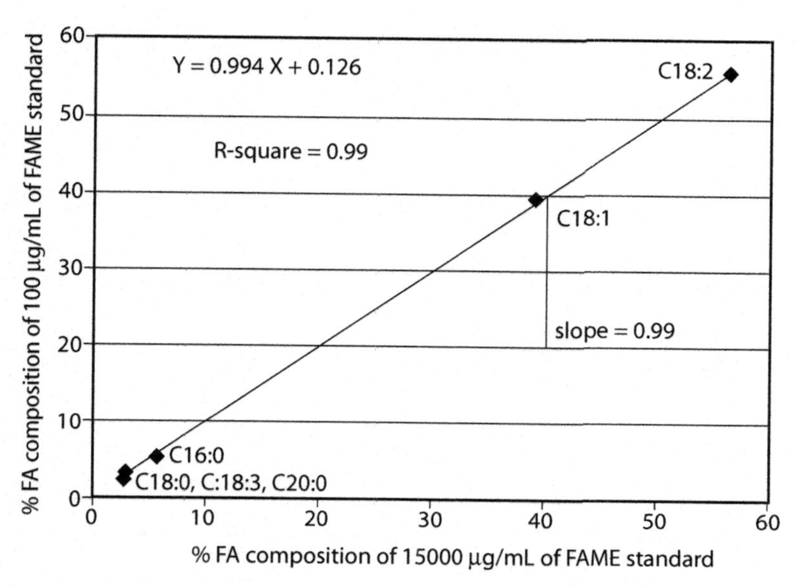

Fig. 12.27. Correlation using orthogonal regression between fatty acid composition of 1 mg of soy oil determined by the micro-method and 100 mg soy oil determined by the conventional macro-method.

A validation test set of 23 photo-isomerized oil samples was used to validate the developed models. The levels of CLA as measured by FTIR method were compared to the actual values as measured by GC-FID fame analysis. The measured CLA levels (by GC-FID) were plotted against predicted CLA levels (using the developed models) by JMP 7.0.2 software to determine the linear correlation of the data based on the equation of the line, coefficient of determination (R^2p), and the root-mean square error of prediction (RMSEP). The statistics of these linear correlations are tabulated in Table 12-D. The predicted and measured CLA levels were highly correlated, with correlations of determination (R^2p) ranging from 0.94 to 0.96 for total CLA, *trans-*, *trans*-CLA isomers, total mono *trans* isomers, *trans*-10, *cis*-12 CLA, *trans*-9, *cis*-11 CLA and *cis*-10, *trans*-12 CLA, and *cis*-9, *trans*-11 CLA, respectively. The entire process of prediction, including the ATR-FTIR scan of the oil sample, takes about 10 min, which is rapid when compared to GC-FID.

Kadamne et al. (2011) developed validated ATR-FTIR calibration curves to measure CLA isomers in CLA-rich potato chips. A representative sample of potato chips was analyzed by ATR-FTIR to collect spectra. Oil was solvent extracted and GC-FID analyzed for actual fatty acid quantities. Models were developed for different isomers of CLA using PLS regression analysis. The calibration models were cross validated with samples not used to develop the models. Table 12-E shows the correlation coefficients and corresponding root mean square error of prediction.

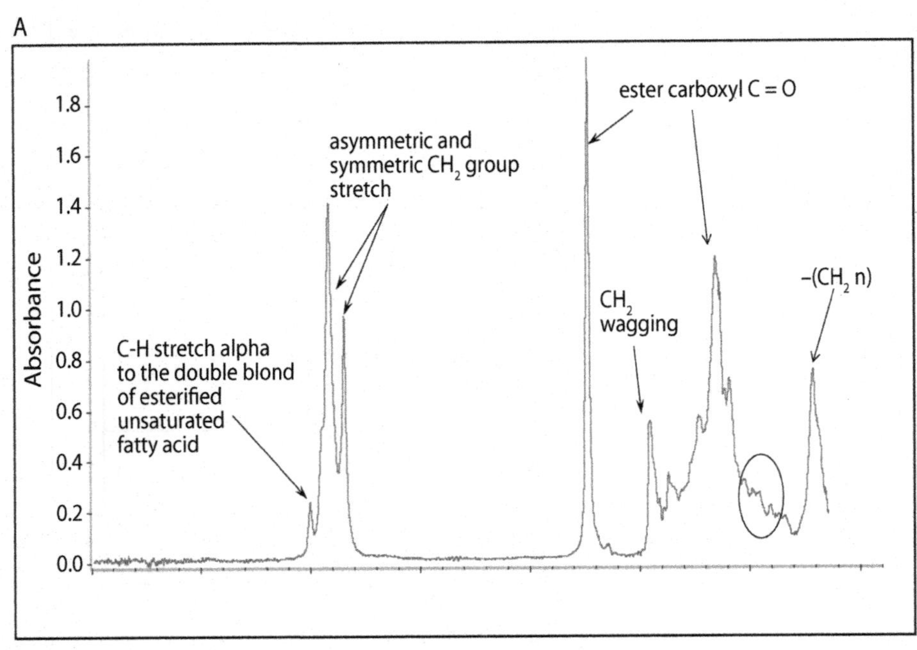

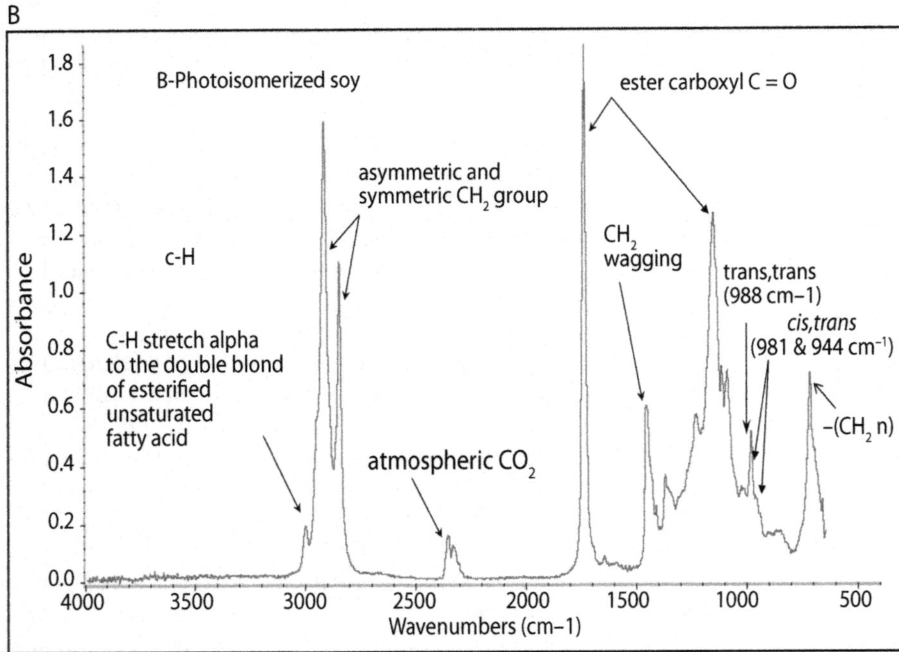

Fig. 12.28. ATR-FTIR spectra of 4000–650 cm^{-1} region of A) control soy oil and B) photo-isomerized soy oil containing 10% total CLA.

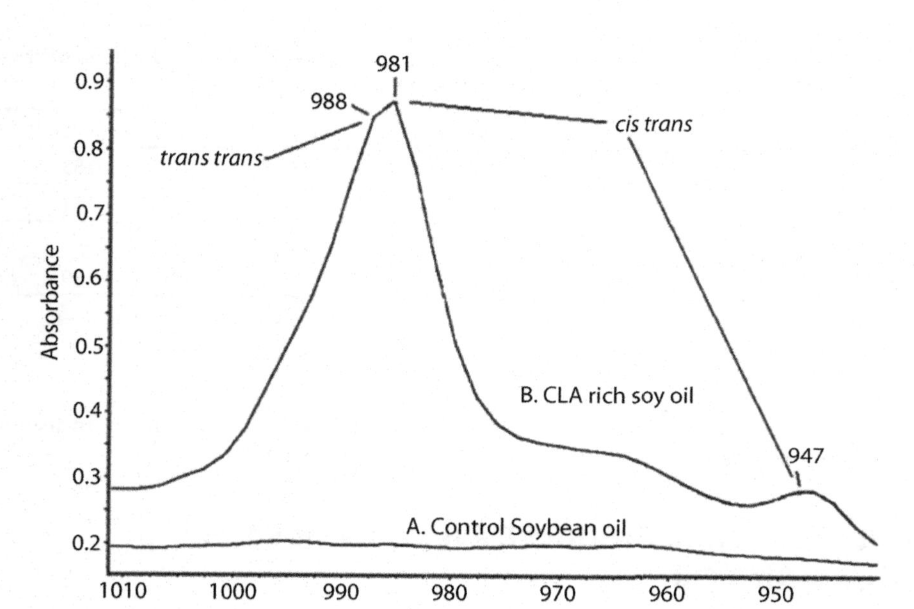

Fig. 12.29. ATR-FTIR spectra of the 1010–900 cm^{-1} region of A) control soy oil and B) photo-isomerized soy oil containing 10% total CLA.

Table 12-D. Statistics of the Linear Plot of the Measured versus Predicted Values of the CLA Isomers Based on the Developed Models.

Model	Equation	R^2p	RMSEP
trans-, trans-CLA	y = 0.99x – 0.038	0.97	0.61
total mono *trans-CLA*	y = 1.02x – 0.042	0.98	0.21
t-10, *c*-12 CLA	y = 1.01x – 0.008	0.98	0.05
t-9, *c*-11 & *c*-10, *t*-12 CLA	y = 1.03x – 0.020	0.97	0.12
c-9, *t*-11 CLA	y = 1.01x – 0.004	0.98	0.06

R^2p, coefficient of determination of prediction; equation, equation for the linear fit of the data; RMSEP, root mean square error of prediction.

Table 12-E. Statistics of the Linear Plot of the Measured versus Predicted Values of the CLA Isomers in CLA Enriched Potato Chips Based on the Developed Models.

Model	Equation	R^2	RMSEP
trans-, trans-CLA	y = 1.02 x – 0.02	0.97	0.85
t-10, *c*-12 CLA	y = 0.72 x + 0.13	0.95	0.11
t-9, *c*-11 & *c*-10, *t*-12 CLA	y = 0.86 x +0.01	0.98	0.01
c-9, *t*-11 CLA	y = 0.86 x + 0.17	0.98	0.09

R^2p, coefficient of determination of prediction; equation, equation for the linear fit of the data; RMSEP, root mean square error of prediction.

The method can be adapted and proves to be a viable solution to quickly measure CLA isomers in food product.

Nutritional and Product Development Value of CLA-Rich Oil

Soy oil accounts for nearly 80% of vegetable oil produced in the U.S., and provides for a viable raw material source for CLA production. The photoisomerization process converts a portion of soy oil linoleic acid to CLA, and thus creates soy oil with a substantially improved health-promoting value and allows CLA intake in traditional food without increased dietary saturated fat. Such CLA rich oil can be used as a food ingredient or a processing medium to develop CLA-enriched food products.

Jain and Proctor (2007a) used refined CLA-rich soy oil as a processing medium to fry potato chips (Fig. 12.30). Chips prepared with RBD commercial soy oil and CLA-rich soy both contained ~39% of oil (w/w) and did not differ in the overall appearance. Oil from CLA-rich potato chips was extracted and analyzed for its fatty acid profile. Table 12-F shows the fatty acid distribution of RBD soy oil, photo-isomerized soy oil, and the oil extracted from fried potato chips made with CLA-rich soy oil. Commercial RBD soy oil had 52.5% linoleic acid, and no CLA isomers were detected. The photo-processed soy oil had ~21% CLA isomers accounted for by a decreased LA content. The oil extracted from CLA-rich potato chips had statistically similar composition to the oil used to make the chips. This suggested that CLA isomers survived frying conditions and were all present in the final product. Extracted oil was analyzed for peroxide value and showed minimal increase, suggesting stability of CLA oil under frying conditions.

An ounce of potato chips prepared with CLA-rich soy oil contained 2.37 g CLA as compared with 0.1 g CLA in a 3-oz. serving of steak filet and 0.06 g CLA in an 8-oz. serving of whole milk. The calculations were based on the fat content of potato chips, steak filet, and whole milk, that is, 39, 20, and 4%, respectively. The results indicate that the *cis*-9, *trans*-11 CLA content of potato chips was approximately 35

Fig. 12.30. Potato chips prepared with CLA-rich soy oil and refined, bleached, and deodorized soy oil.

times higher than the concentration found in a steak filet. The total chip CLA concentration was 80 times higher than found in beef and whole milk.

A nutritional study by Gilbert et al. (2011) showed efficacy of such photo-processed CLA-rich soy oil. CLA-rich soy oil was fed as a dietary supplement and monitored for effect of CLA on body composition, dyslipidemia, hepatic steatosis, and markers of glucose control and liver function of obese *fa/fa* Zucker rats. The

Table 12-F. Linoleic, Linolenic, and Conjugated Linoleic Acid (CLA) Concentration in (1) Refined, Bleached, and Deodorized (RBD) Soy Oil, (2) Photoisomerized Soy Oil, and (3) Oil Extracted from Fried Potato Chips Made with CLA-rich Soy Oil.

	Soy Oil		
Fatty Acid[a]	**RBD Oil**	**Photoisomerized Oil**	**Oil extracted from Chips**
Linoleic Acid	52.2[a]	31.46[b]	30.87[b]
Linolenic Acid	6.5[a]	4.73[b]	4.64[b]
c-9, *t*-11 CLA	0.0[a]	1.78[b]	1.79[b]
t-9, *c*-11/*c*-10, *t*-12 CLA	0.0[a]	2.74[b]	2.81[b]
t-10, *c*-12 CLA	0.0[a]	1.75[b]	1.81[b]
t, *t*-CLA[c]	0.0[a]	14.73[b]	14.55[b]

[a]Fatty acid concentration based on total oil content, weight-by-weight basis.
[b]Values are means of duplicate analyses of two replications, n=4. Means with different letters within rows differ significantly, $p < 0.05$.
[c]Consists of *trans*-8, *trans*-10 CLA, *trans*-9, *trans*-11 CLA, and *trans*-10, *trans*-12 CLA.

Table 12-G. Effects of CLA-Enriched Soybean Oil on Serum and Liver Lipids in Obese Zucker Rats.

	L-Ctrl		O-Ctrl		O-CLA	
	Mean	SE	Mean	SE	Mean	SE
Cholesterol (mmol/l)	2.16[c]	0.09	13.32[a]	1.53	7.86[b]	1.14
HDL-C (mmol/l)	0.58[b]	0.02	1.8[a]	0.1	1.7[a]	0.17
LDL-C (mmol/l)	0.13[c]	0.01	1.39[a]	0.18	0.7[b]	0.15
Triglycerides (mmol/l)	0.97[b]	0.09	1.95[a]	0.31	1.87[a]	0.23
Liver lipid %	9.7[c]	0.4	32.9[a]	1.3	20.1[b]	1.8
Liver cholesterol %	21.9[b]	0.4	23.7[a]	0.3	23.4[a]	0.6

Data represent the mean values and standard error (SE; n = 12/group). Values in a row without common superscripts are significantly different ($P < 0.05$). Cholesterol, HDL-C, LDL-C, and triglycerides were measured in serum. Liver lipid percentage and liver cholesterol percentage were measured in the liver. Liver lipid percentage refers to the percentage of total liver weight found to be lipid. Liver cholesterol percentage refers to the percentage of liver lipids found to be cholesterol. Abbreviations used: L-Ctrl = Lean + control diet, O-Ctrl = Obese + control diet, O-CLA = Obese + Conjugated Linoleic Acid Diet (0.5%).

CLA-rich soy oil consisted of 75% *trans-*, *trans*-CLA isomers, and thus the nutritional results can be attributed mostly to these isomers. The CLA-enriched diet lowered serum cholesterol and low density lipoprotein-cholesterol levels by 41 and 50%, respectively as compared to control rats (Table 12-G). The CLA diet also decreased liver lipids and total liver weight in these obese rats. CLA-enriched soy oil significantly upregulated peroxisome proliferator-activated receptor-γ gene expression in the heart, and this may, in part, be responsible for the positive effects.

Conclusions

A simple rapid means to photoisomerize linoleic acid to CLA in oil triglycerides developed in the last few years, using only UV light and an iodine catalyst, has enabled the production of large quantities of CLA in food oils that can be readily used to produce food products. This CLA has been shown in animal studies to be effective in reducing indicators of heart disease and diabetes. However, there are challenges to commercial production. These include iodine removal prior to food use, and increase in the scale of production and industrial investment to achieve commercialization.

While experimental means of removing iodine are effective, current proprietory research is addressing how this can be done effectively in a large-scale operation with minimal oil loss. However, an increase in the scale of production has not been achieved beyond the pilot plant. Industrial production would require larger or multiple production units in series, which would require the accompanying engineering problems to be addressed. In addition to the technical issues, there is a need for industrial investment to move the technology from an academic research project to commercial viability. Although the challenges are significant, they are not insurmountable, and we are optimistic about developments in the coming years.

References

Adlof, R.O. In Preparation of unlabeled and isotope—Labeled conjugated linoleic and related fatty acid isomers in advances in conjugated linoleic acid research; Yurawecz, M.P., Mossoba M.M., Kramer J.K.G., Pariza M.W. and Nelson G.J. Vol 1. AOCS Press: Champaign, Illinois. 1999; pp. 21–38.

Ando, A.; Ogawa, J.; Kishino, S.; Shimzu, S. CLA production from ricinoleic acid by lactic acid bacteria. J. Am. Oil Chem. Soc. 2003, 80, 8889–8894.

Arcos, J.A.; Otero, C.; Hill, C.G. Jr. Rapid enzymatic production of acylglycerols from conjugated linoleic acid and glycerol in a solvent free system. Biotech. Letters 1998, 20, 617–621.

Beaulieu, A.; Drackley, J.; Merchen, N. Concentrations of conjugated linoleic acid (*cis*-9, *trans*-11-octadecadienoic acid) are not increased in tissue lipids of cattle fed a high-concentrate diet supplemented with soybean oil. J. Anim. Sci. 2002, 80, 847–61.

Bernas, A.; Kumar, N.; Mäki-Arvela, P.; Holmbom, B.; Salmi, T.; Murzin, D.Y. Heterogeneous catalytic production of conjugated linoleic acid. Organ. Process Research and Development 2004, 8, 341–352.

Booth, R.G.; Kon, S.K.; Dann, W.J.; Moore, T. A study of seasonal variation in butter fat II. A seasonal spectroscopic variation in the fatty acid fraction. Biochem. J. 1935, 29, 133–137.

Bordeaux, O.; Christie, W.W.; Gunstone, F.D.; Sebedio J.L. Large-scale synthesis of methyl *cis*-9,*trans*-11-Octadecadienoate from methyl ricinoleate. J. Am. Oil Chem. Soc. 1997, 74, 1011–1015.

Canaguier, R.; Cecchi, G.; Ucciani, E.; Chevalier, J.L. Isomerisation photochemique des acides gras polyinsatures catalysee par I'Iode. Fr. Corps Gras. 1984, 31, 401–409.

Canaguier, R.; Chevalier, J.L.; Cecchi, G.; Ucciani, E. Photoisomerization solaire des huilles vegetales catalysee par I'Iode. Fr. Corps. Gras. 1986, 33, 57–162.

Collomb, M.; Sieber, R.; Butikofer, U. CLA isomers in milk fat from cows fed diets with high levels of unsaturated fatty acids. Lipids 2004, 39, 355–364.

Corl, B.A.; Barbano, D.E., Ip, C. *cis*-9,-11 derived endogenously from *trans*-11 18:1 reduces cancer risk in rats. J. Nut. 2003, 3, 2893–2900.

Gammill, W.; Jain, V.; Proctor, A. A comparative study of high-linoleic acid vegetable oils for the production of conjugated linoleic acid. J. Agric. Food Chem. 2010, 58(5), 2952–2957.

Gangidi, R.R.; Proctor, A. Photochemical production of conjugated linoleic acid from soybean oil. Lipids 2004, 39, 577–582.

Gilbert, W.; Gadang, V.; Devareddy, L.; Proctor, A.; Jain, V. *Trans–trans* Conjugated Linoleic Acid Enriched Soybean Oil Reduces Fatty Liver and Lowers Serum Cholesterol in Obese Zucker Rats. Lipids 2011, 46(10), 961–968.

Ha, Y.L.; Grimm, N.K.; Pariza, M.W. Anticarcinogens from fried ground beef: Heat altered derivatives of linoleic acid. Carcinogenesis 1987, 8, 1881–1887.

Haugen, M.; Vikse, R.; Alexander, J. CLA and adverse health effects: a review of the relevant literature. http://matportalen.no/Filer/fil_CLA_Norwegian_Institute_for_Public_Health.pdf. (accessed 10/10/2003)

Hilditch, T.P.; Jasperson, H. Milk fats from cows fed on fresh pasture and on ensiled green fodder: Observations in the component fatty acids. J. Soc. Chem. Ind. 1941, 60, 305–310.

Ip, C.; Chin, S.F.; Scimeca, J.A.; Pariza, M.W. Mammary cancer prevention by conjugated dienoic derivatives of linoleic acid. Cancer Res. 1991, 51, 6118–6124.

Ip, C.; Singh, M.; Thompson, H.J.; Scimeca, J.A. Conjugated linoleic acid suppresses mammary carcinogenesis and proliferative activity of the mammary gland in the rat. Cancer Res. 1994, 54, 1212–1215.

Iwata, T.; Kamegai, T.; Sato, Y.; Watanabe, K.; Kasai, M. 1999. Method for producing conjugated linoleic acid. U.S. patent 5,986,116.

Jain, V.; Lall, R.; Proctor, A. Pilot-scale production of conjugated linoleic acid-rich soy oil by photoirradiation. J. Food Sci. 2008, 73(4), E183–E193.

Jain, V.; Proctor, A. 2006. Photocatalytic production and processing of conjugated linoleic acid-rich soy oil. J. Agric. Food Chem. 2006, 54(15), 5590–5596.

Jain, V.; Proctor, A. Kinetics of photoirradiation-induced synthesis of soy oil-conjugated linoleic acid isomers. J. Agric. Food Chem. 2007b, 55(3), 889–894.

Jain, V.; Proctor, A. Production of conjugated linoleic acid-rich potato chips. J. Food Sci. 2007a, 72, S75–S78.

Jain, V.; Tokle, T.; Kelkar, S.; Proctor, A. The effect of the degree of processing on soy oil conjugated linoleic acid yields. J. Agric. Food Chem. 2008, 56(17), 8174–8178.

Ju, W.J.; Jung, J.M.Y. Formation of conjugated linoleic acids in soybean oil during hydrogenation with nickel catalyst as affected by sulfur addition. J. Ag. Food. Chem. 2003, 51, 3144–3149.

Julliard, M.; Luciani, A.; Chevalier, J.L.; Cecchi, G.; Ucciani, E. Photosensitized conjugation of methyl 9,12, octodecadienoate. J. Photochem. 1987, 38, 345–355.

Kadamne, J.; Castrodale, C.; Proctor, A. Measurement of Conjugated Linoleic Acid (CLA) in CLA-Rich Potato Chips by ATR-FTIR Spectroscopy. J. Agric. Food Chem. 2011, 59 (6), 2190–2196.

Kadamne, J.; Jain, V.; Saleh, M.; Proctor, A. Measurement of Conjugated Linoleic Acid in CLA-Rich Soy Oil by ATR-FTIR. J. Agric. Food Chem. 2009, 57(22), 10483–10488.

Kepler, C.R.; Tove, S.B. Biohydrogenation of unsaturated fatty acids. J. Biol. Chem. 1967, 242, 5686–5692.

Kepler, C.R.; Tucker, W.P.; Tove, S.B. Intermediates and products of the biohydrogenation of linoleic acid by Butyrivibio fibrisolvens. J. Biol. Chem. 1966, 241, 1350–1354.

Khanal, R.C.; Olson, K.C. Factors affecting conjugated linoleic acid content in milk, meat and egg: A review. Pakistan J. of Nutri. 2004, 3(2), 82–98

Kim, Y.J.; Lee, K.W.; Lee, S.; Kim, H.; Lee, H.J. The production of high-purity conjugated linoleic acid (CLA) using a two-step urea-inclusion crystallization and hydrophilic arginine-CLA complex. J. Food Sci. 2003, 68, 1948–1505.

Lall, R.; Jain, V.; Proctor, A.A rapid, micro FAME preparation method for vegetable oil fatty acid analysis by gas chromatography. J Am. Oil Chem. Soc. 2009b, 86, 309–314.

Lall, R.; Proctor, A.; Jain, V.; Lay, J. Conjugated linoleic acid-rich soy oil triacylglyceride fraction identification. J. Ag. Food. Chem. 2009a, 57, 1727–1734.

Larsen, T.M.; Toubro, S.; Astrup, A. Efficacy and safety of dietary supplements containing CLA for the treatment of obesity: evidence from animal and human studies. J. Lipid Res. 2003, 44, 2234–2241.

Lawson, R.E.; Moss, A.R.; Givens, D.I. The role of dairy products in supplying conjugated Linoleic acid to man's diet: a review. Nutritional Research Reviews. 2001, 14, 153–172.

Lee, K.N.; Kritchevsky, D.; Pariza, M.W. Conjugated linoleic acid and atherosclerosis in rabbits. Atherosclerosis 1994, 108, 19–25.

Lee, S.O.; Kim, C.S.; Cho, S.K.; Choi, J.H.; Ji, G.E.; Oh, D.K. Bioconversion of linoleic acid into conjugated linoleic acid during fermentation and by washed cells of Lactobacillus reuteri. Biotechnology Letters 2003, 25, 935–938.

Lee, S.O.; Hong, G.W.; Oh, D.K. Bioconversion of linoleic acid in conjugated linoleic acid by immobilized Lactobacillus reteri. Biotechnol. 2003, 1081–1084.

Lin, T.Y.; Hung, T.H.; Cheng, T.S.J. Conjugated linoleic acid production by immobilized cells of Lactobacillus delbrueckii ssp. bulgaricus and Lactobacillus acidophilus. Food Chem. 2005, 92, 23–28.

Lin, T.Y.; Lin, C.W.; Wang, Y.J. Production of linoleic acid by enzyme extract of Lacto bacillus acidophilus CCRC 14079. Food Chem. 2003, 83, 27–31.

Martin, S.A.; Jenkins, T.C. Factors affecting conjugated linoleic acid production by mixed ruminal bacteria. The Univ of Georgia, CAES, Dept of Animal and Dairy Sci., 2001/2002 Annual Report.

McInthosh, M. University of Arkansas. Personal Communication 2009.

Moore, T. Spectroscopic changes in fatty acids. 1. Changes in the absorption spectra of various fats induced by treatment with potassium hydroxide. Biochem. J. 1937, 31, 138–154.

Park, Y.; Albright, K. J.; Strokson, J. M.; Cook, M. E.; Pariza, M. W. Effects of conjugated linoleic acid on body composition of mice. Lipids 1997, 32, 853–858.

Reaney, M.J.T.; Dong, L.Y.; Westcott, N.D. In Advances in conjugated linoleic acid research, Volume 1 Ed. 1999.

Risérus, U.; Arner, P.; Brismar, K.; Vessby, B. Treatment with dietary *trans*-10, *cis*-12 conjugated linoleic acid causes isomer-specific insulin resistance in obese men with the metabolic syndrome. Diabetes Care 2002, 25, 1516–1521.

Risérus, U.; Berglund, L.; Vessby, B. Conjugated linoleic acid (CLA) reduced abdominal adipose tissue in obese middle-aged men with signs of the metabolic syndrome: a randomized controlled trial. Int. Jour. Obes. 2001, 25, 1129–1135.

Scholfield, C.R.; Cowan, T.C. Cyclization of linolenic acid by alkali isomerization. J. Am. Oil Chem. Soc. 1959, 36(12), 631–632.

Seko, K.; Kaneko, R.; Kobayashi, K. Photoconjugation of methyl linoleate in the presence of iodine as a sensitizer. Yukagaku 1998, 38, 949–954.

Thom, E.; Wadstein, J.; Gudmundsen, O. Conjugated linoleic acid reduces body fat in healthy exercising humans. J. Int. Med. Res. 2001, 29, 392–396.

Tokle, T.; Jain, V.; Proctor, A. Effect of minor oil constituents on soy oil conjugated linoleic acid production. J. Agric. Food Chem. 2009, 57(19), 8989–8997.

Vahvaselka, M.; Lehtinen, P.; Sippola, S.; Laakso, S. Enrichment of conjugated linoleic acid on oats (Avena sativa L.). J. Ag. Food Chem. 2004, 52, 1749–1752.

Veth, M.; Griinari, M.; Pfeiffer, A.; Bauman, D. Effect of CLA on milk fat synthesis in dairy cows: comparison of inhibition by methyl esters and free acids, and relationships among studies. Lipids 2004, 39, 365–372.

Wang, J.H.; Zhu, B.W.; Song, M.K.; Choi, Y.J. Effect of monensin, fish oil or their combination on in vitro fermentation and conjugated linoleic acid (CLA) production by ruminal bacteria. Animal Feed Sci. and Tech. 2005, 120, 341–349.

Yang, T.S.; Lui, T.T. Optimization of production of conjugated linoleic acid from soybean oil. J. Ag. Food Chem. 2004, 52, 5079–5084.

Yettella, R.; Henbest, B.; Proctor, A. Effect of Antioxidants on Soy Oil Conjugated Linoleic Acid Production and Its Oxidative Stability. J. Agric. Food Chem. 2011, 59 (13), 7377–7384.

Contributors

Chris L. G. Dayton

Chris L. G. Dayton is Director of Fats and Oils Processing for Bunge's Global Innovation group in White Plains, New York. He majored in chemistry while obtaining his Bachelor of Science degree from Purdue University in West Lafayette, Indiana (1986). Chris has been working for over 25 years for Bunge in various research, development and innovation groups. He holds a number of patents on enzymatic degumming, interesterification, and modifications of fats and oils. Chris has served in a number of positions within the American Oil Chemists' Society (AOCS), including At-Large Governing Board Representative and Uniform Methods Committee Chairman. He currently serves the AOCS as the Examination Board Chairman, voting member of the Uniform Methods Committee, and an Associate Editor for the Journal of the AOCS.

Dr. Laurence Eyres

Laurence Eyres, FNZIFST, runs his own contracting/consulting business for the food and dietary supplements industries, specializing in dairy, oils and fats and related lipids, and product and business development. Laurence has had a varied career in the food industry, spanning 35 years. He has held managing director (Sabre Safety) and general manager positions as well as technical and operations director's roles with Abels Ltd., ETA Foods, APV, Bluebird Foods, NZ Dairy Foods and Fonterra Brands. He has also held roles in the Universities, namely Director and Associate Professor of Food Technology & Agribusiness at Massey University (2000), and Business Development Director at University of Auckland. He is a retired board Director of FSANZ and has advisory roles with the NZ Nutrition Foundation and NZ Heart Foundation. He is chairman of the NZIC Oils and Fats specialist group, a role he has held on and off for 28 years.

Walter E. Farr

Walter E. Farr graduated from Mississippi State University in 1960, with a B.S. in Chemistry, Minor in Chemical Engineering. He performed graduate work at UT Medical Center, Memphis, specializing in industrial statistics. He joined Southern Cotton Oil Company, Wesson Oil and Snowdrift Division (later to become Hunt Wesson Foods) in 1960 as a quality control chemist, where he designed and set up the first statistical process control system.

He transferred to the Wesson Oil Refinery in Fullerton CA, where he spent four more years as the refinery superintendent. This was followed by a tour of duty with ADM, Decatur IL; Anderson, Clayton & Co., Houston TX; and Kraft Foods, Memphis TN. He was named Kraft Technology Fellow in 1992, and retired from Kraft in 1993. He then spent 5 years with Owensboro Grain Co., Owensboro KY, and 5 years with Desmet NA, Atlanta, GA. Retiring again, he formed The Farr Group of Companies in Memphis, TN, in 2003.

He joined the American Oil Chemists' Society in 1973, where he was and he has been co-editor of several books, and chapter author in many more. Farr was named Fellow, AOCS in 2007.

Flavio S. Galhardo

Flavio S. Galhardo studied at Sao Paulo University (Chemical Engineering in 1985, MBA in 2000). He initially developed his career in Brazil, in R&D and Operations, and later joined the Bunge Global Innovation group in White Plains, NY. He currently directs business initiatives with strong technology components for Agribusiness, Bioenergy and Food Products.

Kevin Hicks

Kevin Hicks, Ph.D., is the Research Leader of the Sustainable Biofuels and CoProducts Research Unit of the Eastern Regional Research Center (ERRC), ARS, USDA in Wyndmoor (suburban Philadelphia) Pennsylvania. He obtained his Ph.D. from the University of Missouri–Columbia, in Biochemistry. At ERRC, where he has worked for 32 years, Kevin leads a team of 25 scientists and engineers conducting research to develop sustainable biofuels and coproducts from agricultural commodities and byproducts. Dr. Hicks is the author of approximately 200 peer-reviewed and technical publications and 10 patents, as well as numerous presentations to national and international audiences. He has won 11 major awards for his work and has served as an officer in the Carbohydrate Division of the American Chemical Society and in several other professional societies.

Vishal Jain

Vishal has a Ph.D. in Food Science from University of Arkansas, Fayetteville. He has over 10 years of experience in Oils and Fats research with a focus on lipid processing, lipid analysis, and conjugated linoleic acid. He has worked with several multi-national companies. He is currently the Raw Material Scientist with Mars Global Chocolate responsible for projects to support global chocolate growth. Vishal has 15 peer-reviewed publications and over 20 conference presentations. He has held several leadership roles with AOCS and continues to provide leadership and vision.

Dr. David B. Johnston

Dr. David B. Johnston is a Lead Scientist at the USDA-Agricultural Research Service's Eastern Regional Research Center in Wyndmoor, Pennsylvania (suburban Philadelphia) where he leads a team conducting research on value-added coproducts for improving the economics and greenhouse gas emissions of corn and cellulosic fuel ethanol production. Dr. Johnston received his B.S. degree in Microbiology (1990) and Ph.D. degree in Food Biochemistry (1997) from the University of California, Davis.

Dr. ir. Marc Kellens

Marc Kellens is the group technical director and member of the executive committee of the Desmet Ballestra group, headquartered in Brussels Belgium. Desmet is an international technology and process supplier for the oils and fats and oleochemical industry, with worldwide activities in more than 150 countries, which are covered by over 15 offices. Kellens started as manager R&D at Desmet Ballestra over 20 years ago, in 1991. In 1999 he became responsible for the oil refining and fat modification department, and in 2002 for the oilseed preparation and extraction activities. Since 2003, he has been group technical director for technologies and processes developed. He carries R&D in his heart and is today still strongly involved in the new product developments and R&D activities of the group.

He has an engineering degree in food processing and technology and a Ph.D. in oils and fats crystallization, obtained in 1984 and 1990 respectively, at the University of Leuven, Belgium. He is the author of numerous scientific articles and books, and has several patents in oil and oilseed processing technologies.

He is the current president of the European Federation for the Science and Technology of Lipids, a federation of 13 scientific associations in Europe concerned with Lipids, Fats and Oils.

Robert A. Moreau

Robert A. Moreau obtained his B.A. from Boston University and his Ph.D. from the University of South Carolina, where he was mentored by Dr. Anthony Huang. After a two year postdoc with Dr. Paul Stumpf at UC Davis, he joined the staff at the Eastern Regional Research Center (USDA, ARS) in Wyndmoor, Pennsylvania, USA, and he has enjoyed working at ERRC for the last 30 years. Bob's research interests have focused on method development for lipid analysis and their use to solve problems in agriculture and food science, including functional lipids such as phytosterols, tocotrienols, and carotenoids. He is the author of almost 200 publications and four patents and is an AOCS Fellow.

Richard Ozer

Richard has been with Crown Iron Works since 1997 and as Product Sales Manager of Specialty Products since 2000. The Specialty Division is charged with taking Crown's traditional product lines of Extraction and Desolventization and extending them into non-traditional applications such as Algae, Fermentation Products for Omega 3s, SPC, and other vegetable proteins. Currently Richard has taken on White Flake Desolventization and extending this technology to the desolventization of other proteins and temperature sensitive products. Richard obtained his Bachelor's Degree in Mechanical Engineering from Drexel University and Professional Engineers License from the State of Pennsylvania. He has worked for several Equipment Supply Companies such as Bepex and Aljet, where he developed several new products for industry, and is the holder of two patents. He has spoken at AOCS and other conferences including short courses related to Algae.

Andrew Proctor

Andrew Proctor, of the University of Arkansas in Fayetteville, was recognized as a leading professor of lipid chemistry and food science, with a focus on lipid analysis, conjugated linoleic acid, and rice oil co-products. He has distinguished himself as a researcher with more than 100 publications to his credit.

He was also instrumental in establishing an ex--change program between the European Union (E.U.) and United States (U.S.) on renewable resources and clean technology. The program provides grants for international curriculum development and related student exchange.

Proctor has served on close to 30 different AOCS committees or boards, including the Books and Special Publications Committee (2000–2002) and the Governing Board, as a member-at-large (2007–present). He served as chairperson of the Analytical Division Program Committee from 2005–2007 and has been an associate editor or senior associate editor of JAOCS since 1992. In 2008, he organized and chaired the Professional Educators' Common Interest Group—a group that seeks to provide tools for the teaching of fats, oils, and lipid chemistry.

Proctor received the AOCS Herbert J. Dutton Award in 2010 and has been a Fellow of the UK's Royal Society of Chemistry since 2006.

Dr. Leandro Ravetti

Leandro Ravetti graduated as an Agricultural Engineer in Argentina. Leandro leads the Modern Olives technical team, which manages the largest olive groves and olive oil processing plants in Australia. Leandro covered the position of Drafting Leader for the new Australian Standard for Olive Oil from June 2010 until its publication in July 2011. He received the 2011 Standards Award for outstanding contribution to standardization in Australia.

Vincent J. Vavpot

Vince Vavpot received two degrees, the first in Chemical Engineering Technology from the University of Dayton in 1963, followed by his B.S. in Mechanical Engineering from the Fenn College of Engineering of the Cleveland State University in 1968. After achieving the Chemical Engineering Technology degree, Vince worked for the National Cash Register Company as a lab technician in their Research and Development department on materials strength testing. After graduating in Mechanical Engineering, Vince worked as a process engineer for the Dow Chemical Company in their packaging products division from 1968 to 1974. In 1974 Vince started working for the Anderson International Company of Cleveland, Ohio, first as a project engineer for Dairy evaporator and Spray Dryer Systems, and quickly became Product Line Manager for Dairy systems. After 3 years, Vince was promoted to Sales Engineering Manager over all Anderson product lines for the Vegetable Oil, Animal Feeds, Synthetic Rubber, and Steam Specialty product groups. In 1980 Vince accepted the position of Regional Sales Manager for the territory of Asia and Oceania, working with territory representatives, visiting accounts and helping troubleshoot systems for process improvements. Vince has achieved successful developments in process improvements through the use of expander technology for both solvent and mechanical extraction systems, and has acted on a consulting basis for a number of processing clients. Vince is currently Vice President for Anderson's Vegetable Oil and Animal Feeds Business Group.

Maurice A. Williams

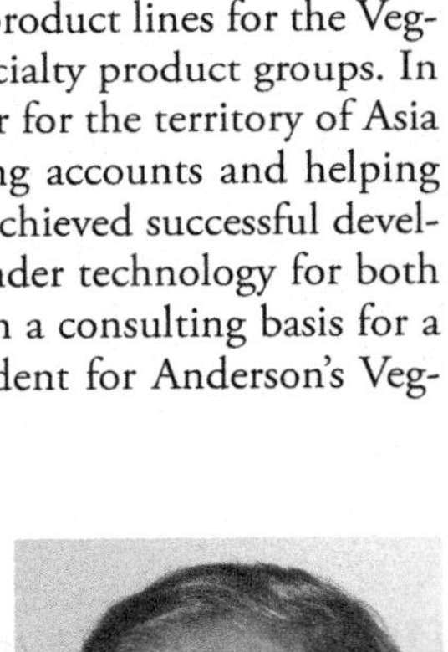

Maurice Williams earned a degree in 1959 in Natural Science with a major in biology and minors in chemistry and philosophy at John Carroll University in Cleveland, OH, following his four-year tour of duty in the U.S. Air Force. He then began his working career at Anderson International Corp in Cleveland, OH, where he worked for a total of 52 years.

He specialized on their mechanical screw presses, expanders, and multiple stage evaporators that were used in the dairy industry and in the rendering industry. He did field service on all of this equipment, especially on expanders making animal feeds and screw presses squeezing oil out of oilseeds.

After retiring in 2003, he continued to work as a consultant for Anderson until November 2011. Prior to his retirement, he held the title of Director of R&D and had spent many years working in various research and development capacities for Anderson International.

Williams spoke frequently at AOCS conventions. He published many journal articles and several book chapters for Wiley and Sons, Marcel Dekker, AOCS Press, and Woodhead Publishing on that equipment. For 31 years he spoke about expanders for feeds and screw pressing vegetable oils in short courses held at Texas A&M University in College Station, TX. His work brought him to many countries servicing Anderson equipment. Williams is a long-time member of AOCS.

Robert J. Williams

Robert J. (Rob) Williams joined Anderson International Corp in 1999 after 20 years in the process industries. Prior to joining Anderson he worked in various process/engineering positions for DOE, Lubrizol, Avery Dennison and UniRoyal Chemical. He has worked for Anderson International Corp as Product Line Manager of Polymer Products, Managing Director of Engineering and Vice President of Engineering, prior to being appointed Senior Scientist in January 2007. Rob took an undergraduate degree in Chemistry from the University of Dayton (1974) and a Master of Science degree in Chemical Engineering from Case Western Reserve University (1978). He lives with his wife in Willoughby Hills, Ohio.

Marie Wong

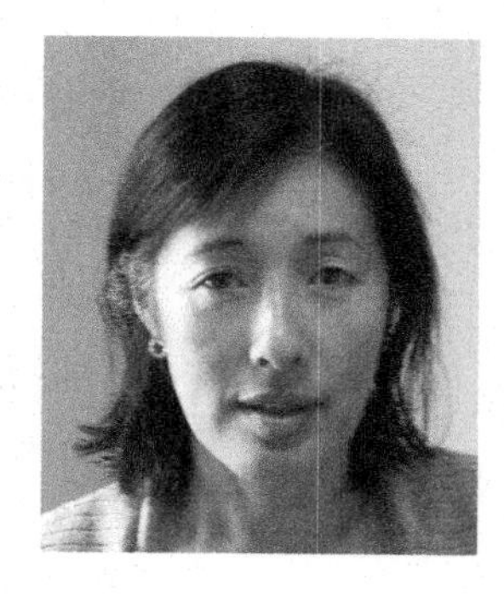

Marie Wong has a Ph.D. in Food Engineering from Massey University. She is currently the Regional Director for the Institute of Food, Nutrition and Human Health at the Auckland campus of Massey University. Marie's research career began with horticultural products processing. For the last 10 years she has been involved with edible oil research, focusing on avocado oil and extra virgin olive oil from New Zealand.

Index

Q

R

S